Large Area Networked Detectors for Particle Astrophysics

Large Area Networked Detectors for Particle Astrophysics

Editors

Pierre Sokolsky
The University of Utah, USA

Gus Sinnis
Los Alamos National Laboratory, USA

World Scientific

EW JERSEY · LONDON · SINGAPORE · BEIJING · SHANGHAI · HONG KONG · TAIPEI · CHENNAI · TOKYO

Published by

World Scientific Publishing Europe Ltd.

57 Shelton Street, Covent Garden, London WC2H 9HE

Head office: 5 Toh Tuck Link, Singapore 596224

USA office: 27 Warren Street, Suite 401-402, Hackensack, NJ 07601

Library of Congress Cataloging-in-Publication Data

Names: Sokolsky, Pierre, editor. | Sinnis, Gus, editor.
Title: Large area networked detectors for particle astrophysics / editors, Pierre Sokolsky,
  The University of Utah, USA, Gus Sinnis, Los Alamos National Laboratory, USA.
Description: New Jersey : World Scientific, [2023] | Includes bibliographical references and index.
Identifiers: LCCN 2022013705 | ISBN 9781800612600 (hardcover) |
  ISBN 9781800612617 (ebook for institutions) | ISBN 9781800612624 (ebook for individuals)
Subjects: LCSH: Nuclear astrophysics. | Nuclear astrophysics--Instruments. |
  Particles (Nuclear physics) | Cosmic rays. | Neutrino astrophysics. | Gamma ray astronomy.
Classification: LCC QB464 .L36 2023 | DDC 523.01/972--dc23/eng20220621
LC record available at https://lccn.loc.gov/2022013705

British Library Cataloguing-in-Publication Data
A catalogue record for this book is available from the British Library.

For any available supplementary material, please visit
https://www.worldscientific.com/worldscibooks/10.1142/Q0372#t=suppl

Desk Editors: Balamurugan Rajendran/Adam Binnie/Shi Ying Koe

Typeset by Stallion Press
Email: enquiries@stallionpress.com

To Michael and Stephanie, with love and appreciation.
Pierre Sokolsky

To my children, Marika and Jimmy, with affection.
Gus Sinnis

Preface

The field of particle astrophysics has its origins with Victor Hess' discovery of cosmic rays in 1912. Today, particle astrophysics is an active field at the intersection of particle physics and astrophysics that uses cosmic messengers to better understand the nature and evolution of the universe. While cosmic rays and gamma rays have been the traditional messengers of particle astrophysics, recent technological advances have added neutrinos and gravitational waves to this list. The resulting multi-messenger astronomy is a powerful tool in the study of some of the most energetic and powerful phenomena in the universe.

In this book, we focus our attention on ultra-high-energy cosmic rays (UHECRs), very-high-energy gamma rays, and very/ultra-high-energy neutrinos. Because these particles display significantly different properties and interactions, they can probe different aspects of the same astrophysical phenomena. Like multi-wavelength observations in astronomy, this provides a much more complete view of the sources of such particles as well as the properties of the intervening space.

UHECRs are the highest energy particles ever detected, and their nature and origin have been the subject of intense scientific effort for decades. However, because they are charged particles, their trajectory is subject to extra-galactic and Galactic magnetic fields. Gamma rays, as neutral particles, point directly back to their sources and can be used to trace potential sources of UHECRs. However, gamma rays with energies above several TeV suffer from absorption in the extragalactic background light and are thus most useful for the study of relatively nearby sources. Neutrinos, feeling only the weak force, can travel through the entire universe unimpeded and point to their source. Their limitation as a probe is due to the smallness of their interaction cross section, requiring massive detectors to produce a sufficient event rate.

Though UHECRs, gamma rays, and neutrinos interact with matter and radiation differently, the detection techniques used have evolved to have significant similarities. In all cases, though for different reasons, large area networked detector systems are used to detect these particles. UHECRs are so rare that detectors dispersed over thousands of square kilometers are needed to detect sufficient numbers of these particles. These arrays measure their energy spectrum and chemical composition, and they search for any anisotropies in their arrival directions. With neutrinos, it is their incredibly small cross section that leads us to instrument a cubic kilometer of ice (or more) to detect them with sufficient statistics to shed light on their sources. With very-high-energy gamma rays, their relatively low flux and the need to differentiate them from the more abundant cosmic-ray background requires detectors with excellent angular resolution spread over tens of thousands of square meters and even a square kilometer to unravel the highest energy sources in our Galaxy.

The story of this field is the movement from the small pioneering arrays such as the Fly's Eye and AGASA for UHECR, Whipple and Milagro for gamma rays, and Amanda for neutrinos, to the huge experiments currently operating such as Auger, Telescope Array, VERITAS, HAWC, and IceCube. This seemingly inevitable increase in scale is primarily driven by the need to increase statistics and push the discovery limits in these three areas. As the reader of this book will see, new discoveries and new questions are driving the need for even larger and more complex detector networks, some proposing to utilize novel detection techniques such as radio and space-based observations.

The three messenger particles, their importance for particle astrophysics, and the current and future detector arrays are discussed in the following three sections. Each section begins with a review of the current physical understanding of the sources, propagation, and interaction properties of the particle. This is followed by chapters focusing on the different networked detector systems currently in use and on the planned upgrades. Finally, there are chapters on new techniques and approaches that will hopefully answer the emergent new questions and finally settle old controversies.

The editors would like to thank all the authors for expeditiously meeting our deadlines and for their hard work at summarizing what is now a vast field of particle astrophysics.

About the Editors

Pierre Sokolsky is an experimental particle astrophysicist. He is a distinguished professor of physics and astronomy (emeritus) at The University of Utah in Salt Lake City. His research interests lie in ultra-high-energy cosmic-ray physics, particularly elucidating the cosmic-ray spectrum and composition. Prof. Sokolsky has played a number of leadership roles in a series of experiments utilizing the atmospheric air fluorescence technique to study ultra-high-energy cosmic rays. These include the Fly's Eye, High Resolution Fly's Eye (HiRes), and most recently, the Telescope Array (TA) observatories. He was the spokesperson for the HiRes and TA experiments. He is a fellow of the American Physical Society and the recipient of the Panofsky Prize in Experimental Particle Physics "for the development of the atmospheric air fluorescence technique". He was a Sloan and Guggenheim Fellow, Chair of the Physics Department, as well as Dean of the College of Science at the University of Utah.

Gus Sinnis is an experimental astroparticle physicist who recently retired from Los Alamos National Laboratory. Prior to retiring, Dr. Sinnis was the Director of the Los Alamos Neutron Science Facility (LANSCE), a national user facility devoted to increasing our understanding of fundamental and applied nuclear physics, the behavior of materials under extreme conditions, and the production of radionuclides for medical imaging and therapeutics. Gus' scientific research focused on exploring the very-high-energy universe with gamma rays and cosmic rays.

He was a lead developer and co-spokesperson of the Milagro experiment, the first ground-based particle detection array to detect a cosmic source of gamma rays, and was one of the primary developers of the HAWC Observatory, which has now detected over 65 sources of very-high-energy gamma rays. Gus was a member of the High Resolution Fly's Eye experiment, which was the first to detect the suppression in the cosmic-ray spectrum above 6×10^{19} eV. Dr. Sinnis was recognized as a fellow of the American Physical Society in 2005 for the development and use of ground-based telescopes to study high-energy gamma rays and cosmic rays from a variety of astrophysical sources.

About the Contributors

Felix Aharonian is a physicist/astronomer with 45 years of experience in high-energy astrophysics.

Since the beginning of his scientific career, Dr. Aharonian has been engaged in observational and theoretical studies of high-energy phenomena in the Universe. His research interests cover a wide range of topical areas, including acceleration, propagation, and radiation processes in astrophysics; the origin of cosmic rays; and the physics and astrophysics of relativistic outflows, such as pulsar winds, AGN jets, and GRBs. Gamma rays as carriers of information about the hadronic and leptonic processes at high energies have been the focus of his research. Dr. Aharonian has also been heavily involved in detection methods and instrumentation; he played a key role in the introduction of the stereoscopic approach to the Imaging Atmospheric Cherenkov Telescope (IACT) technique. It is at present the standard principle of operation of IACT arrays realized, for the first time, by the HEGRA project, which he proposed and initiated in the late 1980s. Dr. Aharonian was among the founding members of other large-scale international projects, particularly the HESS telescope array. Currently, he is actively involved in the LHAASO project. He is a professor at the Dublin Institute for Advanced Studies (DIAS), Ireland.

Douglas R. Bergman is an associate professor in the Department of Physics and Astronomy at the University of Utah, where he studies ultra-high-energy cosmic rays using the Telescope Array. His focus has been on using optical techniques to detect and reconstruct extensive air showers. He received his PhD from Yale University in 1997, studying rare kaon decay at Brookhaven National Lab. He continued studying rare kaon decays as a postdoctoral fellow at Rutgers University before switching to cosmic-ray observations with the High Resolution Fly's Eye experiment. The connection between rare kaon decays and cosmic-ray observations necessitates the use of detailed Monte Carlo simulations, to determine experimental verifications, and other techniques to make sure the simulations accurately reproduce the reality of the experiments.

Amy Connolly earned her PhD from the University of California, Berkeley, in particle physics before turning to particle astrophysics experiments searching for ultra-high-energy neutrinos with the Askaryan technique. Over the past two decades, she has contributed to the balloon-borne ANITA experiment and the in-ice ARA array, which together are the most sensitive to the highest-energy astrophysical neutrinos spanning many decades in energy. Connolly has also investigated the sensitivity of UHE neutrino experiments to physics beyond the Standard Model. She currently leads the ARA collaboration; is the Science principal investigator (PI) for PUEO, the next-generation successor to ANITA; and has played an integral role in the design of the radio component of IceCube-Gen2.

She is also involved with developing novel techniques such as radar detection of UHE neutrinos with RET and using genetic algorithms to design detectors with the student-driven GENETIS project.

Ke Fang is an assistant professor in the Department of Physics at the University of Wisconsin–Madison. Her research focuses on understanding the Universe through its energetic messengers, including ultra-high-energy cosmic rays, gamma rays, and high-energy neutrinos. She runs numerical simulations to study theories of astroparticle sources and analyzes data from HAWC, Fermi-LAT, and IceCube. Ke obtained her PhD in astrophysics from the University of Chicago in 2015. After that, she held a Joint Space-Science Institute (JSI) fellowship jointly at the University of Maryland and the NASA Goddard Space Flight Center from 2015 to 2018, and a NASA Einstein fellowship at Stanford University from 2018 to 2020. She received the Shakti P. Duggal Award in 2021, which recognizes outstanding work by a young scientist in the field of cosmic-ray physics.

Francis Halzen is Vilas Research and Gregory Breit professor of physics at the University of Wisconsin–Madison. He is the principal investigator of IceCube, a cubic-kilometer neutrino telescope buried in the Antarctic ice at the South Pole. IceCube's first observations of high-energy cosmic neutrinos garnered the 2013 *Physics World* Breakthrough of the Year award. In September 2017, IceCube detected a high-energy neutrino from the direction of a blazar called TXS 0506+056. A multi-messenger campaign pinpointed a source of high-energy cosmic rays whose origins have been notoriously difficult to resolve since they were discovered over one hundred years ago. Halzen is the co-author of *Quarks and Leptons*, a textbook on modern particle physics.

Petra Huentemeyer is a professor of physics and the Director of the Earth, Planetary, and Space Sciences Institute at Michigan Technological University and the US spokesperson of the HAWC Observatory in Mexico. She received her PhD from the University of Hamburg, Germany, working on the OPAL experiment at the LEP storage ring at CERN. She then joined the HiRes and Milagro collaborations as a postdoctoral researcher at the University of Utah and Los Alamos National Laboratory. Her current research focuses on investigating cosmic-ray accelerators using very-high-energy gamma-ray emission. Most recently, she has been leading the US effort in the Southern Wide-Field Gamma-Ray Observatory.

John F. Krizmanic is an astroparticle physicist in the High-Energy Cosmic Radiation group within the Laboratory for Astroparticle Physics at NASA/Goddard Space Flight Center (GSFC). He has over 30 years of experience developing science instrumentation for both accelerator- and space-based experiments. He received his PhD in 1989 from Johns Hopkins University with the dissertation *A Search for the Oscillation of Muon Antineutrinos to Electron Antineutrinos using the AGS Wide Band Beam*. He has been at NASA/GSFC since 1994 starting as an NRC fellow. His research interests include space-based high-energy cosmic-ray and neutrino detection with a focus on simulation physics, semiconductor-based detectors for the measurement of charged particles and photons, and diffractive X-ray and g-ray optics. He is the principal investigator (PI) of the nSpaceSim collaboration that is developing an end-to-end simulation package for space-based neutrino experiments; deputy PI of the POEMMA UHECR and VHE neutrino astrophysics probe mission; deputy PI of the TIGERISS mission being developed to measure the flux of galactic cosmic rays from helium to lead with individual element resolution; PI of the Virtual Telescope for X-ray Observations (VTXO), which proposes to use X-ray Phaser Fresnel Lens and precision formation flying SmallSats to form an X-ray telescope with 50 milli-arcsecond–resolution mission, CoI on the ISS-based CALET experiment operating since 2015, the long-duration SuperTIGER & ADAPT balloon experiments, and the JEM-EUSO ultra-long-duration balloon mission.

Frank G. Schroeder is assistant professor at the Department of Physics and Astronomy of the University of Delaware and member of the Bartol Research Institute. He also leads a research team at the Karlsruhe Institute of Technology, where he obtained his doctoral degree in 2011.

His research focus includes detection and analysis techniques to increase the measurement accuracy of energy and mass for ultra-high-energy cosmic rays, especially radio detection of atmospheric particle cascades. A particular research goal is the origin and physics of the most energetic cosmic rays of the Galaxy. For this purpose, he actively contributes to the research conducted by leading astroparticle observatories, such as IceCube and the Pierre Auger Observatory. He has been awarded a Sloan Research Fellowship and the IUPAP Young Scientist Prize in Astroparticle Physics.

Justin A. Vandenbroucke is an associate professor at the University of Wisconsin–Madison, in the Physics Department and Wisconsin IceCube Particle Astrophysics Center, with a joint appointment in the Astronomy Department. His research focuses on gamma-ray instrumentation for the Cherenkov Telescope Array, analysis of data from the IceCube Neutrino Observatory, and citizen science with the Distributed Electronic Cosmic-Ray Observatory.

Contents

© 2023 World Scientific Publishing Europe Ltd.
https://doi.org/10.1142/9781800612617_0001

Chapter 1

Ultra-High-Energy Cosmic Rays: Scientific Motivation

Ke Fang

Department of Physics, Wisconsin IceCube Particle Astrophysics Center, University of Wisconsin–Madison, Madison, WI 53706, USA
kefang@physics.wisc.edu

Ultra-high-energy (E > 10^{18} electron volts) cosmic rays (UHECRs) are the highest energy particles ever detected by human beings. Decades of measurements have revealed that these energetic, charged particles have an extragalactic origin, though the nature of their sources remains unknown. Much has been learned about the energy spectrum, chemical composition, and arrival directions of UHECRs via existing observatories, in particular, the Pierre Auger Observatory in the Southern Hemisphere and the Telescope Array in the Northern Hemisphere. Many mysteries, both old and new, remain unaddressed, such as an excess in the muon content in air showers as well as potential differences in the spectral shapes, composition, and anisotropy patterns of UHECRs from different sky regions. This chapter reviews the science of UHECRs, investigates candidate astrophysical sources, and discusses the role of UHECRs in multi-messenger astrophysics.

1. Overview

Ultra-high-energy cosmic rays (UHECRs) were first detected in 1962 with the array of scintillation detectors at the MIT Volcano Ranch station.[1] Since then, decades of efforts have been devoted to the measurement of UHECRs. See Refs. [2,3] for reviews on historical observations. In this review, we focus on the latest results from the two UHECR experiments currently in operation: the Pierre Auger Observatory and the Telescope Array, which we refer to as Auger and TA, respectively.

Both Auger and TA use two types of instruments for UHECR detection. One is a surface particle detector array on the ground, which samples air shower fronts as they arrive at the Earth's surface. The other is fluorescence telescopes, which measure the light produced as air shower particles excite atmospheric nitrogen.

The Auger is located in the Mendoza Province, Argentina, and has been collecting data since 2004. It consists of a surface array and a fluorescence detector (FD) system. The surface array comprises 1.6×10^3 surface detector (SD) stations, covering an area of about 3000 km^2.[4] The FD system consists of four fluorescence detector stations, each hosting six telescopes that overlook the surface array.[5] The duty cycle of the SD array reaches ∼100% and that of the FDs is 10–15%.[6]

Since 2016, Auger has begun an upgrade called AugerPrime. This dedicated instrumental upgrade includes an installation of a plastic scintillator on top of each existing SD station, enhancement of SD electronics, addition of a radio antenna to each of the SD stations, installation of the underground scintillator muon detector, and measures to increase the duty cycle of the FD.[7-9]

TA is located in West-Central Utah, covering an area of approximately 700 km^2, and has been collecting data since 2007. It consists of 507 scintillation SDs, which are sensitive to muons and electrons,[10] and three FD stations that house a total of 48 telescopes.[11] TA is currently undergoing an upgrade that will increase the SD area fourfold toward a quadrupled TA detector named TA×4.[12,13]

In this chapter, we review recent advances in the study of UHECR physics. In Sections 2–5, we summarize new findings about the UHECR characteristics, including energy spectrum, chemical composition, arrival directions, and muon content in the air showers. We then investigate the origin of these extreme-energy particles in Section 6. In particular, we discuss particle acceleration mechanisms that have the potential to accelerate particles to 10^{18} eV and higher and candidate astrophysical sources that host such accelerators. Finally, we discuss the connection of UHECRs and multi-messenger astrophysics, with an emphasis on the implication of UHECR physics on the detection of cosmogenic neutrinos and UHE photons and on the galactic-to-extragalactic transition. This review focuses on recent developments. Readers should also refer to previous reviews: Refs. 14–17 and references therein.

2. Spectrum

The combined energy spectrum measured by Auger, multiplied by E^3, is shown by the circle markers in Fig. 1. The measurements use several datasets of the Auger Observatory, including events from the 1500 m array,[22] inclined events,[23] hybrid events, events detected by the 750 m array, and the FD events dominated by Cherenkov light.[19]

The data points indicated by square markers in the same figure show the energy spectrum measured by TA using the 22-month TA Low Energy Extension (TALE) measurement below $10^{18.2}$ eV and an 11-year TA SD spectrum[24] above $10^{18.2}$ eV.[20,25]

The UHECR spectrum clearly presents a well-established "ankle" around 4×10^{18} eV and an abrupt suppression above 5×10^{19} eV.

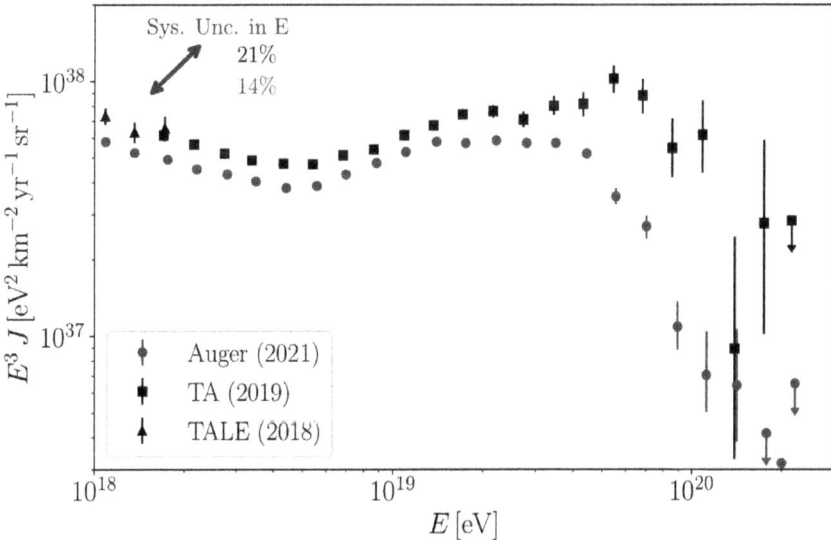

Fig. 1.   Energy spectrum of UHECRs measured by the Pierre Auger Observatory (circle markers)[18,19] and Telescope Array (square markers).[20] The systematic uncertainties in the energy scale (not plotted) are about 14% in Auger[19] and 21% in TA[21] measurements.

The total energy density in cosmic rays above 5×10^{18} eV is measured to be 5.7×10^{53} erg Mpc^{-3}.[18] Just above 10^{19} eV, an "instep" steepening of the spectrum, also referred as the "shoulder", was recently reported by Auger[18,22] and confirmed by TA.[20] The fact that the spectrum hardens from ankle to instep and then softens from instep to 5×10^{19} eV indicates a two-step suppression. The instep feature has no significant dependence on the declination,[18] disfavoring a scenario that the feature is caused by a nearby source. Reference [18] suggests that the spectral features can be reproduced in models with an energy-dependent mass composition.

A joint analysis by the Auger–TA energy spectrum working group finds an agreement in the spectra in the region of the sky accessible to both observatories ($[-15, +24]$ degrees in declination). An alignment of the spectra in the common declination band may be reached by rescaling the energies by $+4.5\%$ for Auger and -4.5% for TA. Such values are well within the systematic uncertainties of both experiments.

In the entire accessible field of view, the Auger spectra in different declination bands are fully consistent.[18,22] TA observed slightly different spectra in the northern and southern parts of the TA sky. The positions where the spectra steepen are different at a 3.5σ confidence level.[20] The difference remains after removing events within $20°$ of the TA "hotspot".[26] More statistics at the highest energy by the next-generation experiments are needed to investigate the significance and cause of these differences in TA and Auger spectra.

3.   Chemical Composition

Understanding the mass composition of UHECRs is important for the development and validation of astrophysical models. Measurements of the depth of shower maximum, $X_{\max}$, and its standard deviation, $\sigma(X_{\max})$, provide crucial information to infer the mass composition of UHECRs. Figure 2 shows the $X_{\max}$ and $\sigma(X_{\max})$ measured by Auger using hybrid data (taken by both FD and SD) and only the SD data.[23,30] It shows that the mean mass is light up to $\sim 10^{18.4}$ eV and gets heavier above that energy. Measurements from the two datasets are consistent, while the SD analysis extends to well beyond $10^{19.5}$ eV due to the higher duty cycle.

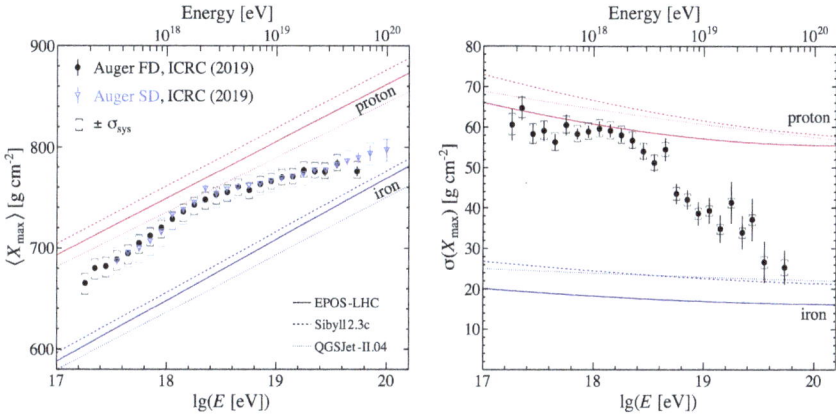

Fig. 2. Measurements of X_{max} (left) and its standard deviation $\sigma(X_{Xmax})$ (right) at the Pierre Auger Observatory, using the hybrid data (black data points) and the Surface Detector data (light-blue data points) from years 2004–2017, compared to the predictions for proton and iron nuclei of the hadronic models EPOS-LHC,[27] Sibyll 2.3c[28] and QGSJetII-04.[29] The error bars indicate the statistical uncertainties, while the brackets indicate the systematic uncertainties of the measurements. The plot is from Ref. [23].

The light composition around 10^{18} eV is independently measured by the Auger Engineering Radio Area (AERA),[31] the LOFAR radio telescope,[32] and the TALE air FD.[33]

Figure 3 shows the X_{max} measured by TA in hybrid mode. By comparing to Monte Carlo X_{max} distributions of unmixed protons, helium, nitrogen, and iron based on the QGSJet II-04 hadronic model,[29] Ref. [34] concluded that QGSJet II-04 protons are compatible with their data at the 95% confidence level. Results from the TA 12-year SD data are consistent with the hybrid measurements.[35] Above 10^{19} eV, TA has insufficient exposure to accurately distinguish the difference between different individual elements.[29]

As pointed out by Ref. [36], a direct comparison of the X_{max} measurements of Auger and TA is not appropriate since the two experiments use different approaches to analyze the measured depths of shower maximum, including different cuts on geometry and shower profiles. The Mass Composition Working Group of Auger and TA presented an analysis technique to compare the measurements of the two experiments. They generated a large number of Monte Carlo air showers for proton, helium, nitrogen, and iron primaries.

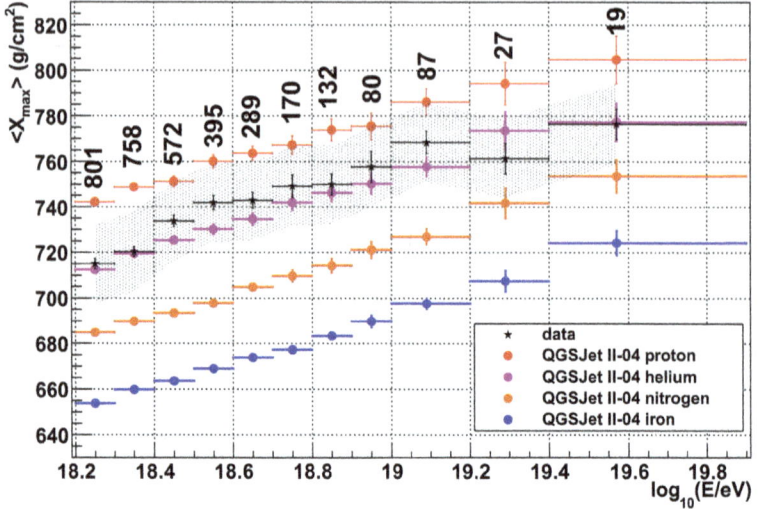

Fig. 3. X_{max} measurement by the Telescope Array using the hybrid data from 8.5 years of observation,[34] compared to the reconstructed Monte Carlo data for proton, helium, nitrogen, and iron nuclei using the QGSJet II-04 hadronic model.[29] The error bars indicate the statistical uncertainties, while the brackets indicate the systematic uncertainties of the measurements.

The combined simulation data, based on Auger's fits of the mass composition using EPOS-LHC and QGSJet II-04 models, were passed through the TA detector simulation. The output from the TA analysis pipeline, which "carries" specifics of the TA acceptance, reconstruction bias and resolution, agrees with the Auger X_{max} distributions within the systematic uncertainties.[36] This result implies that the TA measurement is compatible with both pure protons and the Auger-derived mix composition up to 10^{19} eV.

4. Anisotropy

4.1. *Large scale*

The Auger Observatory detected a dipole with an amplitude of ~7% in the arrival directions of cosmic rays above 8 EeV[37] (see Fig. 4). The post-trial significance of the dipole anisotropy was at the level of 5.2σ in 2017 and increased to 6.6σ in 2021[38] following the increase in statistics. The direction of the dipole,

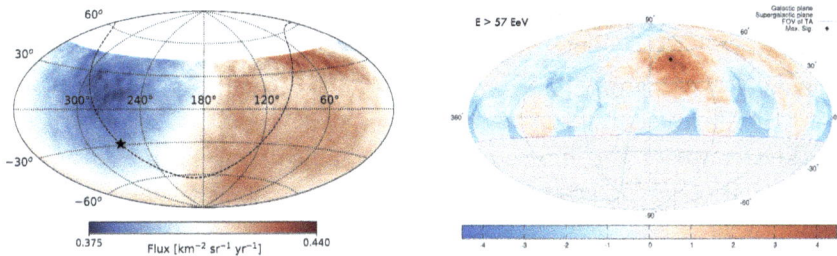

Fig. 4. Left: Sky map of cosmic-ray flux above 8 EeV detected by the Pierre Auger Observatory in equatorial coordinates.[37,38] The map is smoothed with a 45° top-hat function. The star indicates the Galactic center and the dashed line indicates the Galactic plane. A dipolar modulation in right ascension is observed at a statistical significance of 6.6σ. Right: Sky map of direction-dependent excesses and deficit with respect to the isotropic background in pre-trial significance, based on 11 years of TA data above 57 EeV. The plot is from Ref. [40].

(RA = 95 ± 8, Dec = −36 ± 9)[38] is far way from the Galactic center, suggesting that the particles originated from extragalactic sources. The TA searched for the dipole using an energy threshold of 8.8 EeV, which is equivalent to 8 EeV used by Auger.[39,40] They obtained a dipole structure of ∼3% amplitude with a phase similar to that of the Auger dipole. The TA observation is consistent with a fluctuation in the isotropic distribution.[40]

The dipole structure is likely the result of the interplay between the local large-scale matter distribution and deflection in the Galactic magnetic field[41–46] and the intergalactic magnetic field.[47,48] The density field of the local universe may explain the magnitude, direction, and energy dependence of UHECR anisotropies above 8 EeV[46,49–51]

4.2. *Medium scale*

Indications of medium-scale anisotropy have been identified in the arrival directions of UHECRs above 57 EeV in the northern sky.[26] This feature is referred to as the TA hotspot. The hotspot was found by oversampling using 20° radius circles. The post-trial significance is measured to be 3.2σ in 2021.[40] The observation may suffer from ensemble fluctuations due to the discreteness of the emitters.[52]

By using an oversampling search with a 20° circle, TA found an additional excess of events at slightly lower energies.[53] The Perseus–Pisces supercluster, one of the closest superclusters in the

Northern Hemisphere, lies at the location of the new excess of events above $\sim 10^{19.4}$ eV. Monte Carlo studies that generate events from the Perseus–Pisces supercluster reject a chance coincidence with a post-trial significance of 3.5σ.[53]

The cause of the intermediate-scale anisotropy is unknown. Its formation could be related to the structure of the extragalactic and Galactic magnetic field as well as the inhomogeneous distribution of nearby cosmic-ray sources.[54–57]

4.3. *Correlation studies*

Catalog-based searches for anisotropy have been carried out with both Auger and TA data. The searches used jetted and non-jetted active galactic nuclei (AGNs) in the 105-month Swift-BAT catalog,[58] jetted AGNs from the Fermi-LAT 3FHL catalog,[59] starburst galaxies,[60] and objects in the 2MASS catalog up to a K-band magnitude of 11.75.[61] No indications beyond 3σ have been found in these correlation studies.[62,63] The TA analysis concludes that UHECR correlation with the large-scale structure distribution traced by IR emission in the 2MASS catalog can be excluded at 2.4σ level.[40]

On combining the SD data collected at the Auger until 2020 and the TA until 2019, Ref. [64] found a correlation between the arrival directions of UHECRs, including events above 38 EeV by Auger and above 49 EeV by TA, and starburst galaxies[60] with a $\sim 15.5°$ angular scale at a 4.2σ post-trial statistical significance.

The Centaurus A region remains particularly interesting given an excess of the highest energy events from that area of the sky.[62,65] An overdensity search dedicated to a $27°$ circle centered on Centaurus A using events above 41 EeV finds a post-trial p-value of 6×10^{-5}, corresponding to a $3.9\,\sigma$ deviation from isotropy at intermediate angular scales *a priori* focusing on the Centaurus region.[62]

5. Muon Excess and Fundamental Physics

Measurements of the muon content in air showers have been performed by various experiments over the last 20 years. Discrepancies in the number of muons in simulated and observed air showers have been reported by many experiments, such as HiRes-MIA,[66] the Sydney University Giant Air-shower Recorder (SUGAR),[67] and more recently by Auger,[68,69] TA,[70] and NEVOD-DECOR.[71] Using hybrid showers with energies in the range 6–16 EeV and zenith angle $0° - 60°$, Auger finds that the average hadronic shower is 1.33 and 1.61 times larger than predicted using the LHC-tuned models, EPOS-LHC and QGSJetII-04, respectively. It suggests a similar excess in the corresponding muons detected in highly inclined air showers.[69]

Reference [72] carried out a meta-analysis of muon measurements from multiple air shower experiments, covering shower energies between a few PeV and tens of EeV. They find that the measurements agree with simulations based on the models EPOS-LHC and QGSJet II-04 within uncertainties up to $\sim 10^{16}$ eV. Above that energy, a muon excess with respect to simulations is present for all hadronic interaction models. The excess seems to increase with shower energy.

It is not yet known whether the muon excess is due to some incorrectly modeled features of hadron collisions or an indication of the onset of some new phenomenon in hadronic interactions at energies beyond the reach of LHC. Reference [73] shows that shifting the $X_{\max}$ predictions of hadronic interaction models to deeper values and increasing the hadronic signal at extreme zenith angles, the muon excess problem may be alleviated but not removed.

Future analyses that extend to the entire hybrid dataset above the ankle and future upgrades that enable separate measurement of the muon and electromagnetic components of the ground signal would be needed to further understand the muon excess problem.[69]

6. Origin

6.1. *Particle acceleration*

Since the first detection of UHECRs, many theories have been proposed to explain particle acceleration.

Fermi acceleration[74] was proposed in the 1940s as a mechanism for cosmic-ray acceleration. Shock fronts have been found as a plausible site of Fermi acceleration.[75–77] In the shock acceleration model, charged particles stream into magnetic perturbations in the post-shock region, reflect, and are scattered back across the shock by pre-shock Alfvén waves. Repeated reflections steadily accelerate particles to a power-law distribution.

Reacceleration has been proposed to avoid difficulties of accelerating directly to 10^{20} eV energies. One example is the Espresso mechanism. In the Espresso scenario, UHECRs are produced via a one-shot reacceleration of galactic-like cosmic rays in the relativistic jets of AGNs.[78–81]

Particles may also gain energy through electromagnetic processes, such as unipolar induction,[82] magnetic reconnection,[83] and magnetoluminescence.[84] In a unipolar induction, the rotational energy of the highly conducting plasma surrounding the homogeneously magnetized star is converted into electromagnetic energy. Particle-in-cell simulations have confirmed that ion acceleration in the pulsar magnetosphere may reach high efficiency.[85]

6.2.   *Source classes*

Although the actual sources of UHECRs remain a mystery, recent observations have revealed some properties of nature's accelerators.

The UHECR energy generation rate density is evaluated to be $\sim(0.2-2) \times 10^{44}$ erg Mpc^{-3} yr^{-1} at $10^{19.5}$ eV with a nontrivial dependence on the spectral index.[87] Nondetection of significant anisotropy limits the local source density to 10^{-3} Mpc^{-3}.[88,89]

Many source models have been proposed to explain the observed UHECRs. Figure 5 summarizes the source classes that satisfy the Hillas criteria, that is, the size of the acceleration site, R, must be larger than the gyroradius of the cosmic ray:

$$R > \frac{E}{Z e B}. \tag{1}$$

In practice, the acceleration efficiency cannot reach 100%; therefore, only sources well above the red and blue solid lines in Fig. 5 provide plausible acceleration sites. Below the lines, we divide the source classes into three categories based on the mass of the sources.

6.2.1.   *Stellar-mass objects*

Gamma-ray bursts (GRBs): GRBs blast waves may accelerate particles to UHE.[90,91] Protons and heavier nuclei may be accelerated and survive in internal, (external) reverse, and forward shocks.[92–96] Low-luminosity GRBs[97,98] have also been proposed as producers of

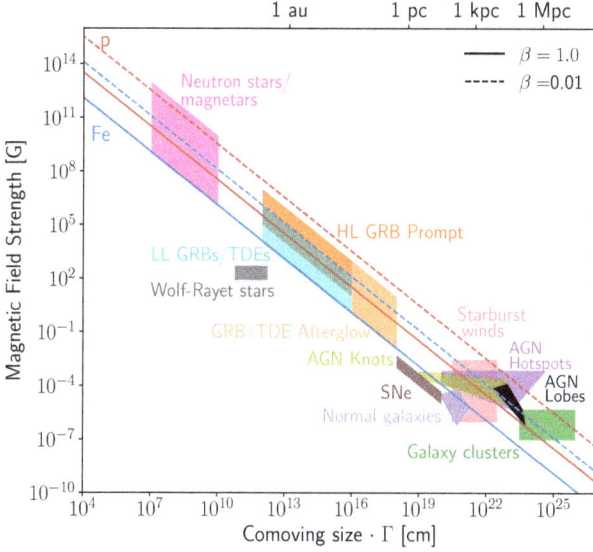

Fig. 5. A modern Hillas diagram[86] (from Ref. [16]). The red and blue lines indicate the magnetic field strength and the size of the particle acceleration region needed to accelerate 10^{20} eV protons ($Z = 1$) and iron nuclei ($Z = 26$), with outflow velocity $\beta = 0.1\,c$ (dashed) and $\beta = c$ (solid), respectively, assuming the maximum achievable efficiency. The colored blocks indicate candidate sources with the uncertainties in their parameters.

UHE nuclei, in particular, in those with Si-rich progenitor and hypernova ejecta. It has been shown that electrons coaccelerated in the UHECR acceleration region of a GRB would result in bright optical-UV emission in tension with observations, if the source is optically thin to synchrotron emission.[99,100] Recently, several TeV gamma-ray emitting GRBs were detected[101–103] confirming that GRBs may produce TeV electrons. Protons, if present, could be accelerated under similar conditions to UHE.

Pulsars and magnetars: Newly born pulsars offer favorable sites for the injection of heavy nuclei and their further acceleration to ultrahigh energies.[82,104–107] Ions may be accelerated in the pulsar magnetosphere[108,109] or at the termination shock of young pulsar winds[110,111] to UHE. The accelerated particles encounter the ejecta material surrounding the neutron star, where they go through hadronuclear and photohadronic interactions. Due to the production of secondary nucleons, the envelope crossing leads to a transition of composition from light to heavy elements at a few EeV and a softer spectral slope than that initially injected.[106] Millisecond magnetars formed from the merger of binary neutron stars have less massive ejecta where UHE nuclei may escape.[112]

Stellar mergers: Reference [113] suggests that UHECRs could be a possible byproduct of the immense energies achieved in black hole mergers. This model requires that relic magnetic fields and disk debris are remaining from the formation of the black holes or their accretion history.

6.2.2.   *Supermassive black holes*

Black hole jets: UHECRs have been proposed to be produced by the jets of radio-loud AGNs,[80,114–121] including BL Lacs,[122,123] and radio galaxies.[124–129] It has also been suggested that radio-quiet AGNs[130] and the black hole magnetosphere[131–133] could also accelerate ions to UHE. Survival of nuclei is shown to be possible in TeV BL Lacs[122] and low-luminosity AGNs[134] with weak photohadronic interaction. Recently the H.E.S.S. Observatory resolved the kilo-parsec-scale jets in Centaurus A at TeV energies,[135] suggesting that UHECRs may be accelerated at kpc distances.[136] Galactic cosmic rays may be picked up and reaccelerated in the AGN jet through transrelativistic shear acceleration.[137]

Tidal disruption events: Tidal disruptions are powerful phenomena that happen when a star approaches sufficiently close to a supermassive black hole and is pulled apart by the black hole's tidal force. The disruption of a star by a black hole can naturally provide protons and heavier nuclei, which can be injected and accelerated to UHE within a jet.[138–140]

6.2.3.   *Galaxies and clusters*

Starburst galaxies and superwinds: Starbursts are galaxies undergoing massive episodes of star formation. The combined effect of stellar winds from hot stars and supernova explosions creates high-temperature winds in the nuclear region of these objects that may produce UHECRs.[55,141,142]

Clusters: Shocks around sheets, filaments, and knots of mass distribution when the gas in void regions accretes onto large-scale structure make them viable candidates to accelerate CRs to ultra-high energies.[143–147]

7. Multi-messenger Connection

7.1. *Cosmogenic neutrinos and photons*

Recent discoveries of high-energy cosmic neutrinos and gravitational waves have led to a new era of multi-messenger astrophysics. Cosmogenic neutrinos and UHE photons are produced when UHECRs propagate through the universe and interact with cosmic background photons. The interaction was first proposed by Refs. [148, 149], and cosmogenic neutrinos are therefore sometimes referred to as GZK neutrinos. These neutral particles may point back to the origin of UHECRs. Since both high-energy cosmic rays and cosmic background photons exist, the existence of high-energy cosmogenic neutrinos is certain. However, their flux has not been measured so far.

The understanding of UHECRs is tightly connected to the search for cosmogenic neutrinos and gamma rays. If the cosmic-ray composition is mostly proton, the expected neutrino flux would be close to the sensitivity of the current-generation neutrino experiments. On the other hand, if UHECRs are mostly heavy nuclei and(or) the proton maximum energy is much lower than 10 EeV, the flux of cosmogenic neutrinos would be rather low. By using the best-fit models of the UHECR spectrum and composition measured by the Auger, Ref. [150] found a cosmogenic neutrino flux that is significantly lower than previous predictions, such as Ref. [151]. The neutrino flux also depends on the source evolution models and the strength of the extragalactic magnetic field.[152] These predictions are crucial to the development of the next-generation neutrino experiments, discussed in later chapters of this book.

7.2. *Galactic-to-extragalactic transition*

The study of UHECRs is closely related to the investigation of the origin of Galactic cosmic rays due to the overlap in their energies and similarity in particle acceleration processes. The transition from

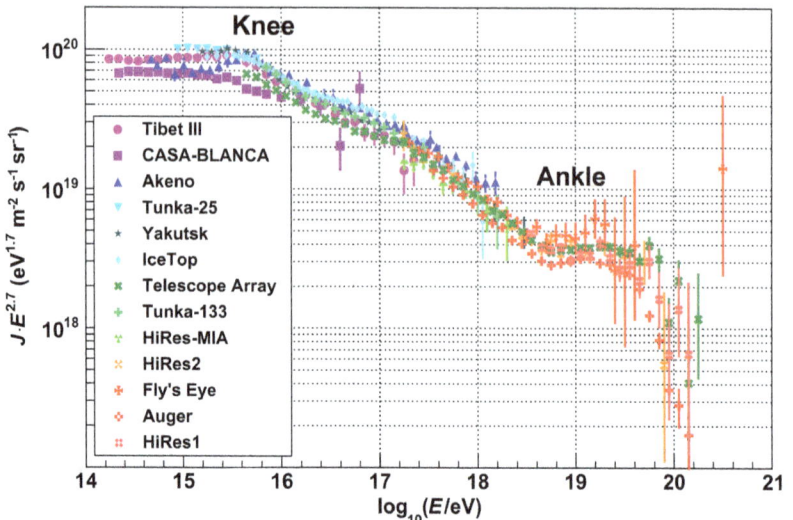

Fig. 6. Cosmic-ray spectrum above 10^{14} eV measured by the listed experiments. Several spectral features can be noticed, including the knee at $\sim$3 PeV, the second knee at 2×10^{17} eV, and the ankle at $\sim$5 EeV. Plot is from Ref. [34].

Galactic to extragalactic cosmic rays remains a mystery. Figure 6 presents the cosmic-ray spectrum above $\sim10^{14}$ eV scaled by $E^{2.7}$. The spectrum generally follows a power-law distribution $dN/dE \propto E^{-s}$, though the spectrum bends at several energies. Specifically, the index changes from $s \approx 2.7$ to $s \approx 3.0$ around 3 PeV. This feature is referred to as the "knee" of the cosmic-ray spectrum. The spectrum becomes hard, $s \approx 2.6$, above the ankle. Between the knee and the ankle, recent observations suggest that a second knee may exist at $\sim 2 \times 10^{17}$ eV. Several scenarios have been proposed to explain these spectral features.

In the most standard scenarios, the knee around 3 PeV is interpreted as the maximum proton energy of Galactic cosmic-ray sources. Heavier elements with the same rigidity are accelerated to Z times higher in energy, where Z is the charge number of nuclei. The iron knee in this scenario appears below 10^{17} eV. Even when extrapolating to the end of the stable elements $Z = 92$, such a model yields transition energy that is still more than an order of magnitude below the ankle at ~ 3 EeV.[153] Another challenge to this scenario is the rareness of "PeVatrons" found by high-energy gamma-ray observations. Supernova remnants (SNRs) have long been suggested as sources of cosmic rays up to the knee, but only a handful of them, out of hundreds of radio-emitting SNRs, have been observed to present hard energy spectra above 10 TeV.[154,155] Other types of sources, such as superbubbles surrounding OB2 stellar clusters, have also been proposed as alternative PeV cosmic-ray accelerators.[156–158]

An alternative scenario is that the transition may occur already in the 10^{17} eV regime. A third population[159] mainly contributes to the energy range between the knee and the ankle.[160] The population could be a new type of source, a subpopulation of the Galactic sources,[161–164] the Galactic counterpart of the extragalactic sources,[165] or additional interaction in the vicinity of the UHECR source.[166]

8. Summary

Decades of efforts have been dedicated to the study of UHECRs. Much has been learned since the discovery of the "oh-my-god" particles, including the shape of the spectrum, the chemical composition, and the arrival directions at the highest energy. A lot remains unknown, such as the origin of UHECRs and the existence of cosmogenic neutrinos from the GZK interactions. Most excitingly, recent observations have introduced new and, sometimes, unexpected questions. These include but are not limited to: the muon puzzle, the difference in the northern- and southern-sky spectra, and the existence of large-scale and intermediate-scale anisotropies. More statistics with better data quality from ongoing upgrades of existing observatories, including the AugerPrime[7] and TA×4,[12] as well as future UHECR experiments, such as the Probe of Extreme Multi-Messenger Astrophysics (POEMMA)[167] and the Global Cosmic-Ray Observatory (GCOS),[168] offers hope for fully unveiling the mysterious Universe at the highest energy.

References

1. J. Linsley, Evidence for a primary cosmic-ray particle with energy 10^{20} ev, *Phys. Rev. Lett.* **10**, 146–148 (February, 1963). doi: 10.1103/PhysRevLett.10.146.

2. M. Nagano and A. A. Watson, Observations and implications of the ultrahigh-energy cosmic rays, *Rev. Mod. Phys.* **72**, 689–732 (July, 2000). doi: 10.1103/RevModPhys.72.689.

3. K.-H. Kampert and A. A. Watson, Extensive air showers and ultra high-energy cosmic rays: A historical review, *Eur. Phys. J. H.* **37**(3), 359–412 (August, 2012). doi: 10.1140/epjh/e2012-30013-x.

4. J. Abraham, P. Abreu, M. Aglietta, E. J. Ahn, D. Allard, *et al.*, Trigger and aperture of the surface detector array of the Pierre Auger Observatory, *Nucl. Instrum. Methods Phys. Res. A.* **613**(1), 29–39 (January, 2010). doi: 10.1016/j.nima.2009.11.018.

5. The Pierre Auger Collaboration, J. Abraham, P. Abreu, M. Aglietta, C. Aguirre, E. J. Ahn, *et al.* The Fluorescence Detector of the Pierre Auger Observatory, *arXiv e-prints.* art. arXiv:0907.4282 (July, 2009).

6. P. Abreu, M. Aglietta, E. J. Ahn, D. Allard, I. Allekotte, others, and Pierre Auger Collaboration, The exposure of the hybrid detector of the Pierre Auger Observatory, *Astropart. Phys.* **34**(6), 368–381 (January, 2011). doi: 10.1016/j.astropartphys.2010.10.001.

7. The Pierre Auger Collaboration, A. Aab, P. Abreu, M. Aglietta, E. J. Ahn, I. A. Samarai, *et al.*, The Pierre Auger Observatory upgrade — Preliminary design report, *arXiv e-prints.* art. arXiv:1604.03637 (April, 2016).

8. P. Abreu, M. Aglietta, J. M. Albury, I. Allekotte, A. Almela, *et al.* AugerPrime upgraded electronics. In *Proceedings of 37th International Cosmic Ray Conference — PoS(ICRC2021)*, Vol. 395, p. 230 (2021). doi: 10.22323/1.395.0230.

9. G. Cataldi, P. Abreu, M. Aglietta, J. M. Albury, I. Allekotte, *et al.* The upgrade of the Pierre Auger Observatory with the Scintillator Surface Detector. In *Proceedings of 37th International Cosmic Ray Conference — PoS(ICRC2021)*, Vol. 395, p. 251 (2021). doi: 10.22323/1.395.0251.

10. T. Abu-Zayyad, R. Aida, M. Allen, R. Anderson, R. Azuma, *et al.*, The surface detector array of the Telescope Array experiment, *Nucl. Instrum. Methods Phys. Res. A.* **689**, 87–97 (October, 2012). doi: 10.1016/j.nima.2012.05.079.

11. H. Tokuno, Y. Tameda, M. Takeda, K. Kadota, D. Ikeda, *et al.*, New air fluorescence detectors employed in the Telescope Array experiment, *Nucl. Instrum. Methods Phys. Res. A.* **676**, 54–65 (June, 2012). doi: 10.1016/j.nima.2012.02.044.

12. Telescope Array Collaboration, R. U. Abbasi, M. Abe, T. Abu-Zayyad, M. Allen, Y. Arai, *et al.*, Surface detectors of the TAx4 experiment, *arXiv e-prints.* art. arXiv:2103.01086 (March, 2021).

13. R. Abbasi, T. Abu-Zayyad, M. Allen, Y. Arai, R. Arimura, *et al.* Current status and prospects of surface detector of the TAx4 experiment. In *Proceedings of 37th International Cosmic Ray Conference — PoS(ICRC2021)*, Vol. 395, p. 203 (2021). doi: 10.22323/1.395.0203.

14. K. Kotera and A. V. Olinto, The astrophysics of ultrahigh-energy cosmic rays, *Ann. Rev. Astron. Astrophys.* **49**(1), 119–153 (September, 2011). doi: 10.1146/annurev-astro-081710-102620.

15. R. Aloisio, Acceleration and propagation of ultra high energy cosmic rays, *arXiv e-prints*. art. arXiv:1707.08471 (July, 2017).

16. R. Alves Batista, J. Biteau, M. Bustamante, K. Dolag, R. Engel, K. Fang, K.-H. Kampert, D. Kostunin, M. Mostafa, K. Murase, F. Oikonomou, A. V. Olinto, M. I. Panasyuk, G. Sigl, A. M. Taylor, and M. Unger, Open questions in cosmic-ray research at ultrahigh energies, *Front. Astron. Space Sci.* **6**, 23 (June, 2019). doi: 10.3389/fspas.2019.00023.

17. L. A. Anchordoqui, Ultra-high-energy cosmic rays, *Phys. Rep.* **801**, 1–93 (April, 2019). doi: 10.1016/j.physrep.2019.01.002.

18. A. Aab, P. Abreu, M. Aglietta, J. M. Albury, I. Allekotte, A. Almela, J. Alvarez Castillo, J. Alvarez-Muñiz, R. Alves Batista, and Pierre Auger collaboration, features of the energy spectrum of cosmic rays above 2.5×10^{18} eV using the Pierre Auger Observatory, *Phys. Rev. Lett.* **125**(12), 121106 (September, 2020). doi: 10.1103/PhysRevLett.125.121106.

19. P. Abreu, M. Aglietta, J. M. Albury, I. Allekotte, A. Almela, J. Alvarez-Muniz, R. Alves Batista, *et al.* Energy spectrum of cosmic rays measured using the Pierre Auger Observatory. In *Proceedings of 37th International Cosmic Ray Conference — PoS(ICRC2021)*, Vol. 395, p. 324 (2021). doi: 10.22323/1.395.0324.

20. D. Ivanov, D. Bergman, G. Furlich, R. Gonzalez, G. Thomson, and Y. Tsunesada. Recent measurement of the Telescope Array energy spectrum and observation of the shoulder feature in the Northern Hemisphere. In *Proceedings of 37th International Cosmic Ray Conference — PoS(ICRC2021)*, Vol. 395, p. 341 (2021). doi: 10.22323/1.395.0341.

21. T. Abu-Zayyad, R. Aida, M. Allen, R. Anderson, R. Azuma, E. Barcikowski, *et al.*, Energy spectrum of ultra-high energy cosmic rays observed with the telescope array using a hybrid technique, *arXiv e-prints*. art. arXiv:1305.7273 (May, 2013).

22. A. Aab, P. Abreu, M. Aglietta, J. M. Albury, I. Allekotte, A. Almela, J. Alvarez Castillo, J. Alvarez-Muñiz, R. Alves Batista, *et al.*, and Pierre Auger Collaboration, Measurement of the cosmic-ray energy spectrum above 2.5×10^{18} eV using the Pierre Auger Observatory,

Phys. Rev. D. **102**(6), 062005 (September, 2020). doi: 10.1103/ PhysRevD.102.062005.

23. The Pierre Auger Collaboration, A. Aab, P. Abreu, M. Aglietta, I. F. M. Albuquerque, J. M. Albury, I. Allekotte, A. Almela, J. Alvarez Castillo, *et al.*, The Pierre Auger Observatory: Contributions to the 36th International Cosmic Ray Conference (ICRC 2019), *arXiv e-prints*. art. arXiv:1909.09073 (September, 2019).

24. R. U. Abbasi, M. Abe, T. Abu-Zayyad, M. Allen, R. Azuma, E. Barcikowski, J. W. Belz, D. R. Bergman, *et al.*, The cosmic ray energy spectrum between 2 PeV and 2 EeV observed with the TALE detector in monocular mode, *Astrophys. J.* **865**(1), 74 (September, 2018). doi: 10.3847/1538-4357/aada05.

25. D. Ivanov. Energy spectrum measured by the telescope array. In *Proceedings of 36th International Cosmic Ray Conference — PoS(ICRC2019)*, Vol. 358, p. 298 (2019). doi: 10.22323/1.358.0298.

26. R. U. Abbasi, M. Abe, T. Abu-Zayyad, M. Allen, R. Anderson, *et al.*, Indications of intermediate-scale anisotropy of cosmic rays with energy greater than 57 EeV in the Northern sky measured with the surface detector of the telescope array experiment, *ApJ. Lett.* **790**(2), L21 (August, 2014). doi: 10.1088/2041-8205/790/2/L21.

27. T. Pierog, I. Karpenko, J. M. Katzy, E. Yatsenko, and K. Werner, EPOS LHC : Test of collective hadronization with LHC data, *arXiv e-prints*. art. arXiv:1306.0121 (June, 2013).

28. F. Riehn, H. Dembinski, R. Engel, A. Fedynitch, T. K. Gaisser, and T. Stanev. The hadronic interaction model Sibyll 2.3c and Feynman scaling. In *35th International Cosmic Ray Conference (ICRC2017)*, Vol. 301, *International Cosmic Ray Conference*, p. 301 (January, 2017).

29. S. Ostapchenko, Monte carlo treatment of hadronic interactions in enhanced pomeron scheme: Qgsjet-ii model, *Phys. Rev. D.* **83**, 014018 (January, 2011). doi: 10.1103/PhysRevD.83.014018.

30. A. Aab, P. Abreu, M. Aglietta, E. J. Ahn, I. Al Samarai, *et al.*, Depth of maximum of air-shower profiles at the pierre auger observatory. i. measurements at energies above $10^{17.8}$ eV, *Phys. Rev. D.* **90**, 122005 (December, 2014). doi: 10.1103/PhysRevD.90.122005.

31. P. Abreu, M. Aglietta, J. M. Albury, I. Allekotte, A. Almela, *et al.* The depth of the shower maximum of air showers measured with AERA. In *Proceedings of 37th International Cosmic Ray Conference — PoS(ICRC2021)*, Vol. 395, p. 387 (2021). doi: 10.22323/1.395.0387.

32. A. Corstanje, S. Buitink, H. Falcke, B. M. Hare, J. Hörandel, T. Huege, G. Krampah, P. Mitra, K. Mulrey, A. Nelles, H. Pandya, J. P. Rachen, O. Scholten, S. Thoudam, G. Trinh, S. ter Veen, and T. Winchen.

Results on mass composition of cosmic rays as measured with LOFAR. In *Proceedings of 37th International Cosmic Ray Conference — PoS(ICRC2021)*, Vol. 395, p. 322 (2021). doi: 10.22323/1.395.0322.

33. R. U. Abbasi, M. Abe, T. Abu-Zayyad, M. Allen, Y. Arai, *et al.*, and Telescope Array Collaboration, The cosmic-ray composition between 2 PeV and 2 EeV observed with the TALE detector in monocular mode, *ApJ.* **909** (2), 178 (March, 2021). doi: 10.3847/1538-4357/abdd30.

34. Telescope Array Collaboration, R. U. Abbasi, M. Abe, T. Abu-Zayyad, M. Allen, R. Azuma, Barcikowski, *et al.*, Depth of ultra high energy cosmic ray induced air shower maxima measured by the telescope array black rock and long ridge FADC fluorescence detectors and surface array in hybrid mode, *Astrophys. J.* **858**(2), 76 (2018). doi: 10.3847/1538-4357/aabad7.

35. Y. Zhezher, Cosmic-ray mass composition with the TA SD 12-year data. In *Proceedings of 37th International Cosmic Ray Conference — PoS(ICRC2021)*, Vol. 395, p. 300 (2021). doi: 10.22323/1.395.0300.

36. W. Hanlon, J. Bellido, J. Belz, S. Blaess, V. de Souza, D. Ikeda, P. Sokolsky, Y. Tsunesada, M. Unger, and A. Yushkov, Report of the working group on the mass composition of ultrahigh energy cosmic rays, *JPS Conf. Proc.* **19**, 011013 (2018). doi: 10.7566/JPSCP.19.011013.

37. Pierre Auger Collaboration, A. Aab, P. Abreu, M. Aglietta, I. A. Samarai, I. F. M. Albuquerque, *et al.*, Observation of a large-scale anisotropy in the arrival directions of cosmic rays above 8×10^{18} eV, *Science* **357**(6357), 1266–1270 (September, 2017). doi: 10.1126/science.aan4338.

38. R. de Almeida, P. Abreu, M. Aglietta, J. M. Albury, I. Allekotte, *et al.* Large-scale and multipolar anisotropies of cosmic rays detected at the Pierre Auger Observatory with energies above 4 EeV. In *Proceedings of 37th International Cosmic Ray Conference — PoS(ICRC2021)*, Vol. 395, p. 335 (2021). doi: 10.22323/1.395.0335.

39. Telescope Array Collaboration, R. U. Abbasi, M. Abe, T. Abu-Zayyad, M. Allen, R. Azuma, *et al.*, Search for large-scale anisotropy on arrival directions of ultra-high-energy cosmic rays observed with the telescope array experiment, *arXiv e-prints*. art. arXiv:2007.00023 (June, 2020).

40. I. Tkachev, T. Fujii, D. Ivanov, C. Jui, K. Kawata, J. Kim, M. Kuznetsov, T. Nonaka, S. Ogio, G. Rubtsov, H. Sagawa, G. Thomson, P. Tinyakov, and S. Troitsky. Telescope Array anisotropy summary. In *Proceedings of 37th International Cosmic Ray Conference — PoS(ICRC2021)*, Vol. 395, p. 392 (2021). doi: 10.22323/1.395.0392.

41. R. Jansson and G. R. Farrar, A new model of the galactic magnetic field, *ApJ.* **757**(1), 14 (September, 2012). doi: 10.1088/0004-637X/757/1/14.

42. N. Globus and T. Piran, The extragalactic ultra-high-energy cosmic-ray dipole, *ApJ. Lett.* **850**(2), L25 (December, 2017). doi: 10.3847/2041-8213/aa991b.

43. G. R. Farrar and M. S. Sutherland, Deflections of UHECRs in the Galactic magnetic field, *JCAP.* **2019**(5), 004 (May, 2019). doi: 10.1088/1475-7516/2019/05/004.

44. M. Erdmann, L. Geiger, D. Schmidt, M. Urban, and M. Wirtz, Origins of extragalactic cosmic ray nuclei by contracting alignment patterns induced in the galactic magnetic field, *Astropart. Phys.* **108**, 74–83 (March, 2019). doi: 10.1016/j.astropartphys.2018.11.004.

45. B. Eichmann and T. Winchen, Galactic magnetic field bias on inferences from UHECR data, *JCAP.* **2020**(4), 047 (April, 2020). doi: 10.1088/1475-7516/2020/04/047.

46. C. Ding, N. Globus, and G. R. Farrar, The imprint of large-scale structure on the ultrahigh-energy cosmic-ray sky, *ApJ. Lett.* **913**(1), L13 (May, 2021). doi: 10.3847/2041-8213/abf11e.

47. R. G. Lang, A. M. Taylor, M. Ahlers, and V. de Souza, Revisiting the distance to the nearest ultrahigh energy cosmic ray source: Effects of extragalactic magnetic fields, *Phys. Rev. D.* **102**(6), 063012 (September, 2020). doi: 10.1103/PhysRevD.102.063012.

48. A. Aramburo Garcia, K. Bondarenko, A. Boyarsky, D. Nelson, A. Pillepich, and A. Sokolenko, Ultra-high energy cosmic rays deflection by the Intergalactic Magnetic Field, *arXiv e-prints.* art. arXiv:2101.07207 (January, 2021).

49. D. Wittkowski and K.-H. Kampert, On the anisotropy in the arrival directions of ultra-high-energy cosmic rays, *ApJ. Lett.* **854**(1), L3 (February, 2018). doi: 10.3847/2041-8213/aaa2f9.

50. S. Mollerach and E. Roulet, Ultrahigh energy cosmic rays from a nearby extragalactic source in the diffusive regime, *Phys. Rev. D.* **99**(10), 103010 (May, 2019). doi: 10.1103/PhysRevD.99.103010.

51. R. G. Lang, A. M. Taylor, and V. de Souza, Ultrahigh-energy cosmic rays dipole and beyond, *Phys. Rev. D.* **103**(6), 063005 (March, 2021). doi: 10.1103/PhysRevD.103.063005.

52. M. Ahlers, L. A. Anchordoqui, and A. M. Taylor, Ensemble fluctuations of the flux and nuclear composition of ultrahigh energy cosmic ray nuclei, *Phys. Rev. D.* **87**(2), 023004 (January, 2013). doi: 10.1103/PhysRevD.87.023004.

53. J. Kim, D. Ivanov, K. Kawata, H. Sagawa, and G. Thomson, Hotspot update, and a new excess of events on the sky seen by the telescope array experiment. In *Proceedings of 37th International Cosmic Ray Conference — PoS(ICRC2021)*, Vol. 395, p. 328 (2021). doi: 10.22323/1.395.0328.

54. R. Alves Batista and G. Sigl, Diffusion of cosmic rays at EeV energies in inhomogeneous extragalactic magnetic fields, *JCAP.* **2014**(11), 031 (November, 2014). doi: 10.1088/1475-7516/2014/11/031.

55. K. Fang, T. Fujii, T. Linden, and A. V. Olinto, Is the ultra-high energy cosmic-ray excess observed by the telescope array correlated with IceCube neutrinos? *ApJ.* **794**(2), 126 (October, 2014). doi: 10.1088/0004-637X/794/2/126.

56. J. A. Carpio and A. M. Gago, Impact of Galactic magnetic field modeling on searches of point sources via ultrahigh energy cosmic ray-neutrino correlations, *Phys. Rev. D.* **93**(2), 023004 (January, 2016). doi: 10.1103/PhysRevD.93.023004.

57. D. Harari, S. Mollerach, and E. Roulet, Angular distribution of cosmic rays from an individual source in a turbulent magnetic field, *Phys. Rev. D.* **93**(6), 063002 (March, 2016). doi: 10.1103/PhysRevD.93.063002.

58. K. Oh, M. Koss, C. B. Markwardt, K. Schawinski, W. H. Baumgartner, S. D. Barthelmy, S. B. Cenko, N. Gehrels, R. Mushotzky, A. Petulante, C. Ricci, A. Lien, and B. Trakhtenbrot, The 105-Month Swift-BAT all-sky hard X-ray survey, *Astrophys. J. Suppl. Ser.* **235**(1), 4 (March, 2018). doi: 10.3847/1538-4365/aaa7fd.

59. M. Ajello, W. B. Atwood, L. Baldini, J. Ballet, G. Barbiellini, *et al.*, 3FHL: The third catalog of hard Fermi-LAT sources, *Astrophys. J. Suppl. Ser.* **232**(2), 18 (October, 2017). doi: 10.3847/1538-4365/aa8221.

60. C. Lunardini, G. S. Vance, K. L. Emig, and R. A. Windhorst, Are starburst galaxies a common source of high energy neutrinos and cosmic rays? *JCAP.* **2019**(10), 073 (October, 2019). doi: 10.1088/1475-7516/2019/10/073.

61. M. F. Skrutskie, R. M. Cutri, R. Stiening, M. D. Weinberg, S. Schneider, J. M. Carpenter, C. Beichman, R. Capps, T. Chester, J. Elias, J. Huchra, J. Liebert, C. Lonsdale, D. G. Monet, S. Price, P. Seitzer, T. Jarrett, J. D. Kirkpatrick, J. E. Gizis, E. Howard, T. Evans, J. Fowler, L. Fullmer, R. Hurt, R. Light, E. L. Kopan, K. A. Marsh, H. L. McCallon, R. Tam, S. Van Dyk, and S. Wheelock, The Two Micron All Sky Survey (2MASS), *Astron. J.* **131**(2), 1163–1183 (February, 2006). doi: 10.1086/498708.

62. P. Abreu, M. Aglietta, J. M. Albury, I. Allekotte, A. Almela, *et al.* The ultra-high-energy cosmic-ray sky above 32 EeV viewed from the Pierre Auger Observatory. In *Proceedings of 37th International Cosmic Ray Conference — PoS(ICRC2021)*, Vol. 395, p. 307 (2021). doi: 10.22323/1.395.0307.

63. P. Motloch, Cross-correlating 2MASS Redshift Survey galaxies with the ultrahigh energy cosmic ray flux from Pierre Auger Observatory, *Phys. Rev. D.* **102**(4), 043014 (August, 2020). doi: 10.1103/PhysRevD.102.043014.

64. L. A. Anchordoqui, T. Bister, J. Biteau, L. Caccianiga, R. de Almeida, O. Deligny, *et al.* UHECR arrival directions in the latest data from the original Auger and TA surface detectors and nearby galaxies. In *Proceedings of 37th International Cosmic Ray Conference — PoS(ICRC2021)*, Vol. 395, p. 308 (2021). doi: 10.22323/1.395.0308.

65. Pierre Auger Collaboration, J. Abraham, P. Abreu, M. Aglietta, C. Aguirre, D. Allard, *et al.*, Correlation of the highest-energy cosmic rays with nearby extragalactic objects, *Science* **318**(5852), 938 (November, 2007). doi: 10.1126/science.1151124.

66. T. Abu-Zayyad, K. Belov, D. J. Bird, J. Boyer, Z. Cao, M. Catanese, G. F. Chen, R. W. Clay, C. E. Covault, J. W. Cronin, H. Y. Dai, B. R. Dawson, J. W. Elbert, B. E. Fick, L. F. Fortson, J. W. Fowler, K. G. Gibbs, M. A. Glasmacher, K. D. Green, Y. Ho, A. Huang, C. C. Jui, M. J. Kidd, D. B. Kieda, B. C. Knapp, S. Ko, C. G. Larsen, W. Lee, E. C. Loh, E. J. Mannel, J. Matthews, J. N. Matthews, B. J. Newport, D. F. Nitz, R. A. Ong, K. M. Simpson, J. D. Smith, D. Sinclair, P. Sokolsky, P. Sommers, C. Song, J. K. Tang, S. B. Thomas, J. C. van der Velde, L. R. Wiencke, C. R. Wilkinson, S. Yoshida, and X. Z. Zhang, Evidence for changing of cosmic ray composition between 10^{17} and 10^{18} eV from multicomponent measurements, *Phys. Rev. Lett.* **84**(19), 4276–4279 (May, 2000). doi: 10.1103/PhysRevLett.84.4276.

67. J. A. Bellido, R. W. Clay, N. N. Kalmykov, I. S. Karpikov, G. I. Rubtsov, S. V. Troitsky, and J. Ulrichs, Muon content of extensive air showers: Comparison of the energy spectra obtained by the sydney university giant air-shower recorder and by the pierre auger observatory, *Phys. Rev. D.* **98**, 023014 (July, 2018). doi: 10.1103/PhysRevD.98.023014.

68. A. Aab, P. Abreu, M. Aglietta, E. J. Ahn, I. Al Samarai, *et al.*, Muons in air showers at the pierre auger observatory: Mean number in highly inclined events, *Phys. Rev. D.* **91**, 032003 (February, 2015). doi: 10.1103/PhysRevD.91.032003.

69. A. Aab, P. Abreu, M. Aglietta, E. J. Ahn, I. Al Samarai, *et al.*, Testing hadronic interactions at ultrahigh energies with air showers measured by the Pierre Auger Observatory, *Phys. Rev. Lett.* **117**, 192001 (October, 2016). doi: 10.1103/PhysRevLett.117.192001.

70. Telescope Array Collaboration, R. U. Abbasi, M. Abe, T. Abu-Zayyad, M. Allen, R. Azuma, *et al.*, Study of muons from ultrahigh energy cosmic ray air showers measured with the Telescope Array experiment, *Phys. Rev. D.* **98**(2), 022002 (2018). doi: 10.1103/PhysRevD. 98.022002.

71. R. Kokoulin, R. P. Kokoulin, N. S. Barbashina, A. G. Bogdanov, S. S. Khokhlov, V. V. Kindin, K. G. Kompaniets, G. Mannocchi, A. A. Petrukhin, V. V. Shutenko, G. Trinchero, I. I. Yashin, E. A. Yurina, and E. A. Zadeba. Muon excess in ultra-high energy inclined EAS according to the NEVOD-DECOR data. In *Proceedings of 37th International Cosmic Ray Conference — PoS(ICRC2021)*, Vol. 395, p. 381 (2021). doi: 10.22323/1.395.0381.

72. D. Soldin, Update on the combined analysis of Muon measurements from nine air shower experiments. In *Proceedings of 37th International Cosmic Ray Conference — PoS(ICRC2021)*, Vol. 395, p. 349 (2021). doi: 10.22323/1.395.0349.

73. P. Abreu, M. Aglietta, J. M. Albury, I. Allekotte, A. Almela, *et al.* Adjustments to model predictions of depth of shower maximum and signals at ground level using hybrid events of the Pierre Auger Observatory. In *Proceedings of 37th International Cosmic Ray Conference — PoS(ICRC2021)*, Vol. 395, p. 310 (2021). doi: 10.22323/1.395.0310.

74. E. Fermi, On the origin of the cosmic radiation, *Phys. Rev.* **75**, 1169–1174 (April, 1949). doi: 10.1103/PhysRev.75.1169.

75. R. D. Blandford and J. P. Ostriker, Particle acceleration by astrophysical shocks, *ApJ. Lett.* **221**, L29–L32 (April, 1978). doi: 10.1086/182658.

76. A. R. Bell, The acceleration of cosmic rays in shock fronts — I., *MNRAS.* **182**, 147–156 (January, 1978). doi: 10.1093/mnras/182.2. 147.

77. R. Blandford and D. Eichler, Particle acceleration at astrophysical shocks: A theory of cosmic ray origin, *Phys. Rep.* **154**(1), 1–75 (October, 1987). doi: 10.1016/0370-1573(87)90134-7.

78. D. Caprioli, "Espresso" acceleration of ultra-high-energy cosmic rays, *ApJ. Lett.* **811**(2), L38 (October, 2015). doi: 10.1088/2041-8205/811/2/L38.

79. D. Caprioli, An original mechanism for the acceleration of ultra-high-energy cosmic rays, *Nucl. Part. Phys. Proc.* **297–299**, 226–233 (April, 2018). doi: 10.1016/j.nuclphysbps.2018.07.032.

80. R. Mbarek and D. Caprioli, Bottom-up acceleration of ultra-high-energy cosmic rays in the jets of active Galactic nuclei, *ApJ.* **886**(1), 8 (November, 2019). doi: 10.3847/1538-4357/ab4a08.

81. R. Mbarek and D. Caprioli, Espresso and stochastic acceleration of ultra-high-energy cosmic rays in relativistic jets, *arXiv e-prints.* art. arXiv:2105.05262 (May, 2021).

82. J. Arons, Magnetars in the metagalaxy: An origin for ultra-high-energy cosmic rays in the nearby universe, *ApJ.* **589**(2), 871–892 (June, 2003). doi: 10.1086/374776.

83. J. F. Drake, M. Swisdak, H. Che, and M. A. Shay, Electron acceleration from contracting magnetic islands during reconnection, *Nature.* **443**(7111), 553–556 (October, 2006). doi: 10.1038/nature05116.

84. R. Blandford, Y. Yuan, M. Hoshino, and L. Sironi, Magnetoluminescence, *Space Sci. Rev.* **207**(1–4), 291–317 (July, 2017). doi: 10.1007/s11214-017-0376-2.

85. A. A. Philippov and A. Spitkovsky, Ab-Initio pulsar magnetosphere: Particle acceleration in oblique rotators and high-energy emission modeling, *Astrophys. J.* **855**(2), 94 (2018). doi: 10.3847/1538-4357/aaabbc.

86. A. M. Hillas, The origin of ultra-high-energy cosmic rays, *Ann. Rev. Astron. Astrophys.* **22**, 425–444 (January, 1984). doi: 10.1146/annurev.aa.22.090184.002233.

87. Y. Jiang, B. T. Zhang, and K. Murase, Energetics of ultrahigh-energy cosmic-ray nuclei, *arXiv e-prints.* art. arXiv:2012.03122 (December, 2020).

88. R.-Y. Liu, A. M. Taylor, M. Lemoine, X.-Y. Wang, and E. Waxman, Constraints on the source of ultra-high-energy cosmic rays using anisotropy versus chemical composition, *ApJ.* **776**(2), 88 (October, 2013). doi: 10.1088/0004-637X/776/2/88.

89. H. Takami, K. Murase, and C. D. Dermer, Isotropy constraints on powerful sources of ultrahigh-energy cosmic rays at 10^{19} eV, *ApJ.* **817**(1), 59 (January, 2016). doi: 10.3847/0004-637X/817/1/59.

90. M. Vietri, The acceleration of ultra–high-energy cosmic rays in gamma-ray bursts, *ApJ.* **453**, 883 (November, 1995). doi: 10.1086/176448.

91. E. Waxman, Gamma-ray bursts: Potential sources of ultra high energy cosmic-rays, *Nucl. Phys. B Proc. Suppl.* **151**, 46–53 (January, 2006). doi: 10.1016/j.nuclphysbps.2005.07.008.

92. K. Murase, K. Ioka, S. Nagataki, and T. Nakamura, High-energy cosmic-ray nuclei from high- and low-luminosity gamma-ray bursts and implications for multimessenger astronomy, *Phys. Rev. D.* **78**(2), 023005 (July, 2008). doi: 10.1103/PhysRevD.78.023005.

93. S. Horiuchi, K. Murase, K. Ioka, and P. Mészáros, The survival of nuclei in jets associated with core-collapse supernovae and gamma-ray bursts, *ApJ.* **753**(1), 69 (July, 2012). doi: 10.1088/0004-637X/753/1/69.

94. N. Globus, D. Allard, R. Mochkovitch, and E. Parizot, UHECR acceleration at GRB internal shocks, *MNRAS.* **451**(1), 751–790 (July, 2015). doi: 10.1093/mnras/stv893.

95. J. Heinze, D. Biehl, A. Fedynitch, D. Boncioli, A. Rudolph, and W. Winter, Systematic parameter space study for the UHECR origin from GRBs in models with multiple internal shocks, *MNRAS.* **498**(4), 5990–6004 (November, 2020). doi: 10.1093/mnras/staa2751.

96. A. Rudolph, Ž. Bošnjak, A. Palladino, I. Sadeh, and W. Winter, Multi-wavelength radiation models for low-luminosity GRBs, and the implications for UHECRs, *arXiv e-prints.* art. arXiv:2107.04612 (July, 2021).

97. B. T. Zhang, K. Murase, S. S. Kimura, S. Horiuchi, and P. Mészáros, Low-luminosity gamma-ray bursts as the sources of ultrahigh-energy cosmic ray nuclei, *Phys. Rev. D.* **97**(8), 083010 (April, 2018). doi: 10.1103/PhysRevD.97.083010.

98. D. Boncioli, D. Biehl, and W. Winter, On the common origin of cosmic rays across the ankle and diffuse neutrinos at the highest energies from low-luminosity gamma-ray bursts, *ApJ.* **872**(1), 110 (February, 2019). doi: 10.3847/1538-4357/aafda7.

99. F. Samuelsson, D. Bégué, F. Ryde, and A. Pe'er, The limited contribution of low- and high-luminosity gamma-ray bursts to ultra-high-energy cosmic rays, *ApJ.* **876**(2), 93 (May, 2019). doi: 10.3847/1538-4357/ab153c.

100. F. Samuelsson, D. Bégué, F. Ryde, A. Pe'er, and K. Murase, Constraining low-luminosity gamma-ray bursts as ultra-high-energy cosmic ray sources using GRB 060218 as a proxy, *ApJ.* **902**(2), 148 (October, 2020). doi: 10.3847/1538-4357/abb60c.

101. MAGIC Collaboration, V. A. Acciari, S. Ansoldi, L. A. Antonelli, A. Arbet Engels, D. Baack, *et al.*, Teraelectronvolt emission from the γ-ray burst GRB 190114C, *Nature.* **575**(7783), 455–458 (November, 2019). doi: 10.1038/s41586-019-1750-x.

102. H. Abdalla, R. Adam, F. Aharonian, F. Ait Benkhali, E. O. Angüner, *et al.*, A very-high-energy component deep in the γ-ray burst afterglow, *Nature.* **575**(7783), 464–467 (November, 2019). doi: 10.1038/s41586-019-1743-9.

103. H. E. S. S. Collaboration, H. Abdalla, F. Aharonian, F. Ait Benkhali, E. O. Angüner, *et al.*, Revealing x-ray and gamma ray temporal and spectral similarities in the GRB 190829A afterglow, *Science.* **372**(6546), 1081–1085 (June, 2021). doi: 10.1126/science.abe8560.

104. P. Blasi, R. I. Epstein, and A. V. Olinto, Ultra-high-energy cosmic rays from young neutron star winds, *ApJ. Lett.* **533**(2), L123–L126 (April, 2000). doi: 10.1086/312626.

105. K. Murase, P. Mészáros, and B. Zhang, Probing the birth of fast rotating magnetars through high-energy neutrinos, *Phys. Rev. D.* **79**(10), 103001 (May, 2009). doi: 10.1103/PhysRevD.79.103001.

106. K. Fang, K. Kotera, and A. V. Olinto, Newly born pulsars as sources of ultrahigh energy cosmic rays, *ApJ.* **750**(2), 118 (May, 2012). doi: 10.1088/0004-637X/750/2/118.

107. A. L. Piro and J. A. Kollmeier, Ultrahigh-energy cosmic rays from the "En Caul" birth of magnetars, *ApJ.* **826**(1), 97 (July, 2016). doi: 10.3847/0004-637X/826/1/97.

108. A. A. Philippov and A. Spitkovsky, Ab-initio pulsar magnetosphere: Particle acceleration in oblique rotators and high-energy emission modeling, *ApJ.* **855**(2), 94 (March, 2018). doi: 10.3847/1538-4357/aaabbc.

109. C. Guépin, B. Cerutti, and K. Kotera, Proton acceleration in pulsar magnetospheres, *A&A.* **635**, A138 (March, 2020). doi: 10.1051/0004-6361/201936816.

110. M. Lemoine, K. Kotera, and J. Pétri, On ultra-high energy cosmic ray acceleration at the termination shock of young pulsar winds, *JCAP.* **2015**(7), 016 (July, 2015). doi: 10.1088/1475-7516/2015/07/016.

111. K. Kotera, E. Amato, and P. Blasi, The fate of ultrahigh energy nuclei in the immediate environment of young fast-rotating pulsars, *JCAP.* **2015**(8), 026 (August, 2015). doi: 10.1088/1475-7516/2015/08/026.

112. K. Fang and B. D. Metzger, High-energy neutrinos from millisecond magnetars formed from the merger of binary neutron stars, *ApJ.* **849**(2), 153 (November, 2017). doi: 10.3847/1538-4357/aa8b6a.

113. K. Kotera and J. Silk, Ultrahigh-energy cosmic rays and black hole mergers, *ApJ. Lett.* **823**(2), L29 (June, 2016). doi: 10.3847/2041-8205/823/2/L29.

114. R. J. Protheroe and A. P. Szabo, High energy cosmic rays from active galactic nuclei, *Phys. Rev. Lett.* **69**(20), 2885–2888 (November, 1992). doi: 10.1103/PhysRevLett.69.2885.

115. G. Sigl, D. N. Schramm, and P. Bhattacharjee, On the origin of highest energy cosmic rays, *Astropart. Phys.* **2**(4), 401–414 (October, 1994). doi: 10.1016/0927-6505(94)90029-9.

116. E. Waxman and J. Bahcall, High energy neutrinos from astrophysical sources: An upper bound, *Phys. Rev. D.* **59**(2), 023002 (December, 1998). doi: 10.1103/PhysRevD.59.023002.

117. A. Meli, J. K. Becker, and J. J. Quenby, On the origin of ultra high energy cosmic rays: Subluminal and superluminal relativistic shocks, *A&A.* **492**(2), 323–336 (December, 2008). doi: 10.1051/0004-6361: 20078681.

118. G. R. Farrar and A. Gruzinov, Giant AGN flares and cosmic ray bursts, *ApJ.* **693**(1), 329–332 (March, 2009). doi: 10.1088/0004-637X/ 693/1/329.

119. C. H. Coimbra-Araújo and R. C. Anjos, Luminosity of ultrahigh energy cosmic rays and bounds on magnetic luminosity of radio-loud active galactic nuclei, *Phys. Rev. D.* **92**(10), 103001 (November, 2015). doi: 10.1103/PhysRevD.92.103001.

120. T. K. Fowler, H. Li, and R. Anantua, A Quasi-static hyper-resistive model of ultra-high-energy cosmic-ray acceleration by magnetically collimated jets created by active galactic nuclei, *ApJ.* **885**(1), 4 (November, 2019). doi: 10.3847/1538-4357/ab44bc.

121. X. Rodrigues, J. Heinze, A. Palladino, A. van Vliet, and W. Winter, Active Galactic nuclei jets as the origin of ultrahigh-energy cosmic rays and perspectives for the detection of astrophysical source neutrinos at EeV energies, *Phys. Rev. Lett.* **126**(19), 191101 (May, 2021). doi: 10.1103/PhysRevLett.126.191101.

122. K. Murase, C. D. Dermer, H. Takami, and G. Migliori, Blazars as ultra-high-energy cosmic-ray sources: Implications for TeV gamma-ray observations, *ApJ.* **749**(1), 63 (April, 2012). doi: 10.1088/0004-637X/ 749/1/63.

123. Y. Inoue and Y. T. Tanaka, Baryon loading efficiency and particle acceleration efficiency of relativistic jets: Cases for low luminosity BL Lacs, *ApJ.* **828**(1), 13 (September, 2016). doi: 10.3847/0004-637X/ 828/1/13.

124. C. D. Dermer, S. Razzaque, J. D. Finke, and A. Atoyan, Ultra-high-energy cosmic rays from black hole jets of radio galaxies, *New J. Phys.* **11**(6), 065016 (June, 2009). doi: 10.1088/1367-2630/11/6/065016.

125. N. Fraija and A. Marinelli, Neutrino, γ-ray, and cosmic-ray fluxes from the core of the closest radio galaxies, *ApJ.* **830**(2), 81 (October, 2016). doi: 10.3847/0004-637X/830/2/81.

126. J. H. Matthews, A. R. Bell, K. M. Blundell, and A. T. Araudo, Fornax A, Centaurus A, and other radio galaxies as sources of ultrahigh energy cosmic rays, *MNRAS.* **479**(1), L76–L80 (September, 2018). doi: 10. 1093/mnrasl/sly099.

127. B. Eichmann, J. P. Rachen, L. Merten, A. van Vliet, and J. Becker Tjus, Ultra-high-energy cosmic rays from radio galaxies, *JCAP.* **2018**(2), 036 (February, 2018). doi: 10.1088/1475-7516/2018/02/036.

128. N. Fraija, E. Aguilar-Ruiz, A. Galván-Gámez, A. Marinelli, and J. A. de Diego, Study of the PeV neutrino, γ-rays, and UHECRs around the lobes of Centaurus A, *MNRAS.* **481**(4), 4461–4471 (Dec., 2018). doi: 10.1093/mnras/sty2561.

129. J. H. Matthews and A. M. Taylor, Particle acceleration in radio galaxies with flickering jets: GeV electrons to ultrahigh energy cosmic rays, *MNRAS.* **503**(4), 5948–5964 (June, 2021). doi: 10.1093/mnras/stab758.

130. A. Pe'Er, K. Murase, and P. Mészáros, Radio-quiet active galactic nuclei as possible sources of ultrahigh-energy cosmic rays, *Phys. Rev. D.* **80**(12), 123018 (December, 2009). doi: 10.1103/PhysRevD.80.123018.

131. A. Y. Neronov, D. V. Semikoz, and I. I. Tkachev, Ultra-high energy cosmic ray production in the polar cap regions of black hole magnetospheres, *New J. Phys.* **11**(6), 065015 (June, 2009). doi: 10.1088/1367-2630/11/6/065015.

132. M. Y. Piotrovich, Y. N. Gnedin, S. D. Buliga, and T. M. Natsvlishvili, Ultra high energy cosmic ray generation in black hole magnetosphere: Testing by polarimetric observations, *Astrophys. Space Sci.* **353**(2), 625–631 (October, 2014). doi: 10.1007/s10509-014-2043-3.

133. A. Tursunov, Z. Stuchlík, M. Kološ, N. Dadhich, and B. Ahmedov, Supermassive black holes as possible sources of ultrahigh-energy cosmic rays, *ApJ.* **895**(1), 14 (May, 2020). doi: 10.3847/1538-4357/ab8ae9.

134. I. Duţan and L. I. Caramete, Ultra-high-energy cosmic rays from low-luminosity active galactic nuclei, *Astropart. Phys.* **62**, 206–216 (March, 2015). doi: 10.1016/j.astropartphys.2014.09.007.

135. H. E. S. S. Collaboration, H. Abdalla, R. Adam, F. Aharonian, F. Ait Benkhali, E. O. Angüner, *et al.*, Resolving acceleration to very high energies along the jet of Centaurus A, *Nature.* **582**(7812), 356–359 (June, 2020). doi: 10.1038/s41586-020-2354-1.

136. P. Banik, A. Bhadra, and A. Bhattacharyya, Interpreting correlated observations of cosmic rays and gamma-rays from Centaurus A with a proton blazar inspired model, *MNRAS.* **500**(1), 1087–1094 (January, 2021). doi: 10.1093/mnras/staa3343.

137. S. S. Kimura, K. Murase, and B. T. Zhang, Ultrahigh-energy cosmic-ray nuclei from black hole jets: Recycling galactic cosmic rays through shear acceleration, *Phys. Rev. D.* **97**(2), 023026 (January, 2018). doi: 10.1103/PhysRevD.97.023026.

138. B. T. Zhang, K. Murase, F. Oikonomou, and Z. Li, High-energy cosmic ray nuclei from tidal disruption events: Origin, survival, and implications, *Phys. Rev. D.* **96**(6), 063007 (September, 2017). doi: 10.1103/PhysRevD.96.063007.

139. R. Alves Batista and J. Silk, Ultrahigh-energy cosmic rays from tidally-ignited white dwarfs, *Phys. Rev. D.* **96**(10), 103003 (November, 2017). doi: 10.1103/PhysRevD.96.103003.

140. C. Guépin, K. Kotera, E. Barausse, K. Fang, and K. Murase, Ultra-high-energy cosmic rays and neutrinos from tidal disruptions by massive black holes, *A&A.* **616**, A179 (September, 2018). doi: 10.1051/0004-6361/201732392.

141. G. E. Romero, A. L. Müller, and M. Roth, Particle acceleration in the superwinds of starburst galaxies, *A&A.* **616**, A57 (August, 2018). doi: 10.1051/0004-6361/201832666.

142. L. A. Anchordoqui, Acceleration of ultrahigh-energy cosmic rays in starburst superwinds, *Phys. Rev. D.* **97**(6), 063010 (March, 2018). doi: 10.1103/PhysRevD.97.063010.

143. C. A. Norman, D. B. Melrose, and A. Achterberg, The origin of cosmic rays above 10 18.5 eV, *ApJ.* **454**, 60 (November, 1995). doi: 10.1086/176465.

144. H. Kang, D. Ryu, and T. W. Jones, Cluster accretion shocks as possible acceleration sites for ultra–high-energy protons below the Greisen cutoff, *ApJ.* **456**, 422 (January, 1996). doi: 10.1086/176666.

145. D. Ryu, H. Kang, E. Hallman, and T. W. Jones, Cosmological shock waves and their role in the large-scale structure of the universe, *ApJ.* **593**(2), 599–610 (August, 2003). doi: 10.1086/376723.

146. S. Das, H. Kang, D. Ryu, and J. Cho, Propagation of ultra-high-energy protons through the magnetized cosmic web, *ApJ.* **682**(1), 29–38 (July, 2008). doi: 10.1086/588278.

147. J. Kim, D. Ryu, H. Kang, S. Kim, and S.-C. Rey, Filaments of galaxies as a clue to the origin of ultrahigh-energy cosmic rays, *Sci. Adv.* **5**(1), eaau8227 (January, 2019). doi: 10.1126/sciadv.aau8227.

148. K. Greisen, End to the cosmic-ray spectrum? *Phys. Rev. Lett.* **16**(17), 748–750 (April, 1966). doi: 10.1103/PhysRevLett.16.748.

149. G. T. Zatsepin and V. A. Kuz'min, Upper limit of the spectrum of cosmic rays, *Soviet J. Exp. Theor. Phys. Lett.* **4**, 78 (August, 1966).

150. R. Alves Batista, R. M. de Almeida, B. Lago, and K. Kotera, Cosmogenic photon and neutrino fluxes in the Auger era, *JCAP.* **2019**(1), 002 (January, 2019). doi: 10.1088/1475-7516/2019/01/002.

151. K. Kotera, D. Allard, and A. V. Olinto, Cosmogenic neutrinos: Parameter space and detectabilty from PeV to ZeV, *JCAP.* **2010**(10), 013 (October, 2010). doi: 10.1088/1475-7516/2010/10/013.

152. D. Wittkowski and K.-H. Kampert, On the flux of high-energy cosmogenic neutrinos and the influence of the extragalactic magnetic field, *MNRAS.* **488**(1), L119–L122 (September, 2019). doi: 10.1093/mnrasl/slz083.

153. J. R. Hörandel, On the knee in the energy spectrum of cosmic rays, *Astropar. Phys.* **19**(2), 193–220 (May, 2003). doi: 10.1016/ S0927-6505(02)00198-6.

154. H. E. S. S. Collaboration, H. Abdalla, A. Abramowski, F. Aharonian, F. Ait Benkhali, E. O. Angüner, *et al.*, Population study of Galactic supernova remnants at very high γ-ray energies with H.E.S.S., *A&A.* **612**, A3 (April, 2018). doi: 10.1051/0004-6361/201732125.

155. H. Fleischhack. A survey of TeV emission from Galactic supernova remnants with HAWC. In *36th International Cosmic Ray Conference (ICRC2019)*, Vol. 36, *International Cosmic Ray Conference*, p. 674 (July, 2019).

156. A. M. Bykov, Nonthermal particles and photons in starburst regions and superbubbles, *A&A Rev.* **22**, 77 (November, 2014). doi: 10.1007/ s00159-014-0077-8.

157. F. Aharonian, R. Yang, and E. de Oña Wilhelmi, Massive stars as major factories of Galactic cosmic rays, *Nat. Astron.* **3**, 561–567 (March, 2019). doi: 10.1038/s41550-019-0724-0.

158. A. U. Abeysekara, A. Albert, R. Alfaro, C. Alvarez, J. R. A. Camacho, *et al.*, HAWC observations of the acceleration of very-high-energy cosmic rays in the Cygnus Cocoon, *Nature Astronomy* **5**, 465–471 (March, 2021). doi: 10.1038/s41550-021-01318-y.

159. A. M. Hillas, The cosmic-ray knee and ensuing spectrum seen as a consequence of Bell's self-magnetized SNR shock acceleration process, *J. Phys. Conf. Ser.* **47**, 168–177 (2006). doi: 10.1088/1742-6596/47/ 1/021.

160. S. Gabici, C. Evoli, D. Gaggero, P. Lipari, P. Mertsch, E. Orlando, A. Strong, and A. Vittino, The origin of Galactic cosmic rays: Challenges to the standard paradigm, *Int. J. Mod. Phys. D.* **28**(15), 1930022-339 (January, 2019). doi: 10.1142/S0218271819300222.

161. V. Ptuskin, V. Zirakashvili, and E.-S. Seo, Spectrum of Galactic cosmic rays accelerated in supernova remnants, *ApJ.* **718**(1), 31–36 (July, 2010). doi: 10.1088/0004-637X/718/1/31.

162. E. Parizot, Cosmic ray origin: Lessons from ultra-high-energy cosmic rays and the Galactic/Extragalactic transition, *Nucl. Phys. B Proc. Suppl.* **256**, 197–212 (November, 2014). doi: 10.1016/j.nuclphysbps. 2014.10.023.

163. R. E. Lingenfelter, Cosmic rays from supernova remnants and superbubbles, *Adv. Space Res.* **62**(10), 2750–2763 (November, 2018). doi: 10.1016/j.asr.2017.04.006.

164. V. Tatischeff and S. Gabici, Particle acceleration by supernova shocks and spallogenic nucleosynthesis of light elements, *Ann. Rev. Nucl. Part. Sci.* **68**(1), 377–404 (October, 2018). doi: 10.1146/ annurev-nucl-101917-021151.

165. K. Fang, K. Kotera, and A. V. Olinto, Ultrahigh energy cosmic ray nuclei from extragalactic pulsars and the effect of their Galactic counterparts, *JCAP*. **2013**(3), 010 (March, 2013). doi: 10.1088/1475-7516/2013/03/010.

166. M. Unger, G. R. Farrar, and L. A. Anchordoqui, Origin of the ankle in the ultrahigh energy cosmic ray spectrum, and of the extragalactic protons below it, *Phys. Rev. D*. **92**(12), 123001 (December, 2015). doi: 10.1103/PhysRevD.92.123001.

167. A. V. Olinto, J. Krizmanic, and P. Collaboration. The roadmap to the POEMMA mission. In *Proceedings of 37th International Cosmic Ray Conference — PoS(ICRC2021)*, Vol. 395, p. 976 (2021). doi: 10.22323/1.395.0976.

168. J. R. Hörandel. GCOS — the global cosmic ray observatory. In *Proceedings of 37th International Cosmic Ray Conference — PoS(ICRC2021)*, Vol. 395, p. 310 (2021). doi: 10.22323/1.395.0310. http://particle.astro.ru.nl/ps/gcos-icrc21.pdf.

https://doi.org/10.1142/9781800612617_0002

Chapter 2

Ground Arrays: Pierre Auger Observatory and Telescope Array

Douglas R. Bergman

Department of Physics & Astronomy and High Energy Astrophysics Institution, The University of Utah, 115 S. 1400 E. Rm 201, Salt Lake City, UT 84112, USA
bergman@physics.utah.edu

Ground arrays detect cosmic-ray–induced extensive air showers over areas of hundreds to thousands of square kilometers. This chapter covers particle detectors on the ground and optical detectors observing signals generated in the atmosphere near the ground. There are two, currently active, large area detectors: the Pierre Auger Observatory in the Southern Hemisphere and the Telescope Array in the Northern Hemisphere.

1. Extensive Air Shower Phenomenology

Cosmic rays with energies above about $1\,\mathrm{PeV}$ ($10^{15}\,\mathrm{eV}$) have a flux which is too low to be measured directly by detectors placed above the Earth's atmosphere.[1] However, these high-energy cosmic rays create a cascade of particles, an extensive air shower (EAS), when they encounter the atmosphere. At the maximum development of an air shower, there will be approximately one shower particle for each GeV of the primary cosmic-ray energy.[2] This results in air showers with millions and billions of particles. The shower may be directly observed on the ground by particle detectors observing signals in coincidence over a large area. The shower may also be observed by collecting the light produced as shower particles excite or disturb air molecules as they pass through the air, fluorescence radiation from excitations of molecular nitrogen or Cherenkov radiation for shower particles traveling faster than the speed of light in the medium. Cosmic rays with energies of tens of EeV and higher produce shower footprints on the ground which extend for 10 km or more in diameter and produce fluorescence radiation which may be observed up to 50 km or more away.

Since these ultra-high-energy cosmic rays (UHECRs, those with $E > 10^{18}$ eV) cannot be observed directly, the properties of the primary cosmic ray must be inferred from measurements and thus the phenomenology of the EAS. The initial interaction of the cosmic ray with an atmospheric molecule happens at a very high center-of-mass energy. The modeling of these interactions,[3,4] the multiplicity distribution, and the inelasticity of the interaction cannot be tested directly against measurements at colliders, leading to significant modeling uncertainty in the development of the early stages of air showers. This modeling uncertainty manifests as uncertainty in the expected lateral distribution of showers and the expected number of muons produced in the shower. There are however phenomenological methods to constrain these uncertainties.

After the first few interactions, shower particles are typically of an energy where their interactions are well understood and the great multiplicity of particles allows for a reliable statistical treatment of the shower. All showers will thus share some universal properties,[5–7] the most important being that the total number of particles in the shower will be proportional to the primary cosmic-ray energy. This implies that the measurement of the total amount of fluorescence light produced by a shower provides a calorimetric measurement of the primary cosmic-ray energy. Only a small correction needs to be made to account for muons and neutrinos (subdominant components of the shower in all cases) which do not contribute fully to producing fluorescence light.

The longitudinal development of showers is often characterized phenomenologically as a function of the amount of atmosphere passed through by the Gaisser–Hillas equation:[2]

$$N(X) = N_{\max} \left(\frac{X - X_0}{X_{\max} - X_0} \right)^{\frac{X_{\max} - X_0}{\Lambda}} \exp\left(\frac{X_{\max} - X}{\Lambda} \right), \quad (1)$$

where the number of shower particles N grows as a function of X (depth, measured in g cm^{-2}). The maximum size of the shower, $N_{\max}$, is reached at a depth $X_{\max}$. There is a "beginning" depth of X_0 (which can be unphysically negative) and an attenuation length for the shower Λ. The shower grows (as a power law) as high-energy particles collide with air molecules producing many more lower-energy particles, then shrinks (exponentially) as lower-energy particles are absorbed in the atmosphere. An example of a Gaisser–Hillas shower profile is shown in Fig. 1(left). Typical values of $X_{\max}$ range from 600 to 1100 g cm^{-2} for UHECRs, compared to a vertical atmospheric depth at sea level of just over 1000 g cm^{-2}. There is considerable variation in $X_{\max}$ between showers of the same species and energy due to fluctuations in the first interaction depth and the multiplicity and inelasticity of the first interaction. Thus, some cosmic rays of a given energy will be attenuated significantly (of order e^{-1}), while others may not even reach their maximum. This makes measuring the total particle count at the ground a poor estimator of the primary cosmic-ray energy.

Showers spread laterally as they develop due to the transverse momentum involved in the collisions.[8] The size of the spread is gauged by the Moliere radius, r_M, which is about 80 m for air at sea level and about 90 m for air a few interaction lengths higher in the atmosphere. The spread is also characterized by a shower age parameter, $s = X/(X - 2X_{\max})$; $s = 0$ at the beginning of a shower, $s = 1$ at the maximum development of a shower, and typically the

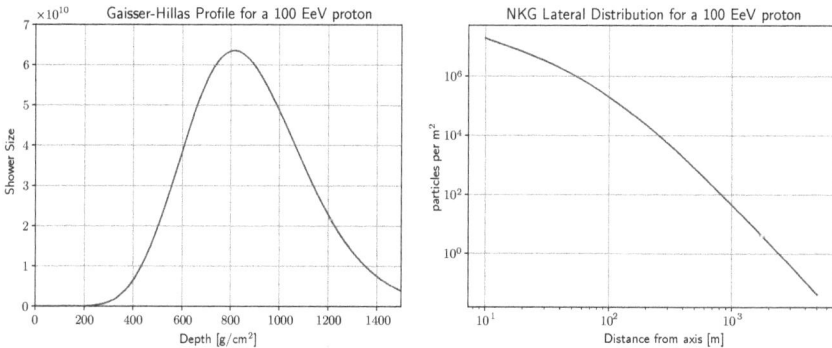

Fig. 1. Left: An example Gaisser–Hillas shower profile for a 100 EeV proton shower with $N_{\max} = 63 \times 10^9$, $X_{\max} = 815.5$ g cm^{-2}, $X_0 = -65.9$ g cm^{-2}, and $\Lambda = 64.0$ g cm^{-2}. Right: An example NKG lateral profile for the same shower taken at age, $s = 1$, and using $r_M = 90$ m.

shower will have been reduced to less than one particle by $s = 2$. The spread of a shower is phenomenologically represented by the NKG function[9,10]

$$\rho(r) = \frac{1}{2\pi r_M^2} \left(\frac{r}{r_M}\right)^{s-2} \left(1 + \frac{r}{r_M}\right)^{s-4.5} \frac{\Gamma(4.5 - s)}{\Gamma(s)\Gamma(4.5 - 2s)}, \qquad (2)$$

so there is a central distribution which has a $\frac{1}{r} \times r^{s-1}$ dependence out to about r_M and then a much steeper drop-off. An example shower lateral distribution is shown in Fig. 1(b). As s increases, the shower becomes wider. The effect of shower attenuation for observations of showers with $s > 1$ can thus be offset by the fact that there will be more particles pushed farther away from the shower axis. Indeed, there will be an angle-dependent distance from the shower axis where attenuation due to fluctuations in X_{max} are almost entirely offset by the spread of the shower. In practical measurements, the lateral distribution is measured with considerable uncertainties so that, in fact, the distance at which to measure the shower particle flux depends just on the arrangement and spacing of shower counters.[11] This distance tends to be slightly smaller than the inter-counter spacing, so 800–1000 m.

Given the uncertainty of modeling the first interaction in the development of a shower, including uncertainty in the transverse momentum distributions, it is very difficult to predict from first principles the expected particle flux at 800–1000 m from the shower core. Measurements of the flux at this point is typically in the tail of the lateral distribution and may well also be in the tail of the longitudinal distribution. However, if one can compare the flux of particles on the ground, 800–1000 m from the shower core, to a calorimetric measurement of the shower energy using fluorescence light, one can form a calibration of the surface detector signal. This calibration will still have statistical uncertainties due to shower-to-shower fluctuations in X_{max} and the lateral spread of the shower, but the systematic uncertainty in the energy scale will be much smaller than what one would have from a purely model-driven calculation. The minimum energy uncertainty is typically of the order of 10%.

The depth of shower maximum, X_{max}, grows logarithmically with the energy of the primary cosmic ray.[12] This is due to the fact that higher energies give rise to more generations of interactions, and thus the shower goes deeper into the atmosphere before being absorbed.

This fact can be turned into a composition measurement if one notes that cosmic rays consisting of multiple nucleons can be treated as giving rise to showers which are a superposition of the showers from each of the nucleons independently. Each of these nucleons has only a fraction of the primary cosmic-ray energy, and so the showers go less deep in the atmosphere. Also, since there is a superposition of several showers, the shower-to-shower fluctuations in X_{max} are statistically averaged out, reducing the spread in expected X_{max} values for the whole shower. The actual distribution of X_{max} values expected at a given energy depends crucially on the high-energy interaction model used to simulate the early stages of a shower. However, the spread of X_{max} values at a given energy is much less model dependent. A measurement of the shower longitudinal profile using fluorescence detection is the surest way to estimate the primary cosmic-ray species. There are, however, many surface detector measurements which can indicate the distance of the detector from X_{max}.

Finally, the direction of the shower axis gives some limited information on the origin of a particular cosmic ray. Since cosmic rays are charged particles, their paths through the universe and within our galaxy are bent by magnetic fields. They do not necessarily point back to their origin. However, at very high energies, above 30 EeV or so, this deflection may be small, of the order of a few degrees. The arrival direction of an air shower is inferred from the pattern of timings of the surface counters in an array. The detailed shape of the shower front is useful to reduce the uncertainty in this measurement. For an event observed by fluorescence telescopes (FTs) in conjunction with a surface array, the pattern of hit pixels in the telescope very accurately determines a shower-detector plane. The angle of the shower within the shower-detector plane is well determined by the measured times of the pixels combined with one or more times from hit surface detector stations.

An example of a surface array event measured by the Telescope Array is shown in Fig. 2.[13] The surface detector station signals and timings are shown on the left of Fig. 2. These data, both signal size and timing, were fit to find the location of shower axis hitting the ground, the direction of the shower axis, and the lateral distribution of the shower. The timing of the signals, showing the effect of shower front curvature, is shown on the top right of Fig. 2, while the lateral

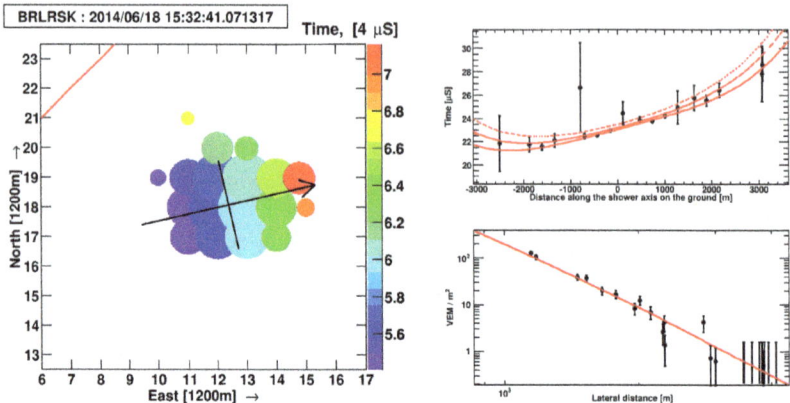

Fig. 2. Example event from the Telescope Array. The estimated energy is 176 EeV. Left: Surface plot of active surface detector stations. The size of the circles is logarithmically proportional to the total signal, while the color represents the time of the signal. The arrow represents the direction of the shower as projected on the ground and the cross point represents the location of the central axis of the shower. Right: *Top:* The time profile of the hit surface detector stations along the ground axis. The solid line is the fit expectation for counters directly on the ground axis, while the dashed and dotted lines represent the fit timing expectation for counters 1.5 and 2 km, respectively, away from the ground axis. *Bottom:* The lateral profile of signals. The fit is for the AGASA-modified NKG function. The signal inferred from the fit at 800 m from the core is the energy indicator.

distribution is shown on the bottom right. The inferred signal at 800 m is used as an energy indicator.

2. The Pierre Auger Observatory

The Pierre Auger Observatory is the largest cosmic-ray observatory in the world, with an area of 3000 km^2.[14] It is a hybrid detector, having both particle detectors on the ground and FTs looking out over the atmosphere above. The configuration and layout of the Auger detectors is shown in Fig. 3.

2.1. *Water-Cherenkov surface detectors*

The primary surface array of the Pierre Auger Observatory consists of 1660 water-Cherenkov particle detectors, generically referred to

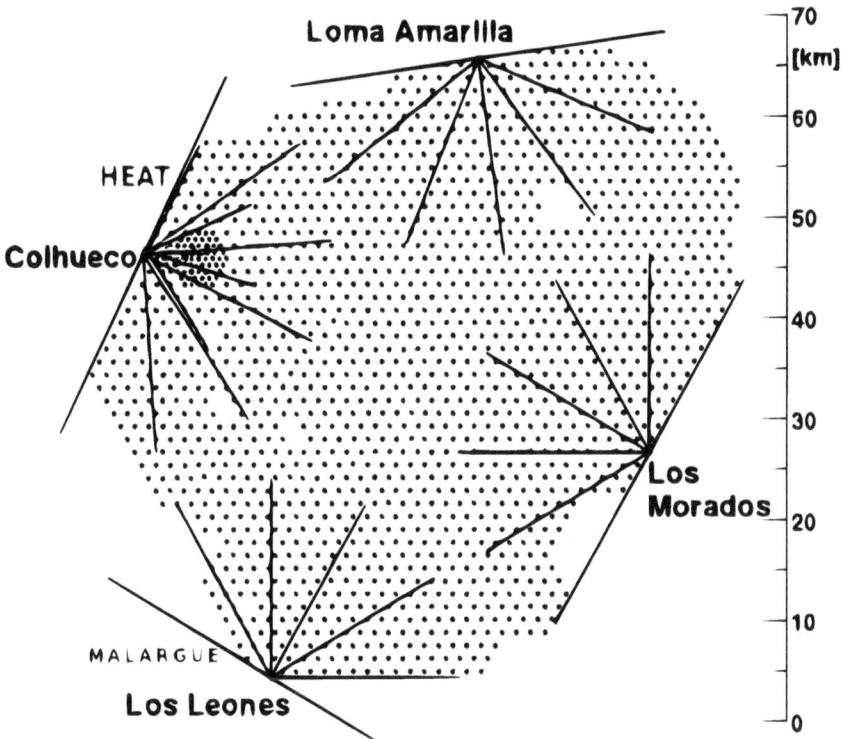

Fig. 3.   A schematic of the Pierre Auger Observatory where each black dot is a water-Cherenkov detector. Locations of the fluorescence telescopes are shown along the perimeter of the surface detector array, where the lines indicate the individual telescope fields of view. The fields of view of the HEAT telescopes are also indicated.[15] Reproduced with permission from Elsevier.

as surface detectors (SDs). The SDs are placed on a triangular grid with an inter-counter spacing of 1500 m. Each detector consists of an opaque tank holding 12 m^3 of ultra-pure water which serves as the medium for generating the Cherenkov radiation used for particle detection. The cylindrical tanks are 3.6 m in diameter and the water is 1.2 m deep. A photograph of one Auger water tank with labels is shown in Fig. 4.

The water tanks are instrumented with three, nine-inch, hemispherical photomultiplier tubes (PMTs), which are set into the top surface of the water in the tanks. The three PMTs are spaced to maximally view Cherenkov light from high-energy particles going through the water. High-energy electrons coming from above will create small

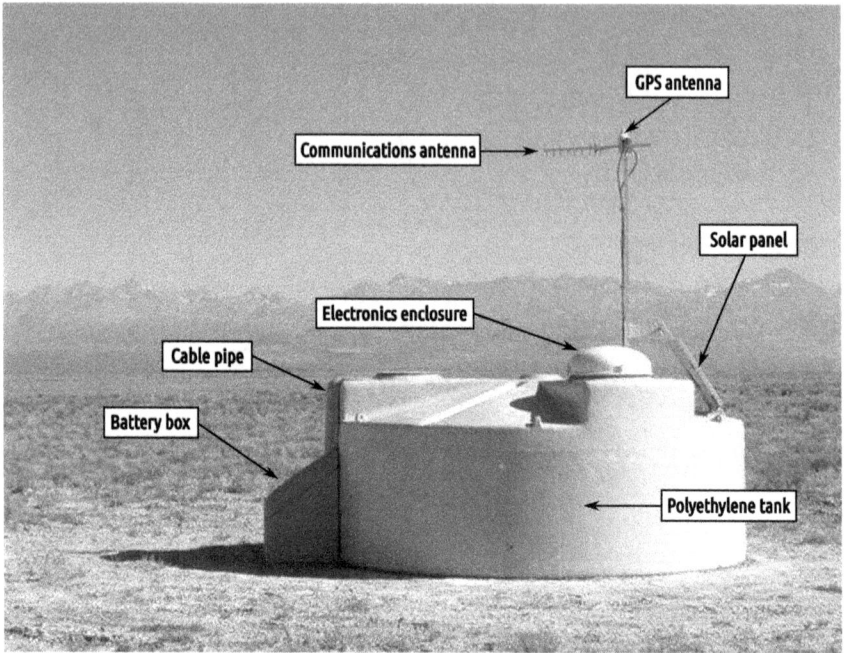

Fig. 4.   A schematic view of an Auger water-Cherenkov detector showing its various components.[14] Reproduced with permission from Elsevier.

showers in the water, and all the light from the shower will be created in the top several cm of the tank. A high-energy muon coming from above will traverse the tank and create Cherenkov radiation throughout the depth of the water. This means that the signal from a muon of a given energy will be much larger than the signal from an electron of the same energy. The Auger SDs are excellent muon detectors. In fact, any triggered signal seen in the counter is compared to that of what is expected from a muon going through the counter vertically, a vertical equivalent muon (VEM).

The 1.2 m height gives the SDs considerable cross section from the side. This implies that the SD array is sensitive to showers arriving at very large zenith angles ($\theta > 80°$).

The Auger SD station has two local trigger modes for physics analyses.[16] The first just requires that each of the three PMTs sees a 1.75-VEM signal. The threshold is set so that individual atmospheric muons (muons from low-energy cosmic-ray showers) do not cause a high rate of triggers. The other trigger is a time-over-threshold (ToT) trigger which uses the flash ADC output from the PMTs (running at 40 MHz). A 0.2-VEM signal is required in 13 bins within a 120-bin (3 μs) window. The first trigger is effective for large signals coming in a short amount of time, such as highly inclined, "old" showers consisting primarily of muons coming in a very thin shower front. The second trigger is effective for vertical, "young" showers which still have a lot of electrons arriving in a thick shell, especially for detectors far from the shower core.

The rate of local triggers has to be kept to 20–25 Hz because each SD communicates with the central data acquisition system (CDAS) by a wireless network with a limited data bandwidth. The CDAS collects the trigger signals from each of the SDs and forms a higher-level trigger. When a high-level trigger is formed, the FADC traces from each of the participating SDs must be sent to the CDAS, and this takes about half the available bandwidth. The main trigger is formed when three ToT signals from three adjacent SDs are coincident within $\pm 25\,\mu$s or when four high-threshold triggers within a hexagonal ring of four counter spacings about one of the SDs.

2.2. *Fluorescence telescopes*

The Pierre Auger Observatory has four sets of FTs placed on the periphery of the surface array and viewing the atmosphere over the surface array.[14] Each of the four sites has six telescopes housed in a climate-controlled building. The telescope design is that of an f/1 Schmidt camera with the curved focal plane instrumented with 440 PMTs arranged in a 22 × 20 array. The segmented primary mirror of the telescope has a radius of 1.7 m. Each of the PMT pixels views a 1.5° diameter spot on the sky. The simplified corrector and stop of the Schmidt camera also forms the window of the building. A schematic diagram of one Auger FT is shown in Fig. 5. Each telescope has an aperture of 30° × 30°, and the apertures are arranged so that the telescopes look from 1.5° 31.5° above the horizon and 180° in azimuth. The entrance window also contains a UV filter which limits the incoming radiation to the range 310–390 nm.

The PMT pixels of the telescopes are eight-stage bialkaline-photocathode PMTs with a peak quantum efficiency of about 25%. The AC-connected PMTs are run at a voltage to give a gain of 5×10^4. The signal from each PMT is sampled by a 10 MHz FADC, giving a

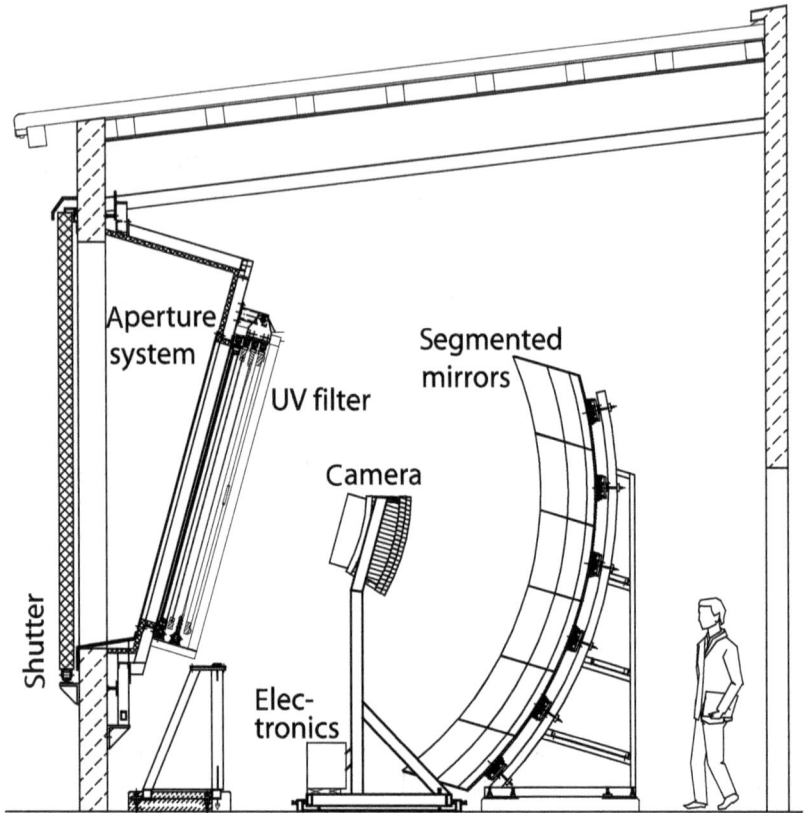

Fig. 5. A schematic view of an Auger fluorescence telescope showing its various components.[15] Reproduced with permission from Elsevier.

measurement every 100 ns. The light collection gaps between PMTs are filled with a simplified Winston cone:[17] a triangular "Mercedes" in the node between three PMTs. The point-spread function for the telescope is about 0.5° in diameter,[18] so the spot size is considerably smaller than the pixel size.

The trigger for the FTs works in three stages. Each pixel has a threshold discriminator which is adjusted dynamically to keep the rate of single-PMT triggers at 100 Hz. The second level of trigger searches for an arrangement of five adjacent PMTs in a "track" in time coincidence. The rate of triggers for each telescope in this mode varies from 0.1 to 10 Hz. A third-level trigger rejects events that are not track-like, caused by lightning and random triggers.

2.3.   *Hybrid energy measurements*

All current UHECR observatories use a hybrid technique, simultaneous fluorescence and surface detection of events, to establish the energy scale of the SD energy estimator. Auger was the first observatory to do this analysis as their primary means of calibrating the SD, though the hybrid technique had been pioneered by the HiRes Prototype/MIA collaboration.[19]

The method Auger uses to measure the geometry of the shower in hybrid mode is to take the SD with the largest signal[20,21] and combine it with the timing fit of the FT data. The addition of the one SD detector dramatically breaks the degeneracy between R_P and ψ in the monocular timing fit.

Once an accurate geometry for a hybrid event has been determined, the shower profile is reconstructed using the measured properties of the atmosphere, the known photosensitivity of the FT pixels, and the measured fluorescence yield of air. Auger uses the AirFLY[22] fluorescence yield measurement. The transparency of the atmosphere is measured by shooting UV lasers into the air and measuring the scattered light in the FTs as a function of position, angle, and the amount of atmosphere traversed by the laser light before and after scattering.

The shower profile determined by Auger is measured in dE/dX, the amount of energy deposited in the atmosphere per unit slant depth (usually measured in $g\,cm^{-2}$). The profile is fit to a functional form such as the Gaisser–Hillas (Eq. (1)), and this is then integrated to find the total energy deposited by the shower in the atmosphere. This is the calorimetric energy of the shower. A small correction needs to be made to account for the energy of the shower that is not deposited in the atmosphere, coming from muons and neutrinos. This correction becomes smaller as the energy of the shower increases. For UHECR air showers, the correction is less than 10% but does depend on the composition, with heavier nuclei having larger amounts of missing energy.

Finally, Auger compares the SD parameter $S(1000)$, the size of the signal at 1000 m from the shower core, to the energy from the FD, E_{FD}, on a shower-by-shower basis.[23] This comparison depends on the angle of the shower since there is an attenuation factor to account for as showers become more inclined and go through more air.

Auger uses the constant-intensity-cut method to compare $S(1000)$ between showers at different zenith angles. The assumption is that the flux of UHECRs is isotropic; therefore, the flux of cosmic rays of a given energy is the same as a function of the zenith angle. However, since showers at larger zenith angles are attenuated, it may appear that the flux as a function of $S(1000)$ is smaller at large zenith angles. One can then correct by adjusting $S(1000)$ to get the same flux for all zenith angles. Auger corrects to the mean zenith angle of their main dataset ($\theta < 60°$), which is $38°$, resulting in the energy estimator S_{38}. S_{38} versus E_{FD} is fit to a simple scaling relation to give the energy for all SD events. A similar procedure is used for highly inclined events.

2.4. *Auger spectrum and composition measurement*

One of the highlight published results of the Pierre Auger Observatory is a measurement of the UHECR spectrum and a simultaneous fit to an astrophysical model of the composition of nuclei making up that spectrum.[24] This result is shown in Fig. 6. A new feature, known now as the "instep" to differentiate it from the other anatomically named features such as the "ankle" and the "knee", is observed at about 13 EeV. To understand the possible sources of UHECRs, and in particular the instep, a simultaneous fit to the spectrum and X_{max} measurements from fluorescence has been performed. The astrophysical model has uniformly distributed sources over cosmological distances, with each source accelerating a mixture of five representative nuclear species with a single power-law spectrum and a rigidity-dependent cutoff. The fit shows that the Auger spectrum can be understood as cosmic rays being produced with a very hard spectrum up to a rigidity-dependent cutoff which is of the same order of the energy range measured. In this model, the instep feature may be the result of the threshold for photodissociation of helium in propagation.

The combined fit analysis is limited by the statistical power of the FTs for measuring the cosmic-ray composition. Being able to robustly determine both the electromagnetic component and the muon component in the SDs would allow for a composition measurement with statistics of the same order as the spectrum measurement. This understanding is part of the motivation for the AugerPrime update.

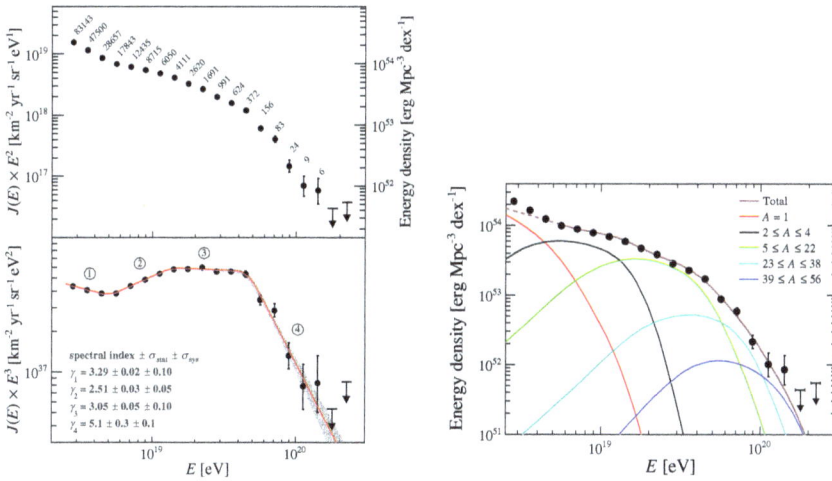

Fig. 6. The Pierre Auger Observatory measurement of a new feature in the cosmic-ray spectrum, and an interpretation of the sources.[24] Left: The energy spectrum of cosmic rays observed by the Pierre Auger Observatory.[24] The top panel shows the measured flux in bins of energy and indicates the number of events in the bin. The bottom panel shows a fit of a broken power law, with three break points, to the data. The feature identified at 13 EeV is a newly discovered feature, now known as the "instep". Right: The result of a simultaneous fit to the spectrum and to the observed distribution of X_{max} to an astrophysical model consisting of sources producing cosmic rays with a single power-law dependence up to a rigidity-dependent cutoff with five representative nuclear species of varying fractions. Reproduced with permission from Creative Commons license.

2.5. *AugerPrime updates*

The Auger observatory has been in the process of enhancing the detector array, as described above, since 2015. The primary aspects of the upgrade include:[25,26]

- surface scintillator detectors (SSDs) installed on top of each existing water-Cherenkov detector (WCD);
- the addition of a small PMT in the WCD to increase the dynamic range of the detector and have fewer saturated counters;
- radio antennas on each SD to capture the radio signals from very inclined showers;
- upgraded SD electronics, with an increase in the data sampling rate from 40 MHz to 120 MHz;

Fig. 7. A schematic view of an upgraded Auger SD station showing the additional scintillator detector and the antenna for radio detection.[27] Reproduced with permission from Creative Commons license.

- a small underground muon detector array overlapping with the infill array;
- extending FD operation into partially moon-lit nights to increase the duty cycle of FD detection.

A schematic depiction of the upgraded SD station is shown in Fig. 7.[27]

The SSDs consist of a $3.8\,\mathrm{m}^2$ plastic scintillator, 1 cm thick. The scintillator is read out using wavelength-shifting fibers collected into one PMT. The SSD sits on top of the existing WCDs overlapping part of the $10\,\mathrm{m}^2$ area. The SSD PMT signal is digitized as another channel in the SD DAQ system. The SSD will be primarily sensitive to the electromagnetic component of the shower signal. This will allow a station-by-station discrimination of the muon flux from the electron and photon flux.

The three main PMTs in the WCD are subject to saturation for counters close to the core of energetic showers. To increase the dynamic range of each WCD counter, a fourth, small PMT is added. The small PMT is placed in the center of the tank, and the small photocathode area provides a smaller gain for signal versus VEM. When the flux of high-energy particles through the detector is very large, the large PMTs will collect enough photons to saturate, but the small PMT will not. In this way, an unsaturated particle flux can be determined over an enlarged range.

The radio signal from an air shower comes from the charge excess in the shower front both from differential absorption of electrons versus positrons and from the deflection of charged particles in opposite directions by the Earth's magnetic field. The 1500 m spacing is too large for useful composition information to be gleaned from the radio signals, except in the case for very inclined showers.

The faster cadence in the FADC will allow for an easier discrimination of single muon signals with the WCDs. This, in conjunction with the SSD, will help to separate the electron and muon components of the shower.

3.   Telescope Array

The Telescope Array is the largest cosmic-ray observatory in the Northern Hemisphere, with an area of 700 km^2. Like the Pierre Auger Observatory, it is a hybrid experiment consisting of SSDs and overlooking FTs. The layout of the TA SDs and FTs is shown in Fig. 8.[28]

3.1. *Scintillator surface detectors*

The SD array consists of 507 scintillator counters, spaced by 1.2 km on a square grid.[29] Each counter has an active area of 3 m^2 and consists of two layers of scintillator, each 1.2 cm thick, separated by a 1 mm thick sheet of stainless steel. A picture of one of the TA SDs is shown in Fig. 9.

The separation of upper and lower layers allows for some distinction between electron, photon, and muon signals. Each scintillator layer is instrumented separately, with wavelength-shifting fibers being laid in grooves on each scintillator panel, before being combined into a bundle and optically connected to a PMT. There are two PMTs per counter, one for the upper layer and one for the lower.

The TA SDs are vertically thin and so present very little effective area to showers coming in from the horizon. Thus, the TA SD array is not used for showers with zenith angles above about 60°.

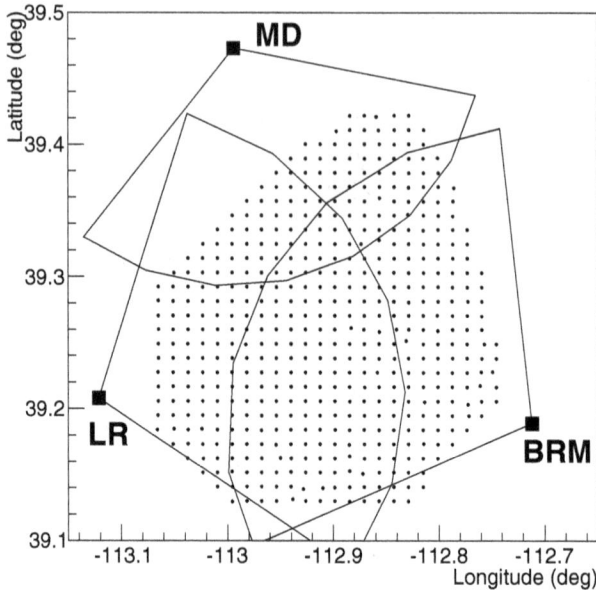

Fig. 8.   The layout of the Telescope Array experiment. The 507 surface detectors are shown as dots, the three fluorescence telescope stations are shown as squares. The approximate field of view and range of the fluorescence telescope stations for 10^{19} eV cosmic rays are also shown. The surface detector spacing is approximately 1200 m.[28] Reproduced with permission from Elsevier.

The triggering for the TA SDs is based on the signal expected for one minimum-ionizing particle (MIP) going vertically through the counter. Electrons and muons are expected to give about the same signal in this case. A Level-0 trigger is generated whenever a signal of more than 0.3 MIPs is observed. These triggers cause a 2.56 µs segment of the FADC trace to be stored locally on the counter. The rate of these triggers is about 750 Hz per counter, which drives the necessary local memory of each counter. A Level-1 trigger is generated whenever a signal of greater than 3 MIPs is observed. Level-1 triggers are communicated to a central DAQ once each second so that a higher-level trigger may be formed. The rate of Level-1 triggers is about 30 Hz per counter.

A Level-2 trigger is formed for air showers at the communication towers if three adjacent SDs have Level-1 triggers within 8 µs. When a Level-2 trigger is formed, the DAQ requests all the waveforms from the Level-0 triggers that are within 30 µs of the Level-2 trigger.

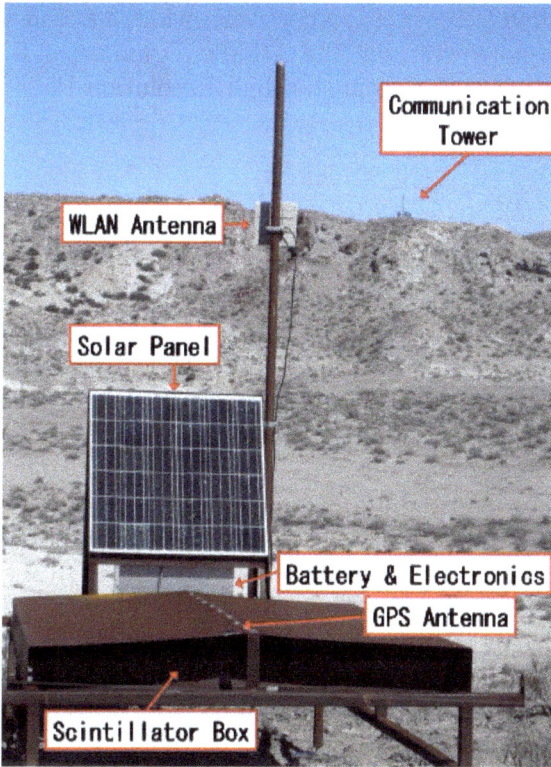

Fig. 9. A picture of deployed surface detector. The scintillator plates are contained in an aluminum box visible under a galvanized-steel cover. The data-acquisition electronics and a battery are located in a box visible behind the solar panel. The WLAN antenna is at the top of the mast behind the counter and is pointed at the communication tower visible on the hill behind the counter.[29] Reproduced with permission from Elsevier.

These waveforms then form an SD event. Since there are three communication towers in charge of three sets of counters, communication between the towers allows for events which fall in the cracks to be collected with full efficiency.

3.2. *Fluorescence telescopes*

The Telescope Array has three sets of FTs overlooking the surface array. These come in two different designs. In the southwest and southeast of the surface array are the Black Rock Mesa (BRM) and Long Ridge (LR) sites, respectively, which have newly designed FTs. In the northwest, at the Middle Drum site, FTs and the data-acquisition system from the High Resolution Fly's Experiment (HiRes) are used.[30]

3.2.1. BRM/LR telescopes

The BRM and LR FT sites consist of 12 telescopes at each site. Each telescope consists of a 3 m diameter, segmented (18 segments), spherical mirror of focal-length 3 m (thus f/1).[31] At the focal plane of the telescope is an array of 256 (16 × 16) PMTs, each of which views about a 1°-diameter circle on the sky. The PMT array is planar. Each telescope has a field of view of about 14° × 16°. The telescopes are pointed in two stacked rings, with the lower ring viewing 3°–17° in elevation and the upper ring viewing 17°–31°. The combined azimuthal coverage is about 106°, with BRM looking to the northwest and LR looking to the northeast. A picture of the BRM site is shown in Fig. 10.

Each PMT in the camera of these telescopes has a Schott BG3 filter attached to the photocathode to limit light outside the UV range from entering the tube. The signal from a PMT is digitized by a 16-bit FADC running at 40 MHz. The signal is summed over four bins to give an effective 18-bit FADC running at 10 MHz. It is worth noting that the PMTs are run at positive HV and the signal is DC coupled to the FADC; thus, the ambient light level can be measured in addition to the short-term fluctuations expected from a shower fluorescence signal.

The signal from the FADC is scanned in a novel way to look for signals over a wide range of time scales. This trigger system is called the signal digitizer/finder (SDF). Rolling averages of the signal in 16-, 32-, 64-, and 128-bin windows are compared to the baseline mean and variance calculated dynamically over the previous milliseconds. When the running average in any of the running sums goes six standard deviations over the mean (standard deviations determined from the running variance), a trigger is formed. Triggers force the storage of 512 bin (51.2 μs waveform).

A higher-level trigger is formed by track finder (TF) module which looks for five coincident SDF triggers in a track-like configuration. There is a list of allowed track templates stored in memory.

Fig. 10.   The Black Rock Mesa fluorescence telescope site. The 12 telescopes look out through the three bays visible. The segmented primary mirrors of several of the telescopes can be seen as well as the backs of the PMT clusters.[31] Reproduced with permission from Elsevier.

There is also a set of limited track templates (containing only four SDF triggers) for use in inter-mirror triggering. Once a TF trigger is formed, *all* the PMT waveforms ($51.2\,\mu$s long) from *all* the mirrors are recorded as part of the event.

3.2.2.   *MD telescopes*

The Middle Drum site contains 14 FTs refurbished from the HiRes experiment, specifically from the HiRes-I site. This fact allows for continuity in photometric calibration between the two experiments and thus also in the overall energy calibration for measuring cosmic-ray air showers.[32,33] Each telescope has a primary mirror consisting of four lobes with an approximate diameter of 2 m and a radius of curvature of 4.74 m. This again is nearly an f/1 design camera. An array of 256 (16 × 16) PMTs is located near but slightly in front of the focal point of the mirror. The location of the imaging plane in front of the exact focal point is to form a nearly uniform spot size of about 1° across the entire plane. The entire field of view of the camera is 14° × 16° (identical to the BRM and LR cameras), and the telescopes are arranged, again like BRM and LR, in two rings to observe from 3°–31° in elevation. A monolithic UV filter is placed over each cluster of PMTs.

The PMTs are digitized by a sample-and-hold system. Each PMT is AC connected to the DAQ system so that ambient light levels do not affect measurements. A variable-threshold discriminator is used as a trigger, and upon a local channel trigger, the charge from the PMT is integrated over a 5.6 μs period and then read out from a QDC. The discriminator thresholds are dynamically adjusted to keep the rate of triggers at 200 Hz. Most of the triggers are noise, and the rate of noise triggers depends on the variance of the baseline voltage on the PMT. The variance will rise if there is more light hitting the PMT, and the adjustable discriminator allows the lowest possible signal to be observed given the noise.

The 16 × 16 array of PMTs is divided into 16 subclusters of PMTs, each 4 × 4 for triggering. Normal operation requires three coincident triggers in a subcluster and two adjacent subclusters triggers to form an event. The event consists of the time and digitized charge of all the tubes with triggers within a 25-μs window.

3.3.   *TA×4*

The Telescope Array has evidence for a source of UHECRs with energies above 57 EeV. The "hot spot" was first noticed in 2013,[34] and it was then estimated that it would require 20 years of data with the Telescope Array to achieve a definitive measurement. A sky map of the arrival directions of cosmic rays with energies above 57 EeV is shown in Fig. 11.

In order to try to get a confirmation of the signal in a much shorter time, Telescope Array took on an expansion project, increasing the exposure by nearly a factor of four, TA×4, in order to get a significant result in on the order of five years. The layout of the new expansion is shown in Fig. 12.[35]

The general design of the array was not changed in this expansion: a scintillator-based surface array overlooked by fluorescence detectors for calibration of the surface signal. The spacing of the SD counters

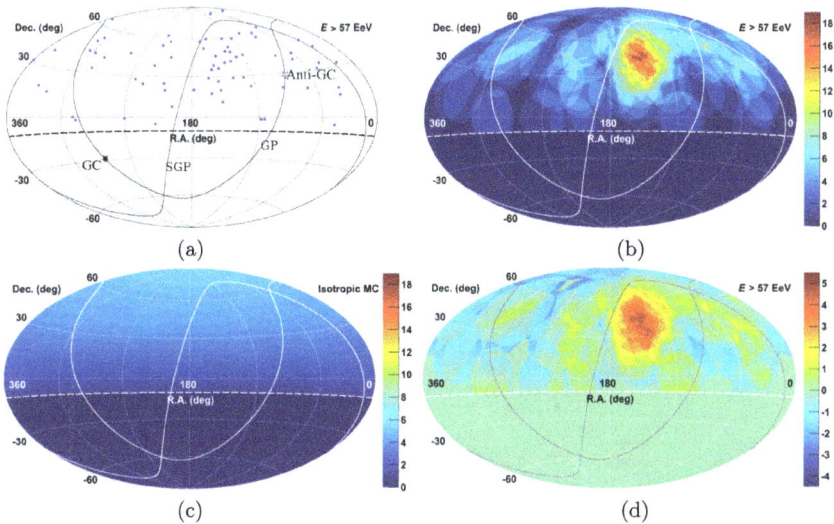

Fig. 11. Aitoff projection of the UHECR maps in equatorial coordinates. The solid curves indicate the galactic plane (GP) and supergalactic plane (SGP). The FoV is defined as the region above the dashed curve at Dec. $= -10°$. (a) The points show the directions of the UHECRs $E > 57$ EeV observed by the TA SD array, and the closed and open stars indicate the Galactic center (GC) and the anti-Galactic center (Anti-GC), respectively; (b) color contours show the number of observed cosmic-ray events summed over a 20°-radius circle; (c) number of background events from the geometrical exposure summed over a 20°-radius circle (the same color scale as (b) is used for comparison); (d) significance map using the Li-Ma method.[34]

was increased to 2.08 km, which would be fully efficient for air showers with energies above 57 EeV. The larger spacing does make the calibration with new FTs a definite requirement.

The FT for the expansion are all refurbished telescopes from the HiRes experiment, and unlike the MD FTs which were from HiRes-I, these telescope are from HiRes-II, which used an FADC data acquisition system.[36] The telescopes are nearly identical in geometry to the telescopes of the MD site at Telescope array, but they are arranged in one ring, viewing 3°–17° only. The reduced elevation coverage is acceptable because the high-energy showers are observed to farther distances and thus do not cover as large a range of the field of view of the telescopes. There are four telescopes covering the northeast lobe situated at the Middle Drum site, and there are eight telescopes covering the southeast lobe located at the Black Rock Mesa site.

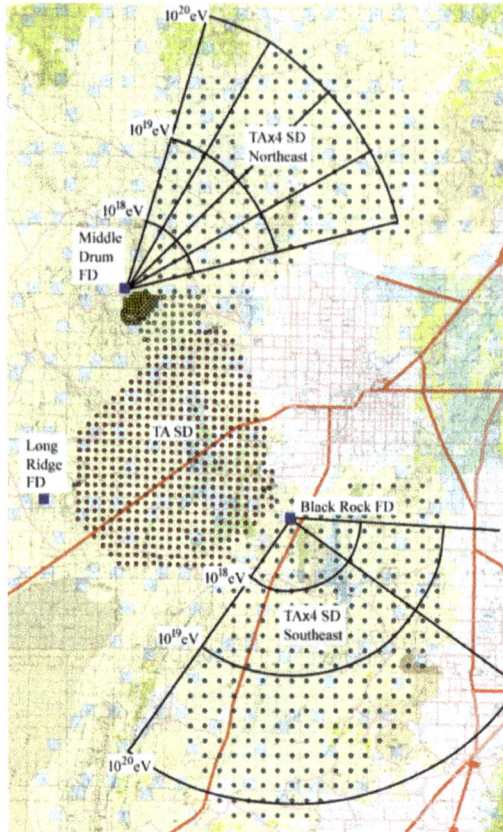

Fig. 12.   An overview of the TA site including TA×4. The green circles in the northeast and southeast correspond to the planned location of TA×4 surface detectors. The red circles in the west show the location of existing TA surface detectors. The two fan shapes show the field of view of TA×4 fluorescence telescopes.[35] Reproduced with permission from Creative Commons license.

The DAQ system for the telescopes is an 8-bit FADC system running at 10 MHz. In addition, there are row and column sums of the PMTs in a cluster, both for triggering and for dealing with saturation of the 8-bit FADC of single channels. The trigger for an event requires three of the row or column sum trigger channels within a given pattern (e.g., three out of five adjacent columns) to be over threshold in coincidence allowing for some time delay between columns. (This trigger can be inefficient for vertical events which only trigger one column of the cluster.) With a mirror trigger, all 256 individual channels are processed to look for non-zero signal and those that have signal are recorded, 100 bins (10 μs) centered around the peak signal.

4.   Ground Array Summary

Large ground arrays detect UHECRs over areas of thousands of square kilometers. Arrays of SDs with kilometer spacing detect the particle footprint of extensive air showers, while FTs observe the emission of fluorescence radiation from the atmosphere as the same air showers pass through. The two currently operating ground arrays are the Pierre Auger Observatory in Argentina with an area of 3000 km^2 and the Telescope Array Observatory in Utah, USA, with an expanded area, TA$\times$4, of 2800 km^2. Both observatories are hybrid detectors combining bother surface arrays and FTs. The detectors measure the energy of cosmic rays; the depth of the shower created by the primary cosmic ray, which is used to estimate the nuclear identity of the primary; and the direction from which the cosmic ray arrived. Using this information, these arrays provide data from which one can attempt to determine the sources of UHECRs and the astrophysical mean by which such large, macroscopic energies can be bestowed upon microscopic particles.

References

1. J. W. Cronin, T. K. Gaisser, and S. P. Swordy, Cosmic rays at the energy frontier, *Sci. Am.* **276**(1), 44–49 (January, 1997). doi: 10.1038/scientificamerican0197-44.
2. T. K. Gaisser and A. M. Hillas, Reliability of the method of constant intensity cuts for reconstructing the average development of vertical showers. In *International Cosmic Ray Conference*, Vol. 8, *International Cosmic Ray Conference*, p. 353 (January, 1977).
3. S. Ostapchenko, Monte Carlo treatment of hadronic interactions in enhanced Pomeron scheme: I. QGSJET-II model, *Phys. Rev. D.* **83**, 014018 (2011). doi: 10.1103/PhysRevD.83.014018.
4. T. Pierog, I. Karpenko, J. M. Katzy, E. Yatsenko, and K. Werner, EPOS LHC: Test of collective hadronization with data measured at the CERN Large Hadron Collider, *Phys. Rev. C.* **92**(3), 034906 (2015). doi: 10.1103/PhysRevC.92.034906.

5. M. Giller, A. Kacperczyk, J. Malinowski, W. Tkaczyk, and G. Wieczorek, Similarity of extensive air showers with respect to the shower age, *J. Phys. G.* **31**, 947–958 (2005). doi: 10.1088/0954-3899/31/8/023.

6. F. Nerling, J. Bluemer, R. Engel, and M. Risse, Universality of electron distributions in high-energy air showers: Description of Cherenkov light production, *Astropart. Phys.* **24**, 421–437 (2006). doi: 10.1016/j.astropartphys.2005.09.002.

7. S. Lafebre, R. Engel, H. Falcke, J. Horandel, T. Huege, J. Kuijpers, and R. Ulrich, Universality of electron-positron distributions in extensive air showers, *Astropart. Phys.* **31**, 243–254 (2009). doi: 10.1016/j.astropartphys.2009.02.002.

8. H. A. Bethe, Moliere's theory of multiple scattering, *Phys. Rev.* **89**, 1256–1266 (1953). doi: 10.1103/PhysRev.89.1256.

9. K. Kamata and J. Nishimura, The lateral and the angular structure functions of electron showers, *Prog. Theor. Phys. Suppl.* **6**, 93–155 (January, 1958). doi: 10.1143/PTPS.6.93.

10. K. Greisen, Cosmic ray showers, *Ann. Rev. Nucl. Part. Sci.* **10**, 63–108 (January, 1960). doi: 10.1146/annurev.ns.10.120160.000431.

11. D. Newton, J. Knapp, and A. A. Watson, The optimum distance at which to determine the size of a giant air shower, *Astropart. Phys.* **26**, 414–419 (2007). doi: 10.1016/j.astropartphys.2006.08.003.

12. J. Matthews, A Heitler model of extensive air showers, *Astropart. Phys.* **22**, 387–397 (2005). doi: 10.1016/j.astropartphys.2004.09.003.

13. D. Ivanov, TA spectrum summary. In *34th International Cosmic Ray Conference (ICRC2015)*, Vol. 34, *International Cosmic Ray Conference*, p. 349 (July, 2015).

14. A. Aab *et al.*, The Pierre Auger cosmic ray observatory, *Nucl. Instrum. Meth. A.* **798**, 172–213 (2015). doi: 10.1016/j.nima.2015.06.058.

15. A. Aab *et al.*, Spectral calibration of the fluorescence telescopes of the Pierre Auger Observatory, *Astrop. Phys.* **95**, 44–56 (October, 2017). doi: 10.1016/j.astropartphys.2017.09.001.

16. J. Abraham *et al.*, Trigger and aperture of the surface detector array of the Pierre Auger Observatory, *Nucl. Instrum. Meth. A.* **613**, 29–39 (2010). doi: 10.1016/j.nima.2009.11.018.

17. W. T. Welford and R. Winston, *High Collection Nonimaging Optics*, Academic Press, Inc., San Diego (1989).

18. J. Abraham *et al.*, The fluorescence detector of the Pierre Auger Observatory, *Nucl. Instrum. Meth. A.* **620**, 227–251 (2010). doi: 10.1016/j.nima.2010.04.023.

19. T. Abu-Zayyad *et al.*, Evidence for changing of cosmic ray composition between 10**17-eV and 10**18-eV from multicomponent measurements, *Phys. Rev. Lett.* **84**, 4276–4279 (2000). doi: 10.1103/PhysRevLett.84.4276.

20. J. Abraham *et al.*, Measurement of the energy spectrum of cosmic rays above 10^{18} eV using the Pierre Auger Observatory, *Phys. Lett. B.* **685**, 239–246 (2010). doi: 10.1016/j.physletb.2010.02.013.

21. M. Mostafa, The hybrid activities of the Pierre Auger Observatory, *Nucl. Phys. B Proc. Suppl.* **165**, 50–58 (2007). doi: 10.1016/j.nuclphysbps.2006.11.009.

22. M. Ave *et al.*, Precise measurement of the absolute fluorescence yield of the 337 nm band in atmospheric gases, *Astropart. Phys.* **42**, 90–102 (2013). doi: 10.1016/j.astropartphys.2012.12.006.

23. A. Aab *et al.*, Measurement of the cosmic-ray energy spectrum above 2.5×10^{18} eV using the Pierre Auger Observatory, *Phys. Rev. D.* **102**(6), 062005 (2020). doi: 10.1103/PhysRevD.102.062005.

24. A. Aab *et al.*, Features of the energy spectrum of cosmic rays above 2.5×10^{18} eV using the Pierre Auger Observatory, *Phys. Rev. Lett.* **125**(12), 121106 (2020). doi: 10.1103/PhysRevLett.125.121106.

25. A. Aab *et al.*, The Pierre Auger Observatory upgrade — preliminary design report, *arXiv:1604.03637* (April, 2016).

26. R. Engel *et al.*, Highlights of the Pierre Auger Observatory. In *37th International Cosmic Ray Conference*, Proceedings of Science (2021).

27. B. Pont. A large radio detector at the Pierre Auger Observatory — measuring the properties of cosmic rays up to the highest energies. In *36th International Cosmic Ray Conference (ICRC2019)*, Vol. 36, *International Cosmic Ray Conference*, p. 395 (July, 2019).

28. T. Abu-Zayyad *et al.*, The energy spectrum of ultra-high-energy cosmic rays measured by the telescope array FADC fluorescence detectors in monocular mode, *Astropart. Phys.* **48**, 16–24 (2013). doi: 10.1016/j.astropartphys.2013.06.007.

29. T. Abu-Zayyad *et al.*, The surface detector array of the Telescope Array experiment, *Nucl. Instrum. Meth. A.* **689**, 87–97 (2013). doi: 10.1016/j.nima.2012.05.079.

30. T. Abu-Zayyad *et al.*, Status of the high resolution Fly's eye detector: Operation and Installation. In *26th International Cosmic Ray Conference*, AIP Conf. Proc. (1999).

31. H. Tokuno *et al.*, New air fluorescence detectors employed in the Telescope Array experiment, *Nucl. Instrum. Meth. A.* **676**, 54–65 (2012). doi: 10.1016/j.nima.2012.02.044.

32. T. Abu-Zayyad *et al.*, The energy spectrum of telescope array's middle drum detector and the direct comparison to the high resolution fly's eye experiment, *Astropart. Phys.* **39–40**, 109–119 (2012). doi: 10.1016/j.astropartphys.2012.05.012.
33. R. U. Abbasi *et al.*, Measurement of the flux of ultrahigh energy cosmic rays from monocular observations by the High Resolution Fly's Eye experiment, *Phys. Rev. Lett.* **92**, 151101 (2004). doi: 10.1103/PhysRevLett.92.151101.
34. R. U. Abbasi *et al.*, Indications of intermediate-scale anisotropy of cosmic rays with energy greater than 57 EeV in the Northern sky measured with the surface detector of the telescope array experiment, *Astrophys. J. Lett.* **790**, L21 (2014). doi: 10.1088/2041-8205/790/2/L21.
35. E. Kido, Status and prospects of the TAx4 experiment. In *36th International Cosmic Ray Conference (ICRC2019)*, Vol. 36, *International Cosmic Ray Conference*, p. 312 (July, 2019).
36. J. H. V. Girard *et al.*, A fiber-optic-based calibration system for the High Resolution Fly's Eye cosmic ray observatory, *Nucl. Instrum. Meth. A.* **460**, 278–288 (2001). doi: 10.1016/S0168-9002(00)01124-4.

Chapter 3

Digital Antenna Arrays for Ultra-High-Energy Cosmic Particles

Frank G. Schroeder

Bartol Research Institute, Department of Physics and Astronomy,
University of Delaware, Newark, DE, USA
Institute for Astroparticle Physics, Karlsruhe Institute of Technology,
Karlsruhe, Germany
fgs@udel.edu

Digital radio detection for air showers has become a widely used technique at cosmic-ray observatories and complements the landscape of detection methods of ultra-high-energy particles. Radio arrays have been demonstrated to deliver resolutions for the arrival direction, energy, and position of the shower maximum that is comparable to established optical techniques. Continuing the remarkable progress in the last decade, future developments of the radio technique may include huge arrays of autonomous radio stations and improvements of the signal processing and analysis by machine learning techniques. Especially, the combination of radio and muon detectors seems to be a promising way to improve the accuracy for the mass composition beyond the capabilities of current air-shower arrays. Hybrid arrays featuring radio and particle detectors have already started to make important contributions to cosmic-ray science in the energy range of the Galactic-to-extragalactic transition. The next generation of radio arrays will expand that use to the extragalactic cosmic rays of the highest energies. This chapter reviews the state of the field, discusses open questions and ongoing developments, and concludes with a description of the coming generation of ground arrays featuring radio antennas, including the SKA, AugerPrime, the surface array of IceCube-Gen2, GRAND, and GCOS.

1. Introduction

Digital detection and analysis techniques have enabled antenna arrays to achieve a measurement accuracy for the properties of cosmic-ray air showers competitive to the traditional optical techniques;[1,2] however, the radio technique is not restricted to dark and clear nights. At the same time, the theoretical understanding of the radio emission has improved to a level that Monte Carlo codes such as CoREAS[3] or ZHAireS[4] reproduce all tested features of the radio emission[5] and are widely used for the interpretation of radio measurements. Overall, the radio technique has achieved sufficient maturity that several ongoing and planned experiments for ultra-high-energy cosmic rays and neutrinos rely on antennas for either stand-alone radio arrays or as part of a hybrid observatory in combination with particle detectors. While stand-alone detectors can offer huge apertures for a reasonable price tag, hybrid detectors provide an increase in measurement accuracy, in particular, for the mass of the primary particle.

Therefore, on the one hand, the radio technique will play an important role in answering astrophysical questions about the origin of the most energetic Galactic[6] and extragalactic cosmic rays.[7] On the other hand, the added accuracy of a calorimetric measurement of the electromagnetic shower energy and the position of the shower maximum will help to solve the particle physics questions at the highest energies,[8] in particular, when combining muon with radio detection.

This chapter starts with an overview of the basic characteristics of the radio emission of air showers and a summary of the achievements of state-of-the-art digital radio arrays. Then, several ongoing developments are discussed that aim at further increasing the measurement accuracy as well as extending the technological reach of the radio technique to huge scales, including new ideas such as radar detection in ice. Last but not least, a major focus is on describing future plans for ground-based radio arrays, in particular, the SKA,[9] the Pierre Auger Observatory with its AugerPrime upgrade,[10] the enhancement of IceTop[11] and the planned IceCube-Gen2 Surface Array,[12] as well as two ideas for huge arrays: GRAND[13] and GCOS.[14]

2.   Basics of the Radio Technique

Air showers emit radiation at radio frequencies in a forward beam. Two mechanisms have been experimentally confirmed to contribute significantly to such radio emission: geomagnetic and Askaryan emission (Fig. 1). Geomagnetic emission is due to a time-varying transverse current induced by the Lorentz force of the Earth's magnetic field. It is linearly polarized in the $\vec{v} \times \vec{B}$ direction, with $\vec{v}$ being the shower axis and $\vec{B}$ the Earth's magnetic field. Askaryan emission is caused by the time variation of a negative net charge excess accumulating in the shower front as the shower ionizes the traversed atmosphere. It is radially polarized along the shower axis. The nature of the geomagnetic and Askaryan emission depends on the frequency

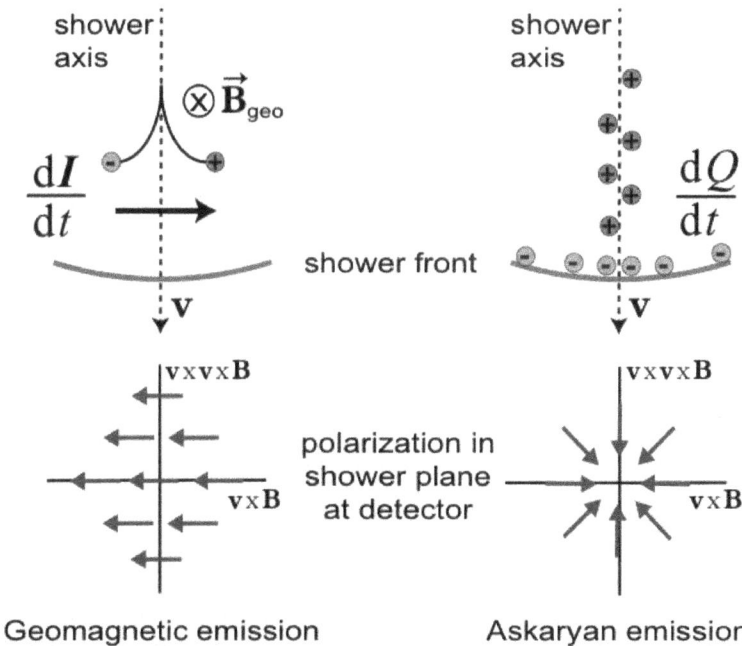

Fig. 1.   Mechanisms for radio emission of air showers. Except for very small geomagnetic angles, the linearly polarized geomagnetic emission (left) due to the time-varying transverse current induced in the shower front is an order of magnitude stronger than the radially polarized Askaryan emission (right) due to the time variation of the net charge accumulated in the shower front.[15]

range, magnetic field, and medium. The situation described here is relevant for typical radio arrays for air-shower detection.[88]

Hence, along the $\vec{v} \times \vec{B}$ axis, the observed emission originates from the interference of the geomagnetic and Askaryan effects, while along the orthogonal $\vec{v} \times \vec{v} \times \vec{B}$, only Askaryan emission is observed. This is why these axes provide a convenient choice of shower coordinates when analyzing radio signals of air showers (Fig. 2). Generally, the Askaryan emission is an order of magnitude weaker in air showers than the geomagnetic one but needs to be taken into account when aiming at high accuracy in the interpretation of radio measurements.[a] Classical Cherenkov-light emission is considered to be negligible

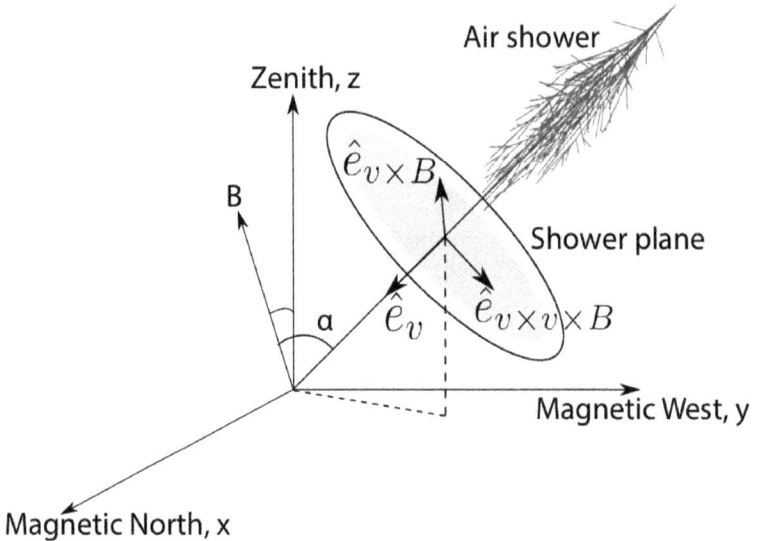

Fig. 2.   Shower coordinate system typically used for the description of radio signals from air showers, with e indicating the unit vectors of the coordinate system, v the shower axis, and B the geomagnetic field.[16]

---

[a]More details regarding the emission mechanisms than summarized in this simple description have been studied and understood, e.g., the Askaryan and geomagnetic emission originate from slightly different stages in the shower development.[17] Furthermore, a phase shift resulting in a circular polarization component[18] and dependencies of the relative geomagnetic and Askaryan emission strengths on the lateral distance, zenith angle,[19] and position of the shower maximum[20] have been found.

at radio frequencies, and searches for other emission mechanisms, including searches for isotropic radio emission, have not yet been successful. Hence, future experiments are planned based on the detailed knowledge of the dominant geomagnetic emission mechanism.

For both emission mechanisms the emission is coherent, which has a number of implications.

First, the radio emission is broad band in the frequency domain and very short in the time domain. Typically, the pulse is a few tens of nanoseconds short. While it is possible to detect air showers by integrating over a corresponding time interval,[21] the leading radio arrays have fine enough sampling to fully resolve the structure of the pulses by sampling at least with the Nyquist frequency. This enables use of Fourier transforms that convert time traces into frequency spectra, which is critical for a number of techniques, such as filtering narrow-band interferences. For experiments with a typical bandwidth of 30−80 MHz on the low side and up to about a GHz on the high side, this implies sampling frequencies of a few $100 \, \mathrm{MSp \, s^{-1}}$ up to a few $\mathrm{GSp \, s^{-1}}$. Even with a marginal fulfillment of the Nyquist sampling frequency, upsampling techniques then enable a number of analysis techniques exploiting the full shape of the pulse.

Second, the radiation energy contained in the radio signal scales approximately quadratically with the energy content of the electromagnetic shower component. This can be exploited to estimate the energy of the primary particle with high accuracy, e.g., by determining the total radiation energy reaching ground by using a model for the radio footprint fitted to the measurements at a handful of antennas.[22] An alternative way is to measure the radio amplitude (instead of the radiation energy) at a fixed reference distance to the shower axis, which then scales approximately linearly with the shower energy.[23,24]

Third, the emission is strongest at the Cherenkov angle (of the order of 1° in air) although generated by mechanisms different than ordinary Cherenkov light. This is because the coherence conditions are fulfilled best for emission at the Cherenkov angle, where a significant amplitude of the radio signal has been observed until frequencies of a few GHz.[26] At frequencies below about 100 MHz, the wavelengths are of the order or larger than the size of the particle disk forming the shower front. Thus, coherence conditions are fulfilled for all angles inside of the Cherenkov cone, i.e., the radio emission is

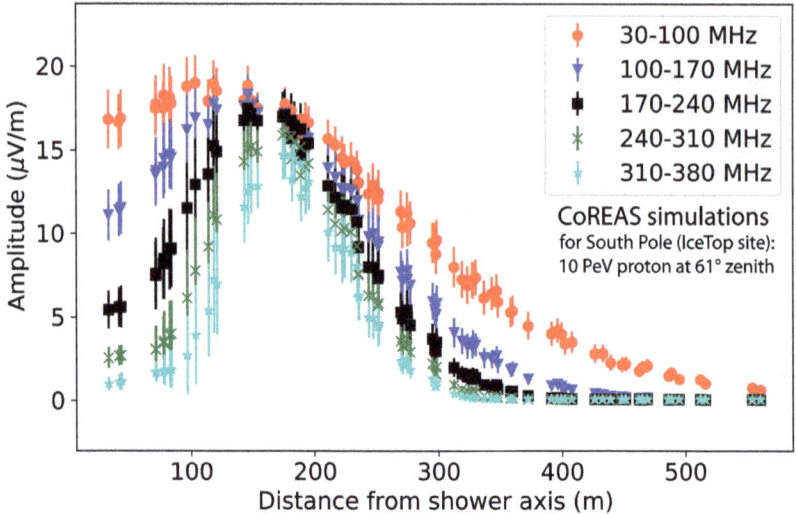

Fig. 3.   Lateral distribution of the radio signal in different frequency bands simulated for a particular proton shower. The size and shape of the lateral distribution depend on shower parameters; for observations below the shower maximum, a Cherenkov ring is visible that becomes more pronounced at higher frequencies.[25]

not only visible at the Cherenkov angle but the complete Cherenkov cone is filled. That implies that at such low frequencies, a significant radio amplitude is detectable in a cone of up to a few degree opening angle around the shower axis (Fig. 3).

From that perspective, it would be an advantage to detect the radio emission at low-MHz frequencies because the emission is strong and the footprint relatively wide. However, the background rises dramatically toward lower frequencies (Fig. 4), and there is no universal optimum frequency band. Lower-frequency bands lead to a higher detection threshold but enable a larger spacing of the antennas because the radio emission extends to wider angles and thus to a larger distance from the shower axis. Avoiding the FM band, many sparse arrays thus operate in a detection band around 30–80 MHz. At higher frequencies, antennas are smaller and easier to deploy and the signal-to-noise ratio is better, but denser arrays are required. Ideal locations for radio arrays consequently are those without significant human-made interferences that can cover a wide frequency range from several tens of MHz up to several hundreds of MHz.

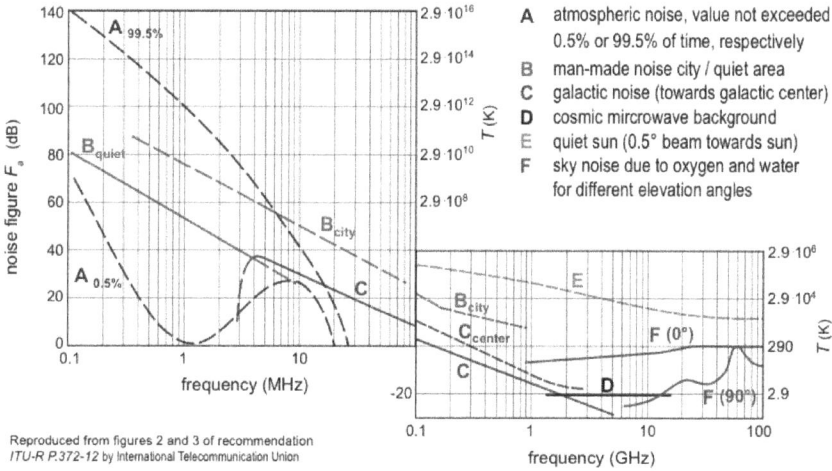

Fig. 4. Typical power of different types of radio backgrounds on Earth over frequency: In remote locations with low anthropogenic background, radio measurements of air showers are typically limited by Galactic noise at lower frequency and by thermal noise of the detection system and/or the environment at higher frequencies.[2]

3.   State of the Art: Digital Radio Arrays for Cosmic-Ray Air Showers

The digital radio technique experienced a revival in the 2000s with prototype experiments, such as LOPES[5] and CODALEMA.[27] By comparing radio measurements to measurements of co-located particle detectors, they delivered a number of proof-of-principle demonstrations and helped to improve knowledge from preceding analog experiments. A second generation of radio arrays for air showers, in particular, LOFAR,[28] AERA,[22] OVRO-LWA,[29] TREND,[30] and Tunka-Rex[31] were able to demonstrate that the radio technique can compete in accuracy with the traditional optical techniques (see the following) and provided demonstrations of self-triggering of air showers.[29,30,32] Yet, a number of technical and analysis challenges still need to be overcome to make the radio technique mature and applicable to forefront cosmic-ray science. This process has started recently in a smooth transition to a third generation of future radio detectors. Before describing the challenges and future plans in detail, this section will give a summary of the state of the art and of the achievements of digital radio arrays.[b] [35,36]

Important for the application of the radio technique on cosmic-ray physics is the accuracy for those shower parameters linked to the properties of the primary particle: the arrival direction, energy, and atmospheric depth of the shower maximum, X_{max}, which is statistically correlated with the mass of the primary particle.

As charged cosmic rays are deflected by (extra)galactic magnetic fields on their way to Earth, a direction resolution of a few degrees is sufficient for most science cases. Indeed, a direction resolution of $o(1°)$ is easily achieved by radio arrays using GPS timing and positioning in combination with a simple plane wavefront fit. Subdegree resolution is of particular interest for photon and neutrino searches (although these have not yet been detected in the energy range covered by radio arrays). Subdegree resolution requires a precise relative timing and positioning of the antennas as well as properly taking into account the hyberbolic shape of the radio wavefront.[37,38] This has been experimentally demonstrated, for example, by the compact LOPES array using a cable-based clock distribution and an interferometric beamforming technique.[5] With the ongoing improvements in satellite-based timing and positioning, it may be possible that subdegree angular resolution will be available in the future with even simpler techniques and for autonomous stations over large baselines. In any case, the state-of-the-art accuracy for the arrival direction is more than sufficient for charged cosmic rays.

Several methods have been developed for the reconstruction of the energy contained in the electromagnetic shower component, which is closely related to the total energy of the shower and thus to the energy of the primary particle. While a rough energy estimation is possible with a single antenna station,[34,40,41] competitive measurement accuracies for the energy have been achieved by measuring the lateral distribution of the radio signal of an air shower at several antenna stations. Details of the approaches vary, and several experiments have demonstrated that they reach a statistical precision for

---

[b]In parallel, there has also been progress regarding radio observation of air showers from mountain tops[33] and balloons.[34] These approaches aim mostly at neutrino detection and are discussed in the corresponding chapter.

individual events of $o(10-20\%)$.[5,22,42] A major systematic uncertainty for both the energy reconstruction of individual events and the absolute energy scale is the gain pattern of the antennas, which needs to be accurately known on an absolute level for all arrival directions. Current experiments use simulations of the gain pattern for the direction dependence and external reference sources[43,44] or the Galactic background[36] to calibrate the absolute scale. The achieved accuracy of $o(15\%)$ is similar to the energy scale uncertainties of the currently leading air-fluorescence technique and might be reduced further to $o(10\%)$ by addressing known systematic uncertainties in the calibration methods. Moreover, the radio technique seems to be well suited to compare the energy scales of different experiments with each other with a relative accuracy of 10% or better.[35,36]

X_{max}, the depth of the shower maximum, is one of the most important shower parameters for the estimation of the mass composition of cosmic rays. Although it has less statistical separation power for mass discrimination than the electron–muon ratio, X_{max} overall is the more accurate parameter. This is because its interpretation is only marginally affected by the deficits that hadronic interaction models show regarding the muon number (Fig. 5).

Various features of the radio signal of air showers are sensitive to X_{max}: the slope of the lateral distribution,[48,49] the shape of the wavefront,[37] the pulse shape at a given lateral distance or, equivalently, the frequency spectrum,[50] and finally the polarization angle at a given position relative to the shower axis, which depends on the relative contributions of the Askaryan and geomagnetic emission.[16,17] Today, the most precise method to reconstruct X_{max} is a special variant of template matching. Monte Carlo simulations of the radio signal for different X_{max} values are compared to the measurements at the various positions of an antenna array.[42,51] When at least $o(5)$ antennas with clear signals contribute to the reconstruction, the precision seems to be competitive or even better than those of the traditional optical techniques of $o(20\,\mathrm{g\,cm^{-2}})$. However, the radio technique has not been studied as thoroughly as the fluorescence technique for unknown systematic uncertainties, and self-consistency checks of the different radio X_{max} methods still need to be performed. Among the contemporary radio arrays, LOFAR[45] and AERA[46] feature the most complete check of systematic uncertainties, but further

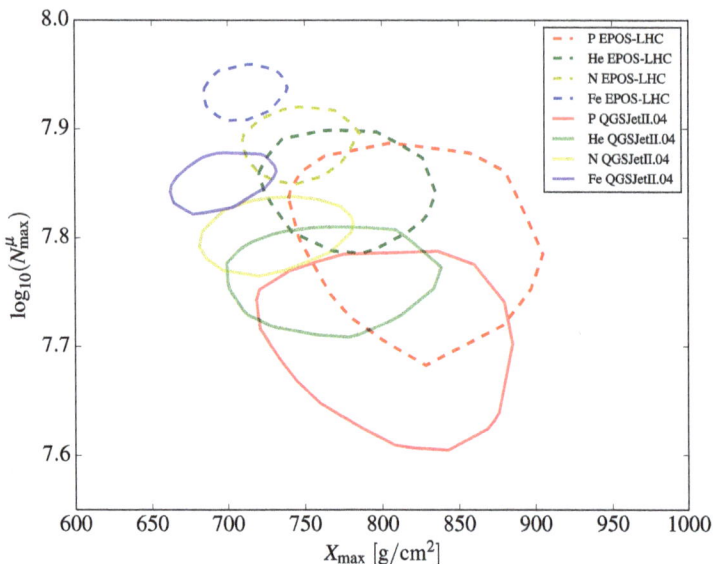

Fig. 5.   One sigma contours of the maximum muon number and X_{max} of $10\,\text{EeV}$ CORSIKA simulations at $\theta = 38°$ (taken from the AugerPrime PDR[39]). Three important conclusions can be drawn from this plot: First, the muon number is the more sensitive parameter for mass separation — at least when the energy is known, for e.g., from calorimetric fluorescence or radio measurements. Second, different hadronic interaction models disagree in the absolute mass scale for the interpretation of the muon number but are similar in the interpretation of X_{max}. Therefore, the muon number is the more precise and X_{max} is the more accurate parameter for the estimation of the mass composition (which may change in the future when the deficiencies of hadronic interaction models regarding the muon content will be fixed). Third, the mass information in the muon content and in X_{max} is partly complementary, which implies that a higher total accuracy for the mass can be achieved when simultaneously measuring both parameters of a shower by a hybrid detector.

studies are needed to understand the tension between LOFAR and the Auger measurements. At the same time, Auger fluorescence and radio measurements agree well with each other (Fig. 6). These insufficiently understood systematic uncertainties are one reason why in the next decade it will remain important to continue optical and radio observations in parallel.

Overall, radio measurements have reached a level of accuracy for all important shower parameters similar to the established techniques. That does not mean that the radio technique could fully

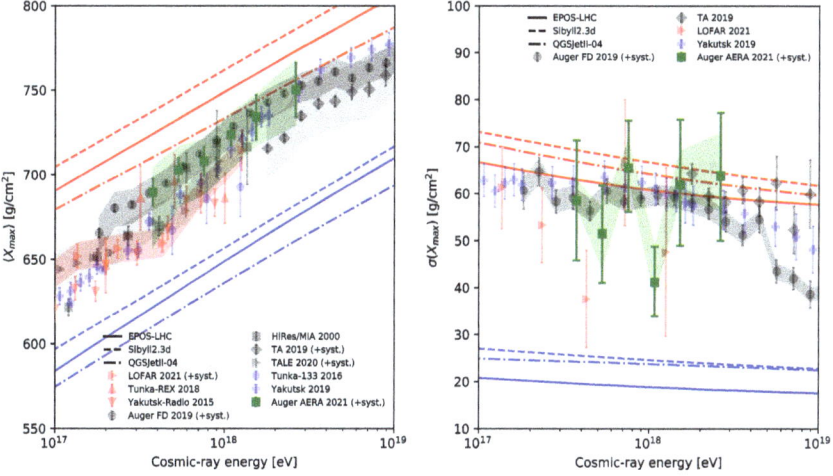

Fig. 6. Measurement of the mean $X_{\max}$ (left) and its variation (right) by various experiments using different techniques: radio (red + green), air-Cherenkov (blue), and air-fluorescence (gray). There is some tension between LOFAR[45] and Auger[46] radio measurements which requires further investigation.[47]

replace existing techniques. On the one hand, the successful demonstrations all come from hybrid arrays, where the radio measurements can be triggered and cross-checked (e.g., for the removal of outliers) by co-located ground arrays of either air-Cherenkov or particle detectors. On the other hand, we will see that the main goal of exceeding the state-of-the-art accuracy for the next generation of cosmic-ray experiments will best be achieved by hybrid arrays combining radio antennas with other detection methods, such as muon detectors.

4. Open Questions and Overcoming Limitations

While the radio technique has achieved a maturity level that makes it applicable to cosmic-ray science in the coming generation of air-shower arrays, there still are a number of open questions. Continuing research and development thus is important to develop the full potential of the radio technique. These developments aim at further increasing the accuracy of the radio technique for the energy and mass composition, at decreasing its detection threshold, and at making it applicable for the largest scales of future experiments.

The following paragraphs provide a personal summary on areas where further developments may be useful to maximize the potential of radio arrays for cosmic-ray physics.

4.1. *Efficiency, aperture, and exposure*

Because the interplay between the two emission mechanisms causes an asymmetric lateral distribution of the radio signal, the detection efficiency of antenna arrays depends on the azimuth and zenith angles of the shower and on the position of the shower core in the array. Thus, calculating the efficiency and aperture is more difficult than for traditional detection methods whose efficiency is almost independent of the azimuth angle. Depending on the antenna spacing of a radio array, full efficiency might be achieved only for a certain region of the sky. This is similar to particle-detector arrays, except that the sky region is not defined by a simple cut in zenith angle but by a more complex cut including the zenith angle and the angular distance to the geomagnetic field (Fig. 7). A further complication can be a varying level of anthropogenic radio background and the slight variations of the Galactic noise over the sidereal day. This may create a time dependence in the efficiency, which needs to be taken into account when estimating the exposure at an energy very close to the detection threshold. Yet, knowing the exposure to a few percent accuracy is important when measuring energy spectra of cosmic rays. Moreover, full efficiency is important when selecting events for anisotropy or mass-composition studies.

There are currently two approaches to solve this problem for radio arrays. First, a computationally expensive approach used by LOFAR:[51] For each measured shower, a set of Monte Carlo simulations is produced to check if almost every shower with the particular energy, core position, and arrival direction would have been detected independent of the primary particle. The advantage of this approach is that it enables the selection of full-efficiency datasets even for very irregular arrays. However, the computational effort makes the method unfeasible for large-scale arrays with high event statistics. Another disadvantage is that it is very challenging to calculate the exposure of the array if the efficiency is only evaluated for those showers measured but not for the full sky.

Second, the efficiency can be modeled as a function of core position and of zenith and azimuth angles.[52] Although developing an

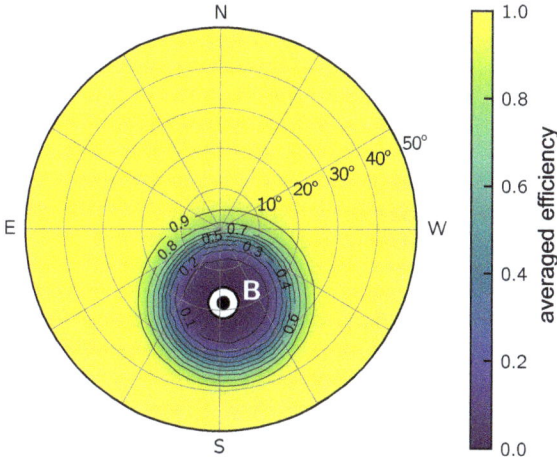

Fig. 7. Modeled detection efficiency of Tunka-Rex for air showers of $10^{17.3}$ eV energy for arrival directions up to $50°$ zenith angle. The detection efficiency depends mostly on the geomagnetic angle. It is approximately, but not completely, symmetric around the geomagnetic field direction because the size of the radio footprint depends on the zenith angle.[52]

accurate model for a particular array geometry and location is cumbersome, once the model is developed, it can easily be applied to the full dataset of measured air showers. As none of the available models to date is perfect, this approach is particularly recommendable if the statistics are high enough to allow for a certain safety margin in the full-efficiency selection. Moreover, having a reasonably accurate model in place enables calculation of the aperture and exposure of the radio array as function of energy. Therefore, it seems advisable to dedicate further effort in developing such efficiency models also for future arrays.

4.2. *Computational methods*

Modern computational methods, often summarized as data science, offer significant potential for cosmic-ray experiments in general, including digital radio arrays. Methods such as template fitting or machine learning can enhance the reconstruction precision for air-shower parameters during data analysis or lower the detection threshold by identifying radio signals at low signal-to-noise ratios. These methods have in common that they rely on simulations as input. State-of-the-art simulation codes such as ZHAireS[53] and CoREAS[54] seem to be fully compatible with measured signals within systematic uncertainties.[5] Yet, it is not clear what their absolute accuracy is, i.e., to what level they can be used as tools to improve experimental accuracies beyond the state of the art. In addition, current codes have certain limitations, for example, regarding atmospheric models. Thus, it is important to continue developments on codes for simulations of the air-shower particle cascade as well as its radio emission. A community effort to improve some of the current limitations is already ongoing as part of the CORSIKA 8 project.[55]

Nonetheless, the currently available simulations are good enough to be successfully applied to the reconstruction of measured radio events. For the shower energy and for X_{max}, the most precise reconstruction method available today[c] currently is a variant of template matching using several dozens of simulated CoREAS showers for each measured air shower. For the template matching, either the peak amplitudes in the individual antennas[51] or the full shape of the pulse envelopes[42] in each antenna is used. In the future, this method might be extended to also include timing and polarization information. Lately, it has also been suggested to match modeled radio pulses instead of the computationally intensive individual simulations.[56] This may help to identify pulses against background and provide an improvement over some traditional reconstruction methods for shower parameters. It remains to be shown whether the pulses can be modeled accurately enough to also compete in precision for X_{max}.

Machine learning, especially artificial neural networks, has recently been applied to lower the detection threshold of radio pulses in individual antennas (Fig. 8). The approaches seem to work remarkably well to classify simulated signals against modeled background,[20,57] and some success has already been achieved with real measurements.[58] It is noteworthy that lowering the detection threshold in individual antennas will not just lower the energy threshold of the array but will also improve the measurement accuracy for shower parameters at higher energies. This is because the reconstruction

---

[c]Template matching in its currently used version provides only a slightly better precision for the reconstructed energy than other methods but a significantly improved precision for X_{max}.

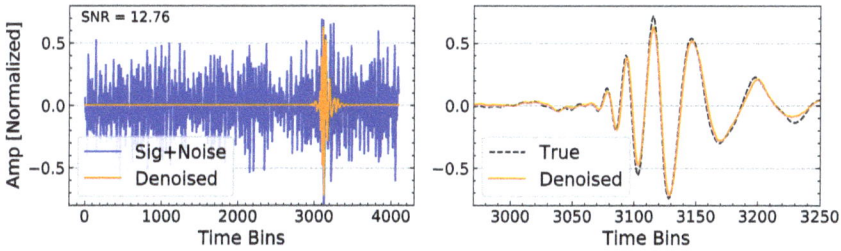

Fig. 8. Example for a simulated radio pulse with modeled Galactic and thermal noise that is denoised by a neural network. The right-hand side plot is a zoom of the left-hand side plot.[20]

precision of X_{max} improves with the number of antennas contributing to the analysis.[42]

There seems to be great potential in the application of deep learning for high-level analysis tasks too. For ground arrays of particle detectors, the benefit for the reconstruction of the shower direction[59] and of X_{max}[60] has already been demonstrated. Therefore, it is promising to test these methods also for radio arrays and finally for combined analyses of radio and particle measurements in hybrid detectors.

4.3. *Radar detection*

Indirect cosmic-ray detection through radar reflection from the ionized air traversed by an air shower was long thought to be a possible alternative to the detection of the radio signals emitted by the air shower itself. However, the TARA experiment determined a firm upper limit on the strength of the reflected radar signal.[61] Thus, even if the radar reflection of air showers eventually could be measured, the method currently is of limited interest due to the technical challenges imposed by the weak signal. Nonetheless, the situation is different for showers in dense media. As indicated by a laboratory experiment using an electron beam at SLAC, showers in dense media may give rise to significant radar reflection.[62] Consequently, the next step will be a practical demonstration confirming the radar detection of real cosmic-ray–induced showers in dense, radio-transparent media, such as ice or salt. This motivates the planned Radar Echo Telescope which aims at the detection of radar signals reflected by cosmic-ray– and neutrino-initiated particle cascades in the Antarctic ice.[63]

4.4. *Interferometry*

Digital interferometry exploits the phase of the radio signal recorded at several antennas. The prerequisites are a relative position accuracy of the antennas that is an order of magnitude better than the wavelengths of the signal and a relative timing accuracy that is an order of magnitude better than the oscillation frequency, i.e., of $o(1\,\text{ns})$ for the typical band of 30–80 MHz[64] and sub-nanosecond for higher frequencies.[65] While the position accuracy can be achieved with satellite positioning systems, such as GPS, their timing accuracy is not yet good enough. Thus, interferometry requires either a cabled setup to achieve the relative timing accuracy and/or some additional technical measures to improve the relative timing, such as a reference beacon,[64] which is challenging when covering arrays of several km of size.[66] To date, digital interferometry has successfully been applied for air-shower detection with local setups featuring several adjacent antennas[34] as well as at the relatively dense LOPES array featuring up to 30 antennas on an area of about 200 m diameter.[5]

One of the benefits of interferometry is a lower detection threshold.[67] In classical radio interferometry, where all antennas see the same signal in the far field, the detection threshold improves with the square root of the number of antennas. For radio detection of air showers, the relation may be slightly worse because the pulse shape is not exactly the same in all antennas as it changes with the distance to the shower axis. The exact improvement will depend on the array configuration and needs to be studied for the specific setup.

Another benefit of interferometric methods, such as cross-correlation beamforming, is an improvement in the reconstruction accuracy, in particular, for the resolution of the arrival direction.[5] As suggested by recent simulation studies,[65,68] interferometry may also enable an accurate reconstruction of X_{max} for large arrays featuring antennas on both sides of the Cherenkov cone visible in the radio footprint. This prediction requires practical demonstration and may further enhance the physics potential of future radio arrays. Certainly, interferometric methods remain of interest for future experiments, in particular, with the continuously improving timing accuracy of remote clocks.

4.5. *Autonomous stations*

Building huge ground arrays for ultra-high-energy cosmic rays requires autonomous station designs without cables connecting the stations. For hybrid arrays of the scale of several thousand square kilometers, such as the AugerPrime upgrade of the Pierre Auger Observatory featuring a radio antenna on top of each particle detector, this problem is solved:[10] The trigger is provided by the particle detectors; each detector features its own power system comprising a solar panel and a battery and its own data acquisition which communicates remotely to the central DAQ via one of a few communication towers at the borders of the array. For radio-only arrays, further improvements of the self-triggering techniques are desirable to enable the operation in environments with human radio interferences as well.

For arrays of significantly larger size, such as the GRAND project which is planned to consist of several patches of $o(10{,}000\,\mathrm{km}^2)$ each, further technical developments are needed. In particular, the maintenance effort needs to be reduced compared to Auger, for example, concerning the lifetime of the batteries and occasional replacements of the low-noise amplifiers in the antennas after lightning damage. Further improvements which will simplify the installation of huge arrays regard the station design itself, which needs to be easily deployable in very remote environments. In summary, the radio technique is mature enough to plan for its application on the largest scales, where some technical developments should be continued over the next years.

5. Future Radio Arrays

The radio technique will play a key role in several experiments planned for future progress in high-energy particle astrophysics. In many cases, antennas will be combined with particle detectors to maximize the accuracy for the mass of the primary particle as a higher accuracy is required to test scenarios for the origin of both the most energetic Galactic and extragalactic cosmic rays.

In particular, the combination of radio and muon detectors seems promising for this purpose. The radio detectors can provide a calorimetric measurement of the size of the electromagnetic shower component and of X_{max}, which yields complementary mass information to the size of the muonic component (cf. Fig. 5). Methods to combine both estimators are under development, and simulation studies indicate that the combination of energy, X_{max}, and muon number will indeed be a powerful tool to separate light and heavy primary particles on a per-event basis even for very inclined showers (Fig. 9).

Moreover, this combination of radio and muon detection is ideal to make further progress in the test of hadronic interaction models, especially, when measuring the muon number over energy. While for arrays of particle detectors the energy estimate of a measured air shower depends at least to some extent on its muon number at ground; this is different for radio arrays which provide an energy

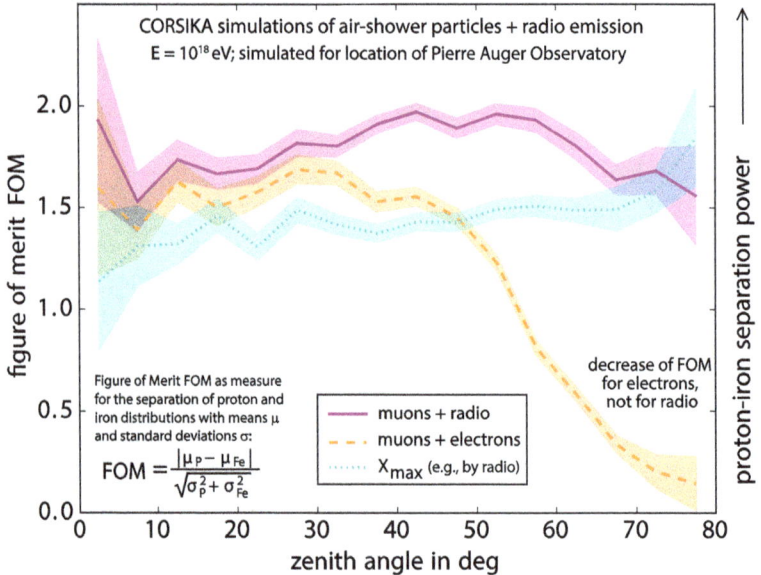

Fig. 9. Figure of merit for the separation of proton and iron nuclei using certain shower observables or their combination: the muon density at 600 m divided by the square root of the radiation energy (purple) or divided by the electron number (yellow) and the depth of the shower maximum X_{max} (blue) over zenith angle. The figure is for the true shower observables while not taking into account any detector effects.[69]

estimate purely of the electromagnetic component (the radio signal by muons is negligible). Therefore, coincident radio and muon measurements of showers provide important complementary information to further understand why hadronic interaction models predict too few muons for air showers in the EeV energy range.[70]

Nonetheless, the combination with muons is not the only way that radio antennas will be utilized to drive the field of ultra-high-energy cosmic-ray physics. The following paragraphs provide an overview on select radio arrays which have cosmic-ray science among their science goals.[d]

5.1. *The Square Kilometer Array (SKA)*

The low-frequency array of the Square Kilometer Array (SKA-low) in Australia will be built primarily for radio astronomy in the band 50–350 MHz. Nonetheless, the design enables the detection of cosmic-ray air showers on ground[71] and potentially also the observation of ultra-high-energy cosmic rays interacting in the lunar regolith.[72]

The inner core of SKA-low features several ten thousand antennas on an area of about $0.5\,\mathrm{km}^2$. This high density of antennas will measure the radio footprint of air showers in extreme detail, which enables a highly precise measurement of X_{max} as well as of the width parameter L of the shower profile.[9] The simultaneous measurement of X_{max} and L can be used to scrutinize hadronic interaction models because they predict significantly different values of L for the same value of X_{max} (Fig. 10). After this problem of the hadronic interaction models has been understood and solved, the same data can be used to measure the mass composition with much higher precision than with X_{max} alone.

The positive impact of the SKA on the field reaches beyond the direct science potential of the SKA. With its broad frequency band, low noise level, and wide sky coverage, the SKA-low antenna (SKALA) is ideally suited for cosmic-ray measurements. Therefore, as a kind of spin-off, SKALA-v2 antennas[73] are used in other projects too, in particular, for the surface enhancement of IceTop[74] and

---

[d]For radio arrays aiming primarily at neutrino detection and for balloon-borne and space experiments, see the corresponding chapters in this book.

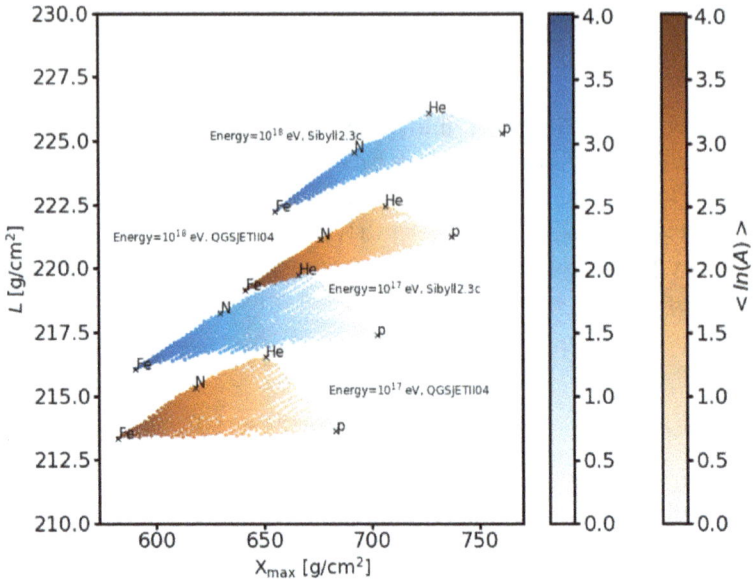

Fig. 10.   Relation of the depth of maximum, X_{max}, and the width, L, of the shower profile for different primary particles at two different energies and for two different hadronic interaction models.[9] The blue triangles are for Sibyll 2.3c simulations and the red triangles for QGSJET II.04, each for an energy of 10^{17} eV (lower triangles) and 10^{18} eV (upper triangles). The corners of all triangles are made by Fe at the bottom left, He at the top, and protons at the right.

a project aiming at cross-calibrating the energy scale of various air-shower arrays by co-locating a few SKALA antennas for coincident measurements.[75] Using exactly the same antenna type for the cross-calibration will reduce one of the main systematic uncertainties in earlier approaches to compare energy scales with radio antennas, which already achieved a low uncertainty of about 10%.[35]

5.2.   *AugerPrime: Upgrade of the Pierre Auger Observatory*

The Pierre Auger Observatory is currently receiving an upgrade of several of its components. This AugerPrime upgrade includes enhancements of the water-Cherenkov detectors of the surface array. These will be enhanced by improved electronics, by an additional photomultiplier to increase the dynamic range, and by a scintillation

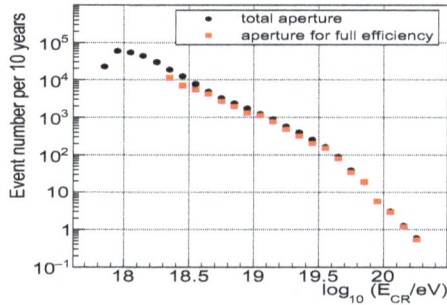

Fig. 11.   Left: Photo of a water-Cherenkov detector of the surface array of the Pierre Auger Observatory enhanced by a scintillation panel and by a SALLA-type radio antenna as part of the AugerPrime upgrade. Right: expected event rate of the radio upgrade.[76]

panel on top of each detector, where the combination of the scintillation and water-Cherenkov signal can be used to estimate the electron and muon numbers. By this method, the upgraded array is capable of statistically separating showers initiated by light and heavy primary particles, which is a prerequisite to search for mass-dependent anisotropies.[39]

Complementing the other upgrades, a SALLA-type antenna[77] for the frequency band of 30–80 MHz will be installed on top of each scintillator (Fig. 11). Because of the large detector spacing of 1.5 km in the 3000 km^2 main array, this radio upgrade will efficiently detect inclined showers with zenith angles greater than 60° at the highest energies. The X_{max} precision still needs to be determined, but independently of this, the radio upgrade will feature per-event mass sensitivity by combining the measured radiation energy with the muon signal measured in the water-Cherenkov tanks, where only muons arrive at ground for these highly inclined showers. Therefore, the radio upgrade is an excellent complement to the scintillator upgrade which provides per-event mass sensitivity for less inclined showers.

5.3.   *IceCube-Gen2 surface array*

The IceCube Neutrino Observatory with its surface array IceTop[79] was completed in 2011 and has made a number of remarkable discoveries (see chapters about high-energy neutrinos and Ref. [80] for results on cosmic rays). On the one hand, the surface array supports the neutrino mission of IceCube by serving as a veto and calibration device. On the other hand, the combination of surface and deep detectors provides a unique instrument for cosmic-ray science, where the deep detector measures high-energy muons produced in air showers detected at the surface. Consequently, it is foreseen to equip also the eight-times-larger deep optical array of IceCube-Gen2 with a surface array, which will cover the energy range from below the knee to approximately the ankle in the cosmic-ray energy spectrum.

The design of the IceCube-Gen2 surface array follows the plans of an upgrade of the existing IceTop detector with elevated scintillation panels and SKALA-v2 radio antennas for the band of 70–350 MHz (Fig. 12).[11,74] The scintillators will provide a measurement of the particles at the surface with a threshold of around 0.5 PeV for vertical showers.[78] They are sensitive primarily to electromagnetic particles, but it has been shown for IceTop that the shower content of GeV muons can be measured by averaging over detectors at large lateral distances. Together with the measurement of TeV muons in the deep

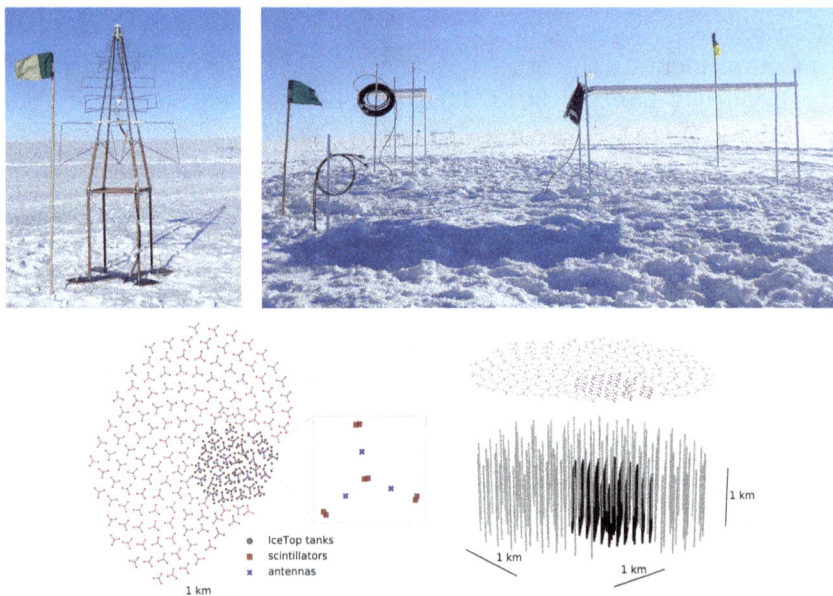

Fig. 12.   Top: Detectors of a prototype station at IceTop at the South Pole comprising three SKALA-v2 radio antennas and four pairs of scintillators. Bottom: Layout of the planned IceCube-Gen2 surface array above the optical array in the ice.[78]

optical detectors, this assembly constitutes a distinct setup for studies of hadronic interactions in air showers, such as the production of conventional and prompt forward muons. Not only will the Gen2 surface array increase the area compared to IceTop but also the zenith range for coincidences with the in-ice detectors, which is why the total aperture for such analyses will be augmented by more than a factor of 30.

Radio antennas will be added to the surface array to maximize the accuracy for the mass composition for the complete energy range of the transition from Galactic to extragalactic sources, which is presumed to occur somewhere between the heavy knee around 100 PeV[81] and the ankle around 5 EeV.[82] Complementary to its neutrino mission driving the design, IceCube-Gen2 will consequently be a leading experiment for the most energetic Galactic cosmic rays.[12]

5.4.   *GRAND and GRANDProto300*

The Giant Radio Array for Neutrino Detection (GRAND) is a staged project that in its final phase will be by far the largest air-shower detector in the world.[13] It will consist of several radio arrays distributed over the globe, each of $o(10,000)$ antennas distributed over $o(10,000\,\mathrm{km}^2)$ and together covering about 200,000 km^2. GRAND will be a highly sensitive detector for ultra-high-energy neutrinos (see corresponding chapter) and will also feature an order-of-magnitude increase in aperture for ultra-high-energy cosmic rays compared to the Pierre Auger Observatory[84] or the TA×4 extension of the Telescope Array.[85]

As the first step of GRAND, a prototype array with 300 antennas on 200 km^2 will be constructed in China, measuring air showers in the frequency band of 50–200 MHz (Fig. 13).[83] This GRAND-Proto300 array will be complemented by a co-located surface array of particle detectors which fulfill two main purposes. First, the particle detectors can be used to cross-check the efficiency and purity of the independent self-trigger of the radio array, including all real-life technical challenges. Second, both arrays together provide hybrid measurements of the electromagnetic and muonic shower components for very inclined showers. While the accuracy for the primary mass may be lower than at IceCube-Gen2 with its multiple detectors, GRAND-Proto300 will certainly have much higher statistics at EeV energies

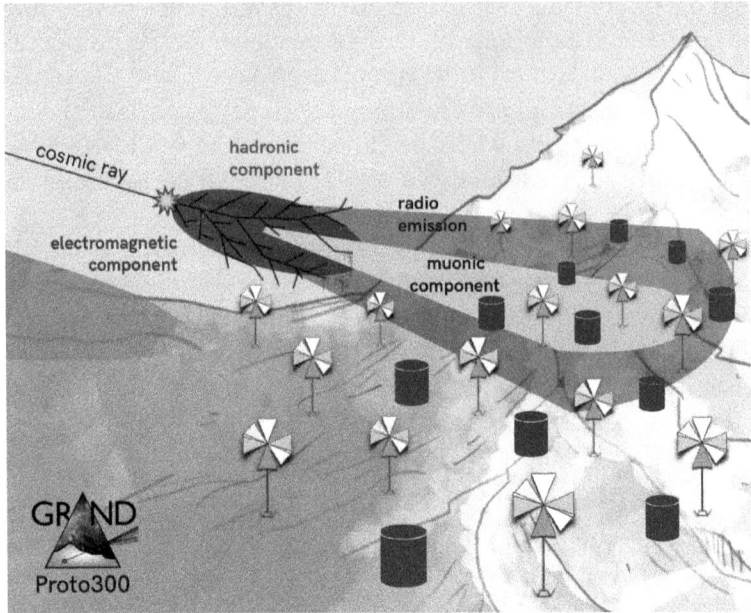

Fig. 13. Sketch of the detection principle of the GRANDProto300 array that will detect inclined air showers with self-triggered radio antennas and particle detectors (courtesy of the GRAND Collaboration[83]).

due to the much larger area. Therefore, GRANDProto300 will be a leading instrument for the search of (mass-dependent) anisotropies in the energy range of the Galactic-to-extragalactic transition.

5.5. *Global cosmic-ray observatory*

Global Cosmic-Ray Observatory (GCOS) is a future project for ground-based cosmic-ray measurements planned as successor to the Pierre Auger Observatory and other ultra-high-energy observatories. The science goals encompass fundamental particle physics, astrophysics with cosmic rays at the highest energies, as well as geophysics and atmospheric physics.[14] Similar to GRAND, GCOS will also combine several sites in the world with a total instrumented area of many 10,000 km^2.[86]

Although the area is smaller than that planned for the final phase of GRAND, the total aperture may be similar because the GCOS design will not be limited to very inclined showers and includes particle detectors providing electron–muon separation. The best design is currently under discussion in the community, with air-fluorescence and radio detectors considered for hybrid instrumentation. Hybrid detection is important since the mass composition of cosmic rays seems to be mixed at all energies,[87] and the best mass separation is provided by the simultaneous measurement of X_{max} and the sizes of the electromagnetic and muonic components. While ultra-high-energy neutrino detectors are discovery instruments aiming at maximum exposure by compromising on measurement accuracy, the situation is different for cosmic-ray instruments, such as GCOS, which require a high accuracy for the energy and mass of the primary particle.

Nonetheless, it is certainly possible that some of the experimental sites will be integrated in both GRAND and GCOS by using GRAND-type radio antennas in addition to the particle-array instrumentation of GCOS. While the priority in the science goals of GCOS (fundamental discoveries with cosmic rays) and GRAND (ultra-high-energy neutrinos) are different and both are needed, there is potential for synergies and collaboration.

6.   Conclusion

Remarkable progress has been achieved in the understanding of the radio emission of air showers and its application to cosmic-ray detection. The radio signal has been understood to $o(10\%)$ of the signal amplitude, and features of its lateral distribution, wavefront shape, and frequency spectrum seem to be consistent between Monte Carlo simulations and experimental measurements. For a hybrid observatory, where radio antennas complement arrays of particle detectors, the radio technique has thus achieved a similar level of maturity as the traditional optical techniques; however, it is not restricted to clear nights. With further improvements planned regarding simulation codes, experiments, and analysis methods, digital radio arrays may eventually become the most accurate technique available for the measurements of the electromagnetic shower component.

Some of the scientific questions in ultra-high-energy cosmic-ray physics require maximum statistics at the highest energies, e.g., the search for ZeV particles or anisotropy studies above the GZK threshold. Huge radio arrays are probably the most economic way to surpass current observatories by an order of magnitude in exposure. Progress has been made for self-triggered radio detection and autonomous station designs. By continuing these ongoing R&D efforts, stand-alone radio arrays such as GRAND appear feasible in the next decade.

Many other science goals in cosmic-ray physics require a high sensitivity to the mass of the primary particle, e.g., testing scenarios for the most energetic Galactic and extragalactic source types or searching for mass-dependent anisotropies. For these science goals, hybrid arrays combining radio antennas with particle detectors seems to be the best option, especially, when the particle detectors can measure the muonic shower content. The technology for such hybrid arrays is ready, and further advances in the analysis techniques will release the full potential of the combination of radio and muon detectors for mass separation. Even with the currently available analysis techniques, radio antennas are already an important asset to achieve the energy and mass resolution required for progress in cosmic-ray science. Consequently, leading cosmic-ray observatories such as Ice-Top and the Pierre Auger Observatory have started enhancements including radio antennas. Moreover, the multi-messenger observatories planned to lead the field in the next decade consider radio antennas as part of a hybrid concept, in particular, IceCube-Gen2 with its surface array for air showers and GCOS.

References

1. T. Huege, Radio detection of cosmic ray air showers in the digital era, *Phys. Rept.* **620**, 1–52 (2016). doi: 10.1016/j.physrep.2016.02.001.
2. F. G. Schröder, Radio detection of Cosmic-Ray Air Showers and High-Energy Neutrinos, *Prog. Part. Nucl. Phys.* **93**, 1–68 (2017). doi: 10.1016/j.ppnp.2016.12.002.
3. T. Huege, M. Ludwig, and C. W. James, Simulating radio emission from air showers with CoREAS, *AIP Conf. Proc.* **1535**(1), 128 (2013). doi: 10.1063/1.4807534.
4. J. Alvarez-Muniz, W. R. Carvalho, Jr., and E. Zas, Monte Carlo simulations of radio pulses in atmospheric showers using ZHAireS, *Astropart. Phys.* **35**, 325–341 (2012). doi: 10.1016/j.astropartphys.2011.10.005.

5. W. D. Apel *et al.*, LOPES collaboration, final results of the LOPES radio interferometer for cosmic-ray air showers, *Eur. Phys. J. C.* **81**(2), 176 (2021). doi: 10.1140/epjc/s10052-021-08912-4.
6. F. G. Schröder *et al.*, High-energy galactic cosmic rays (Astro2020 Science White Paper), *Bull. Am. Astron. Soc.* **51**, 131 (March, 2019).
7. F. Sarazin *et al.*, What is the nature and origin of the highest-energy particles in the universe? *Bull. Am. Astron. Soc.* **51**(3), 93 (2019).
8. Hörandel, J. R. *et al.*, A next-generation cosmic-ray detector to study the physics and properties of the highest-energy particles in nature, *Snowmass2021 Lett. Interest* **CF7**, 117 (2020).
9. S. Buitink, A. Corstanje, H. Falcke, B. M. Hare, J. Hörandel, T. Huege, C. James, G. Krampah, K. Mulrey, P. Mitra, A. Nelles, H. Pandya, J. P. Rachen, O. Scholten, S. ter Veen, S. Thoudam, G. Trinh, and T. Winchen, Performance of SKA as an air shower observatory, *PoS.* **ICRC2021**, 415 (2021). doi: 10.22323/1.395.0415.
10. F. Schlüter for the Pierre Auger Collaboration, Expected performance of the AugerPrime radio detector, *PoS.* **ICRC2021**, 262 (2021).
11. A. Haungs for the IceCube Collaboration, A scintillator and radio enhancement of the IceCube surface detector array, *EPJ Web Conf.* **210**, 06009 (2019). doi: 10.1051/epjconf/201921006009.
12. F. G. Schroeder for the IceCube-Gen2 Collaboration, The surface array planned for IceCube-Gen2, *PoS.* **ICRC2021**, 407 (2021). doi: 10.22323/1.395.0407.
13. J. Álvarez-Muñiz *et al.*, GRAND Collaboration, The Giant Radio Array for Neutrino Detection (GRAND): Science and design, *Sci. China Phys. Mech. Astron.* **63**(1), 219501 (2020). doi: 10.1007/s11433-018-9385-7.
14. J. R. Hörandel on behalf of the GCOS Collaboration, The global cosmic ray Observatory — GCOS, *PoS.* **ICRC2021**, 027 (2021). doi: 10.22323/1.395.0027.
15. F. G. Schröder, Air shower detection by arrays of radio antennas, *EPJ Web Conf.* **208**, 15001 (2019). doi: 10.1051/epjconf/201920815001.
16. E. N. Paudel, A. Coleman, and F. Schroeder, Parametrization of the relative amplitude of geomagnetic and Askaryan radio emission from cosmic-ray air showers using CORSIKA/CoREAS simulations, *PoS.* **ICRC2021**, 429 (2021). doi: 10.22323/1.395.0429.
17. C. Glaser, M. Erdmann, J. R. Hörandel, T. Huege, and J. Schulz, Simulation of radiation energy release in air showers, *JCAP.* **09**, 024 (2016). doi: 10.1088/1475-7516/2016/09/024.
18. O. Scholten *et al.*, LOPES Collaboration, Measurement of the circular polarization in radio emission from extensive air showers confirms emission mechanisms, *Phys. Rev. D.* **94**(10), 103010 (2016). doi: 10.1103/PhysRevD.94.103010.

19. P. Schellart *et al.*, LOPES Collaboration, Polarized radio emission from extensive air showers measured with LOFAR, *JCAP*. **10**, 014 (2014). doi: 10.1088/1475-7516/2014/10/014.

20. A. Rehman, A. Coleman, F. G. Schröder, and D. Kostunin, Classification and denoising of cosmic-ray radio signals using deep learning, *PoS*. **ICRC2021**, 417 (2021). doi: 10.22323/1.395.0417.

21. Q. Gou *et al.*, The GRANDproto35 experiment, *PoS*. **ICRC2017**, 388 (2018). doi: 10.22323/1.301.0388.

22. A. Aab *et al.*, Pierre-Auger Collaboration, Measurement of the radiation energy in the radio signal of extensive air showers as a universal estimator of cosmic-ray energy, *Phys. Rev. Lett.* **116**(24), 241101 (2016). doi: 10.1103/PhysRevLett.116.241101.

23. W. D. Apel *et al.*, LOPES Collaboration, Reconstruction of the energy and depth of maximum of cosmic-ray air-showers from LOPES radio measurements, *Phys. Rev. D.* **90**(6), 062001 (2014). doi: 10.1103/PhysRevD.90.062001.

24. P. A. Bezyazeekov *et al.*, Tunka-Rex Collaboration, Radio measurements of the energy and the depth of the shower maximum of cosmic-ray air showers by Tunka-Rex, *JCAP*. **01**, 052 (2016). doi: 10.1088/1475-7516/2016/01/052.

25. A. Balagopal V., A. Haungs, T. Huege, and F. G. Schroeder, Search for PeVatrons at the Galactic Center using a radio air-shower array at the South Pole, *Eur. Phys. J. C.* **78**(2), 111 (2018). doi: 10.1140/epjc/s10052-018-5537-2. [Erratum: Eur.Phys.J.C 78, 1017 (2018), Erratum: Eur.Phys.J.C 81, 483 (2021)].

26. R. Šmída *et al.*, CROME, First experimental characterization of microwave emission from cosmic ray air showers, *Phys. Rev. Lett.* **113**(22), 221101 (2014). doi: 10.1103/PhysRevLett.113.221101.

27. D. Ardouin *et al.*, CODALEMA Collaboration, Geomagnetic origin of the radio emission from cosmic ray induced air showers observed by CODALEMA, *Astropart. Phys.* **31**, 192–200 (2009). doi: 10.1016/j.astropartphys.2009.01.001.

28. P. Schellart *et al.*, LOFAR, Detecting cosmic rays with the LOFAR radio telescope, *Astron. Astrophys.* **560**, A98 (2013). doi: 10.1051/0004-6361/201322683.

29. R. Monroe *et al.*, OVRO-LWA, Self-triggered radio detection and identification of cosmic air showers with the OVRO-LWA, *Nucl. Instrum. Meth. A.* **953**, 163086 (2020). doi: 10.1016/j.nima.2019.163086.

30. D. Charrier *et al.*, TREND Collaboration, Autonomous radio detection of air showers with the TREND50 antenna array, *Astropart. Phys.* **110**, 15–29 (2019). doi: 10.1016/j.astropartphys.2019.03.002.

31. P. A. Bezyazeekov *et al.*, Tunka-Rex Collaboration, Measurement of cosmic-ray air showers with the Tunka Radio Extension (Tunka-Rex), *Nucl. Instrum. Meth. A.* **802**, 89–96 (2015). doi: 10.1016/j.nima.2015. 08.061.

32. P. Abreu *et al.*, Pierre Auger Collaboration, Results of a self-triggered prototype system for radio-detection of extensive air showers at the Pierre Auger Observatory, *JINST.* **7**, P11023 (2012). doi: 10.1088/ 1748-0221/7/11/P11023.

33. S. Wissel *et al.*, Prospects for high-elevation radio detection of >100 PeV tau neutrinos, *JCAP.* **11**, 065 (2020). doi: 10.1088/ 1475-7516/2020/11/065.

34. H. Schoorlemmer *et al.*, ANITA Collaboration, Energy and flux measurements of ultra-high energy cosmic rays observed during the first ANITA flight, *Astropart. Phys.* **77**, 32–43 (2016). doi: 10.1016/j. astropartphys.2016.01.001.

35. W. D. Apel *et al.*, LOPES and Tunka-Rex Collaborations, A comparison of the cosmic-ray energy scales of Tunka-133 and KASCADE-Grande via their radio extensions Tunka-Rex and LOPES, *Phys. Lett. B.* **763**, 179–185 (2016). doi: 10.1016/j.physletb.2016.10.031.

36. K. Mulrey *et al.*, LOFAR, On the cosmic-ray energy scale of the LOFAR radio telescope, *JCAP.* **11**, 017 (2020). doi: 10.1088/1475-7516/2020/ 11/017.

37. W. D. Apel *et al.*, LOPES Collaboration, The wavefront of the radio signal emitted by cosmic ray air showers, *JCAP.* **09**, 025 (2014). doi: 10.1088/1475-7516/2014/09/025.

38. A. Corstanje *et al.*, LOFAR, The shape of the radio wavefront of extensive air showers as measured with LOFAR, *Astropart. Phys.* **61**, 22–31 (2015). doi: 10.1016/j.astropartphys.2014.06.001.

39. A. Aab *et al.*, Pierre Auger Collaboration, The Pierre Auger Observatory upgrade — preliminary design report (2016). arXiv.org:1604. 03637.

40. R. Hiller for the Tunka-Rex Collaboration, Tunka-Rex: Energy reconstruction with a single antenna station, *EPJ Web Conf.* **135**, 01004 (2017). doi: 10.1051/epjconf/201713501004.

41. C. Welling, C. Glaser, and A. Nelles, Reconstructing the cosmic-ray energy from the radio signal measured in one single station, *JCAP.* **10**, 075 (2019). doi: 10.1088/1475-7516/2019/10/075.

42. P. A. Bezyazeekov *et al.*, Tunka-Rex Collaboration, Reconstruction of cosmic ray air showers with Tunka-Rex data using template fitting of radio pulses, *Phys. Rev. D.* **97**(12), 122004 (2018). doi: 10.1103/ PhysRevD.97.122004.

43. W. D. Apel *et al.*, LOPES Collaboration, Improved absolute calibration of LOPES measurements and its impact on the comparison with REAS 3.11 and CoREAS simulations, *Astropart. Phys.* **75**, 72–74 (2016). doi: 10.1016/j.astropartphys.2015.09.002.

44. A. Aab *et al.*, Pierre Auger Collaboration, Calibration of the logarithmic-periodic dipole antenna (LPDA) radio stations at the Pierre Auger Observatory using an octocopter, *JINST.* **12**(10), T10005 (2017). doi: 10.1088/1748-0221/12/10/T10005.

45. A. Corstanje *et al.*, LOFAR, Results on mass composition of cosmic rays as measured with LOFAR, *PoS.* **ICRC2021**, 322 (2021). doi: 10. 22323/1.395.0322.

46. B. Pont for the Pierre Auger Collaboration, The depth of the shower maximum of air showers measured with AERA, *PoS.* **ICRC2021**, 387 (2021). doi: 10.22323/1.395.0387.

47. B. Pont, Cosmic ray mass composition, PhD Thesis. Radboud University (2021). http://hdl.handle.net/2066/234170.

48. O. Vedeneev, Depth of the maximum of extensive air showers and mass composition of primary cosmic radiation at an energy of 4 x 10**17 eV according to data on radioemission from extensive air showers, *Phys. Atom. Nucl.* **72**, 250–256 (2009). doi: 10.1134/S1063778809020070.

49. W. D. Apel *et al.*, LOPES Collaboration, Experimental evidence for the sensitivity of the air-shower radio signal to the longitudinal shower development, *Phys. Rev. D.* **85**, 071101 (2012). doi: 10.1103/PhysRevD. 85.071101.

50. S. Grebe for the Pierre Auger Collaboration, Spectral index analysis of the data from the Auger Engineering Radio Array, *AIP Conf. Proc.* **1535**(1), 73–77 (2013). doi: 10.1063/1.4807524.

51. A. Corstanje *et al.*, LOFAR, Depth of shower maximum and mass composition of cosmic rays from 50 PeV to 2 EeV measured with the LOFAR radio telescope, *Phys. Rev. D.* **103**(10), 102006 (2021). doi: 10.1103/PhysRevD.103.102006.

52. V. Lenok for the Tunka-Rex Collaboration, Estimation of aperture of the Tunka-Rex radio array for cosmic-ray air-shower measurements, *PoS.* **ICRC2021**, 210 (2021). doi: 10.22323/1.395.0210.

53. J. Alvarez-Muñiz *et al.*, Monte Carlo simulations of radio pulses in atmospheric showers using ZHAireS, *Astropart. Phys.* **35**, 325–341 (2012).

54. T. Huege, M. Ludwig, and C. James, Simulating radio emission from air showers with CoREAS, *AIP Conf. Proceedings.* **1535**, 128–132 (2013).

55. N. Karastathis, R. Prechelt, T. Huege, and J. Ammerman-Yebra, Simulations of radio emission from air showers with CORSIKA 8, *PoS.* **ICRC2021**, 427 (2021). doi: 10.22323/1.395.0427.

56. C. Welling, P. Frank, T. A. Enßlin, and A. Nelles, Reconstructing non-repeating radio pulses with Information Field Theory, *JCAP.* **04**, 071 (2021). doi: 10.1088/1475-7516/2021/04/071.

57. M. Erdmann, F. Schlüter, and R. Smida, Classification and recovery of radio signals from cosmic ray induced air showers with deep learning, *JINST.* **14**(04), P04005 (2019). doi: 10.1088/1748-0221/14/04/P04005.

58. P. A. Bezyazeekov *et al.*, Tunka-Rex Collaboration, Reconstruction of radio signals from air-showers with autoencoder, *MDPI Proc. L.* **2679**, 43–50 (2020).

59. D. Ivanov, O. E. Kalashev, M. Y. Kuznetsov, G. I. Rubtsov, T. Sako, Y. Tsunesada, and Y. V. Zhezher, Using deep learning to enhance event geometry reconstruction for the telescope array surface detector, *Mach. Learn. Sci. Tech.* **2**(1), 015006 (2021). doi: 10.1088/2632-2153/abae74.

60. A. Aab *et al.*, Pierre-Auger Collaboration, Deep-learning based reconstruction of the shower maximum X_{max} using the water-Cherenkov detectors of the Pierre Auger Observatory, *JINST.* **16**(07), P07019 (2021). doi: 10.1088/1748-0221/16/07/P07019.

61. R. U. Abbasi *et al.*, TARA Collaboration, First upper limits on the radar cross section of cosmic-ray induced extensive air showers, *Astropart. Phys.* **87**, 1–17 (2017). doi: 10.1016/j.astropartphys.2016.11.006.

62. S. Prohira *et al.*, Observation of radar echoes from high-energy particle cascades, *Phys. Rev. Lett.* **124**(9), 091101 (2020). doi: 10.1103/PhysRevLett.124.091101.

63. S. Prohira *et al.*, The radar echo telescope for cosmic rays: Pathfinder experiment for a next-generation neutrino observatory, *Phys. Rev. D* **104**, 102006 (2021).

64. F. G. Schröder *et al.*, New method for the time calibration of an interferometric radio antenna array, *Nucl. Instrum. Meth. A.* **615**, 277–284 (2010). doi: 10.1016/j.nima.2010.01.072.

65. F. Schlüter and T. Huege, Expected performance of air-shower measurements with the radio-interferometric technique (February, 2021).

66. A. Aab *et al.*, Pierre Auger Collaboration, Nanosecond-level time synchronization of autonomous radio detector stations for extensive air showers, *JINST.* **11**(01), P01018 (2016). doi: 10.1088/1748-0221/11/01/P01018.

67. A. G. Vieregg, K. Bechtol, and A. Romero-Wolf, A technique for detection of PeV neutrinos using a phased radio array, *JCAP.* **02**, 005 (2016). doi: 10.1088/1475-7516/2016/02/005.

68. H. Schoorlemmer and W. R. Carvalho, Radio interferometry applied to the observation of cosmic-ray induced extensive air showers (June, 2020).

69. E. M. Holt for the Pierre Auger Collaboration, Estimating the mass of cosmic rays by combining radio and muon measurements, *EPJ Web Conf.* **216**, 02002 (2019). doi: 10.1051/epjconf/201921602002.
70. J. R. Hörandel for the Pierre Auger Collaboration, Precision measurements of cosmic rays up to the highest energies with a large radio array at the Pierre Auger Observatory, *EPJ Web Conf.* **210**, 06005 (2019). doi: 10.1051/epjconf/201921006005.
71. T. Huege *et al.*, Ultimate precision in cosmic-ray radio detection — the SKA, *EPJ Web Conf.* **135**, 02003 (2017). doi: 10.1051/epjconf/201713502003.
72. C. W. James *et al.*, Overview of lunar detection of ultra-high energy particles and new plans for the SKA, *EPJ Web Conf.* **135**, 04001 (2017). doi: 10.1051/epjconf/201713504001.
73. E. de Lera Acedo *et al.*, Evolution of skala (skala-2), the log-periodic array antenna for the ska-low instrument, In *2015 International Conference on Electromagnetics in Advanced Applications (ICEAA).* pp. 839–843 (September, 2015). doi: 10.1109/ICEAA.2015.7297231.
74. F. G. Schröder for the IceCube-Gen2 Collaboration, Science Case of a Scintillator and Radio Surface Array at IceCube, *PoS.* **ICRC2019**, 418 (2020). doi: 10.22323/1.358.0418.
75. K. Mulrey, Cross-calibrating the energy scales of cosmic-ray experiments using a portable radio array, *PoS.* **ICRC2021**, 414 (2021). doi: 10.22323/1.395.0414.
76. B. Pont for the Pierre Auger Collaboration, A large radio detector at the Pierre Auger Observatory — measuring the properties of cosmic rays up to the highest energies, *PoS.* **ICRC2019**, 395 (2021). doi: 10.22323/1.358.0395.
77. P. Abreu *et al.*, Pierre Auger Collaboration, Antennas for the detection of radio emission pulses from cosmic-ray, *JINST.* **7**, P10011 (2012). doi: 10.1088/1748-0221/7/10/P10011.
78. A. Leszczynska for the IceCube-Gen2 Collaboration, Simulation study for the future IceCube-Gen2 surface array, *PoS.* **ICRC2021**, 411 (2021). doi: 10.22323/1.395.0411.
79. R. Abbasi *et al.*, IceCube Collaboration, IceTop: The surface component of IceCube, *Nucl. Instrum. Meth. A.* **700**, 188–220 (2013). doi: 10.1016/j.nima.2012.10.067.
80. D. Soldin for the IceCube Collaboration, Recent results of cosmic ray measurements from IceCube and IceTop, *PoS.* **ICRC2019**, 014 (2019). doi: 10.22323/1.358.0014.
81. W. D. Apel *et al.*, KASCADE-Grande Collaboration, Kneelike structure in the spectrum of the heavy component of cosmic rays observed with KASCADE-Grande, *Phys. Rev. Lett.* **107**, 171104 (2011). doi: 10.1103/PhysRevLett.107.171104.

82. A. Aab *et al.*, Pierre Auger Collaboration, Features of the energy spectrum of cosmic rays above 2.5×10^{18} eV using the Pierre Auger Observatory, *Phys. Rev. Lett.* **125**(12), 121106 (2020). doi: 10.1103/ PhysRevLett.125.121106.
83. K. Kotera for the GRAND Collaboration, The Giant Radio Array for Neutrino Detection (GRAND) project, *PoS*. **ICRC2021**, 1181 (2021). doi: 10.22323/1.395.1181.
84. A. Aab *et al.*, Pierre Auger Collaboration, The Pierre Auger Cosmic Ray Observatory, *Nucl. Instrum. Meth. A*. **798**, 172–213 (2015). doi: 10.1016/j.nima.2015.06.058.
85. R. U. Abbasi *et al.*, Telescope Array Collaboration, Surface detectors of the TAx4 experiment (March, 2021).
86. Preliminary Helmholtz Roadmap for Research Infrastructures 2021 (in German). https://www.helmholtz.de/fileadmin/medien_upload/Neu neu_HELMHOLTZ-ROADMAP-2021-Vorlaeufige-Fassung-210623.pdf Accessed: 2021-07-28.
87. F. G. Schröder, News from Cosmic Ray Air Showers (Cosmic Ray Indirect — CRI Rapporteur), *PoS*. **ICRC2019**, 030 (2020). doi: 10.22323/1.358.0030.
88. C. James, Nature of radio-wave radiation from particle cascades, *Phys. Rev. D* **105**, 023014 (2022).

https://doi.org/10.1142/9781800612617_0004

Chapter 4

Space-based Ultra-High-Energy Cosmic-Ray Experiments

John F. Krizmanic

NASA/Goddard Space Flight Center,
Laboratory for Astroparticle Physics,
Greenbelt, MD 20771, USA
john.f.krizmanc@nasa.gov

Space-based experiments, either orbiting the Earth or from scientific balloon altitudes, measure high-energy cosmic rays by measuring from above the atmosphere the optical and radio signals generated by extensive air showers (EASs). These experiments are designed to have a large field of view for observing EASs which translates to monitoring the atmosphere over a large area ($\sim 10^6 \, \text{km}^2$) on the ground. Ultra-high-energy cosmic rays (UHECRs, $E_{CR} \gtrsim 1$ EeV) are measured by using the isotropic near-UV air-fluorescence signal to finely sample the EAS development and to efficiently use the atmosphere as a vast calorimeter. At UHE, these immense EAS particle cascades have sufficient charged particle content to generate the relatively dim fluorescence light that propagates to the space-based instrument. Additionally, the beamed Cherenkov light and geomagnetic radio emission from EASs arrive with $\gtrsim 10$ ns impulses and are measured at small angles away from the cosmic-ray trajectories. In particular, for optical Cherenkov measurements, the energy thresholds can be $\gtrsim 1$ PeV, i.e., for very-high-energy cosmic rays (VHECRs). The instruments that use these EAS optical signals are effectively coarse-imaging telescopes with meter-sized optical collecting areas to minimize the VHECR and UHECR detection energy thresholds, which are in part set by the dark-sky airglow chemiluminescence optical background. Since optical signals need to be measured near astronomical night, these experiments have a mission-averaged measurement live

93

time (duty cycle) around 10–20%, which takes into account effects of the variable viewing conditions, mainly cloud cover, high-altitude ozone thickness, and avoiding viewing areas with high aurora backgrounds. The effects of low-altitude aerosols are small when viewing EASs via fluorescence from above the atmosphere since the majority of the EAS development is above the ∼1 km scale height of the aerosol layer. These effects are further complicated since the monitored atmospheric volume changes are due to motion of the experiment.

In this chapter, the nature of observing the UHECR-induced shower development from orbiting and balloon-borne experiments is detailed, both for missions that have been flown and those currently in development. This will be accomplished by discussing experimental performance in terms of measuring the UHECR spectrum, UHECR nuclear composition, and UHECR arrival direction. The ability for these experiments to also perform VHECR and VHE and UHE cosmic neutrino measurements will also be discussed.

In space, no one can hear you scream ...

— tagline of Ridley Scott's movie *Alien*

1. Introduction

Space-based experiments use optical or radio telescopes that are designed to view from above the atmosphere, the development of extensive air showers (EASs) initiated by ultra-high-energy cosmic rays (UHECRs). Since the EAS particle cascades have an immense number of charged particles, dominated by electrons and positrons even for hadronic UHECRs, they can form signals that can be detected by distant Earth-orbiting or balloon-borne experiments. The relatively coarse imaging requirements needed to accurately sample the $\gtrsim 10\,\mu$s waxing and waning development using the EAS near-UV air-fluorescence signal, ∼km spatial scale near the Earth's surface, allows for the use of large field-of-view (FoV) telescopes and translates to monitoring ∼10^{13} tons of atmosphere for UHECRs. Figure 1 artistically presents the variety of EAS fluorescence and optical Cherenkov instruments that perform UHECR measurements as well as the laser calibration technique and sources of other optical signals, including meteors and transient atmosphere phenomena.[1] Since the majority of EAS development is within ∼30 km altitude

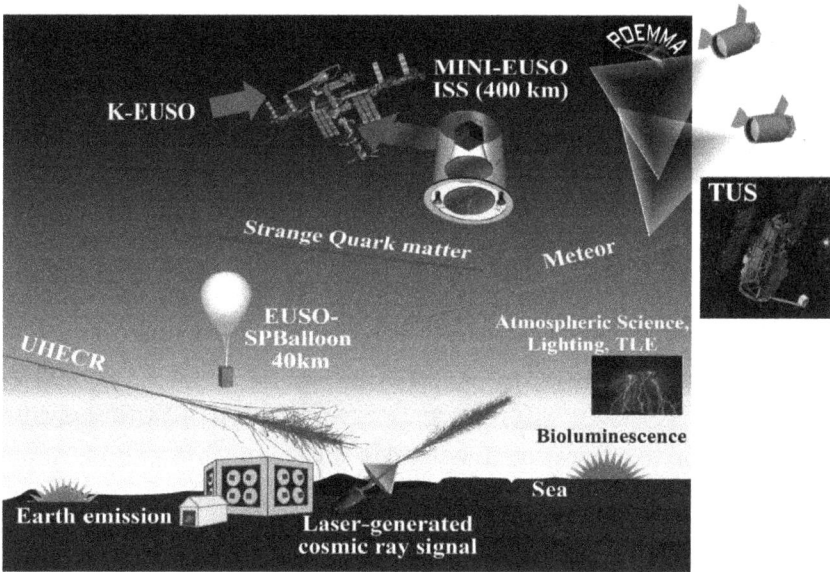

Fig. 1. A composite collection of space-based experiments designed to measure UHECRs via the optical signals from extensive air showers, included are several that have flown (TUS and Mini-EUSO), those that will fly in the near term (EUSO-SPB2), and those that are in the development stage (K-EUSO and POEMMA). UHECR measurements are the primary science goals but the large-FoV's image with 0.1° resolution in the near-UV and optical wavelengths are sensitive to a variety of secondary science signals, but with much longer observational time signatures than UHECRs. These include meteors, lightning and atmospheric transient luminous events, and bioluminescence. Also shown is the existing technique of using upward-pointed laser pulses to mimic EASs for instrument calibration.[1]

above sea level, the large distance from a nominal 525 km low Earth orbit (LEO) to the EASs requires meter-scale optical systems, including refractive systems based on Fresnel lenses and reflective systems, such as Schmidt telescopes. However, even with these large optics, the threshold energy for UHECR detection is $\gtrsim 20$ EeV, which implies space-based UHECR experiments are designed to provide high exposure measurements of the highest energies of the cosmic-ray spectrum. The advantage of using observations from LEO is full-sky sensitivity to cosmic UHECR sources on a time scale of

$\sim$year due to the required operation near astronomical night to measure the optical EAS signal, which yields $\sim$(10–15)% measurement average live time. These also yield all-sky measurement of cosmic UHECR sources under a single experimental framework with well-understood systematic errors on the measurements, especially when two satellites perform precise stereo-fluorescence measurements that properly determine the three-dimensional location of key features of the EAS, including incident angle and the slant depth of EAS maximum, X_{max}, which provides a measure of the identity of the UHECR particle. However, the volumes of atmosphere viewed by an orbiting experiment change quickly. Thus, the UHECR measurements are performed in changing atmospheric conditions, in particular, cloud cover in the FoV. Ancillary devices such as an IR camera measuring cloud top temperature or LIDAR measurements to give more detailed information along the EAS viewing direction provide the needed information. This information will be augmented with that available from other Earth atmosphere monitoring satellites, such as that used in the MERRA-2 global atmospheric properties database.[2]

Experiments flying at scientific balloon altitudes, $\sim$30 km, also have sensitivity to UHECRs via air-fluorescence measurements, but the UHECR exposure is limited due to the smaller atmospheric volume monitored by the telescope and due to shorter mission times. Currently, the maximum float time of a balloon is at most $\sim$100 days assuming an ultra-long duration balloon (ULDB) flight. However, the ANITA long-duration balloon (LDB) experiment has measured the beamed geomagnetic Cherenkov-like EAS radio emission from UHECRs both directly and reflected off the Antarctic ice for $E_{CR} \gtrsim 1$ EeV.[3] Additionally, simulation studies used to develop space-based experiments to detect optical Cherenkov light from EAS generated by cosmic neutrino interactions in the Earth have also shown that a meter-sized telescope pointed above the limb of the Earth will have sensitivity to VHECRs with $E_{CR} \gtrsim 1$ PeV. While these measurements in the radio and optical Cherenkov are currently less precise than air-fluorescence measurements, especially those done with EAS stereo reconstruction, they do offer the opportunity to measure cosmic rays at a much lower $\gtrsim$PeV energy than those measured by space-based instruments.

2. History

The concept of observing UHECRs from space was suggested in a paper by Robert Benson and John Linsley in 1980.[a,4,5] In a 1995 paper, Yoshi Takahasi proposed the concept called *Maximum-energy Auger (AIR)-Shower Satellite (MASS)*[6,7] to employ an Earth-orbiting, wide FoV coarse-imaging telescope to view EAS development using the near-UV fluorescence signal. A sketch of the MASS concept is shown in the left panel of Fig. 2, which was presented at a workshop on space-based UHECR measurements.[8]

Important UHECR information is encoded in the development profile of an EAS: The total emitted optical light is proportional to the energy, sampling with sufficient temporal and spatial resolution, the EAS evolution yields relatively long tracks to provide very good angular resolution for the UHECR primary, and measuring the location of shower maximum point in the development, X_{max}, provides information about the nature of a primary particle, e.g., proton, nucleus, photon, or neutrino.

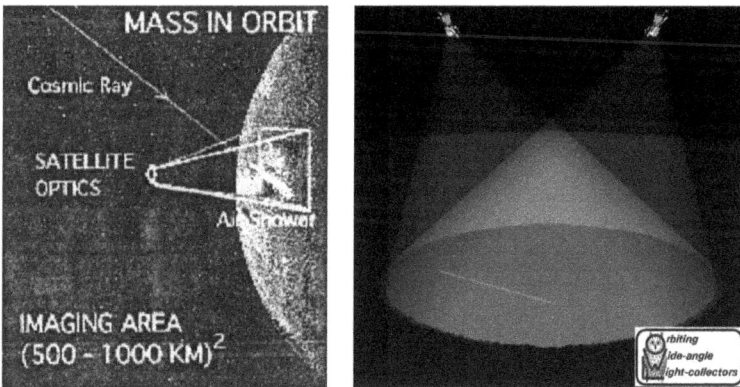

Fig. 2. Left: Artist sketch of the space-based MASS mission concept,[8] Right: Simulation graphic of two OWL spacecraft co-viewing UHECR-induced EAS, the fields of view of the telescopes are highlighted.[9]

---

[a]Folklore has it that the idea of observing UHECR EASs from space was proposed during a coffee break conversation at a scientific workshop between John Linsley and other participants.

The charged particles in an EAS produce ionization that excites air fluorescence that is emitted isotropically as the few hundred meter wide (determined by the Moliére radius, here taken at altitudes below 30 km) and few meter deep "pancake" of EAS particles moves through the atmosphere close to the speed of light. Thus, a 1 km imaging resolution near the ground is more than sufficient to view the vast majority of the EASs in a given snapshot. A μ-second sampling resolution matches well with that needed to accurately sample the waxing and waning of the EAS longitudinal profile to map the EAS development to determine the UHECR physical properties. That is, μs sampling is close to the optimal needed to have a high EAS fluorescence signal to dark-sky airglow background ratio. A km image size on the ground from LEO, i.e., 400–1000 km altitude, is very coarse in terms of the imaging requirement for optical astronomical telescopes.[b] This translates into generous tolerances on the optics, e.g., more similar to a microwave dish than an astronomical telescope.

In the early 2000s, this concept lead to the AIRWATCH concept[7] defining a single orbiting telescope for monocular observation of EASs as well as the Orbiting Wide-angle Light collectors (OWL) study, shown conceptually in right panel of Fig. 2. The two groups worked together in the late 1990s, and Fig. 3 shows a joint AIRWATCH/OWL logo drawn by John Linsley. OWL was designed to use the stereo-fluorescence EAS technique to measure more precisely the properties of the UHECRs and UHE neutrinos due to the superior angular resolution offered by the stereo technique. This provides the ability to achieve degree angular resolution and < 20% energy resolution, both needed to measure the UHECR nuclear composition evolution. The OWL development included using the air-fluorescence technique to search for cosmic neutrinos interacting in the atmosphere and the beamed Cherenkov light from upward-moving EASs from τ-lepton decay source from ν_τ interactions in the Earth.[10] The OWL instrument and mission were developed at the GSFC Integrated Design Center[11] in 2002, and Fig. 4 shows schematics of the OWL telescopes and launch configuration based on the IDC work.

---

[b] At 400 nm, 0.1° resolution using a meter-diameter telescope is $\sim 10^4$ away from the diffraction limit.

Fig. 3. The OWL/AIRWATCH logo drawn by John Linsley presented at an OWL/AIRWATCH meeting in 1998.

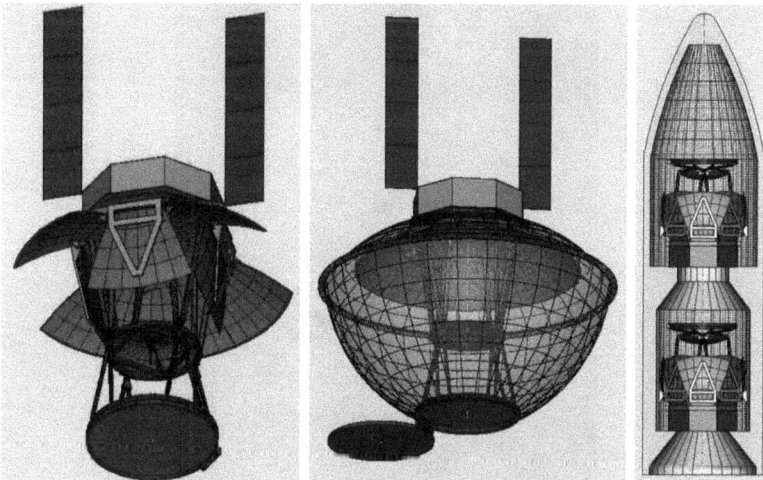

Fig. 4. Left: Schematic of an OWL satellite showing the internal, Schmidt telescope structures during deployment, the lightshield is not shown for clarity. Middle: Schematic of an OWL satellite showing the internal, deployed Schmidt telescope, the lightshield is shown as translucent. Right: The stowed OWL spacecraft in a dual-launch fairing on the launch vehicle.[12]

The AIRWATCH program led to the development of the Extreme Universe Space Observatory (EUSO) program, which included the EUSO on the Japanese Experiment Module (JEM-EUSO) on the International Space Station (ISS) mission concept. JEM-EUSO was to use a single nadir-viewing near-UV telescope using refractive Fresnel optics, to be located on the JEM external facility,[14] shown in Fig. 5. The EUSO program included a number of related balloon-borne missions, including EUSO-balloon[15] with a downward-looking fluorescence telescope using EUSO refractive optics (denoted as EUSO-FT), the Mini-EUSO experiment using a smaller version of the EUSO-FT[16] on the ISS, EUSO-SPB[17] ultra-long-duration balloon (ULDB) payload flown in 2017 with a EUSO-FT, and the EUSO-SPB2 ULDB mission[18] planned to fly in the spring of 2023. The latter includes a downward-looking Schmidt fluorescence telescope and an Earth-limb viewing Schmidt Cherenkov telescope to search for upward tau neutrino-induced EASs and observe Cherenkov light from VHECRs viewed over the limb.

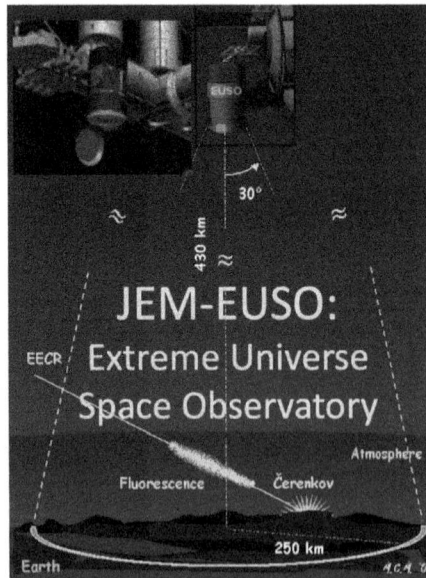

Fig. 5. Artist sketch of the EUSO on the Japanese Experimental Module (JEM-EUSO) mission proposed as an external International Space Station Payload.[13]

3. Space-based UHECR Missions

Somewhat independently, the Tracking Ultraviolet Set-up (TUS) experiment,[19,20] shown in Fig. 6, was built, launched, and operated on the Lomonosov satellite from May 19, 2016 until May 30, 2019. Three years of operation at $\sim$500 km altitude in a sun-synchronous orbit led to an exposure of 12,000 km^2 sr yr for $E_{UHECR} > 300$ EeV.[19] TUS employed an optical system with a $\sim$2 m^2 Fresnel mirror, a FoV of 9°, and imaged the near-UV air-fluorescence emission from EASs in a focal plane with 256 pixels. The design did not include a light shroud and this led to more sensitivity to optical backgrounds. The relatively modest size of the optical light collection led to a UHECR detection energy threshold of $\gtrsim$70 EeV, but TUS did detect six UHECR EAS candidate events whose origin may be due to background.[21,22] Figure 7 shows the location and light curve for an UHECR candidate event observed over Minnesota in 2016. While the UHECR candidate sample is small, TUS has measured a wealth of low-intensity transient atmospheric phenomena,[23] including transient luminous events (TLEs), meteors, and structures in the

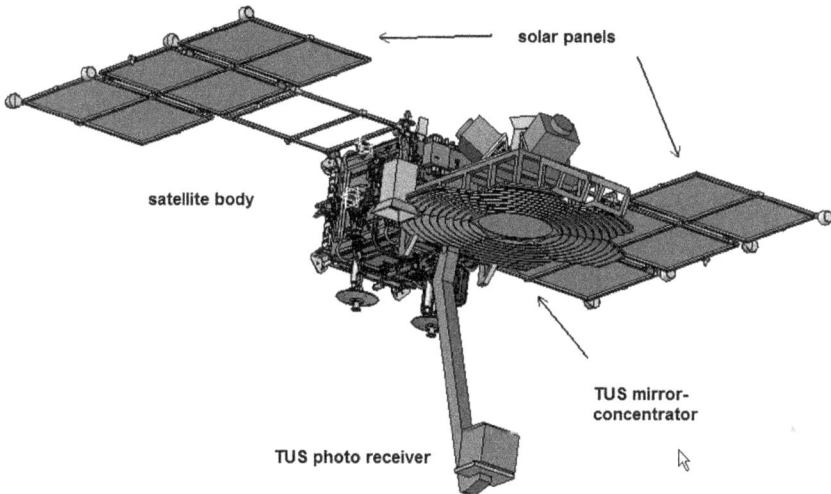

Fig. 6. Schematic of the TUS experiment on the Lomonosov satellite. The Fresnel mirror, with $\sim$2 m^2 area, has a field of view of 9° imaging EASs in a focal plane with 256 pixels. Three years of operation at $\sim$500 km altitude led to an exposure of 12,000 km^2 sr yr for $E_{UHECR} > 300$ EeV.[19]

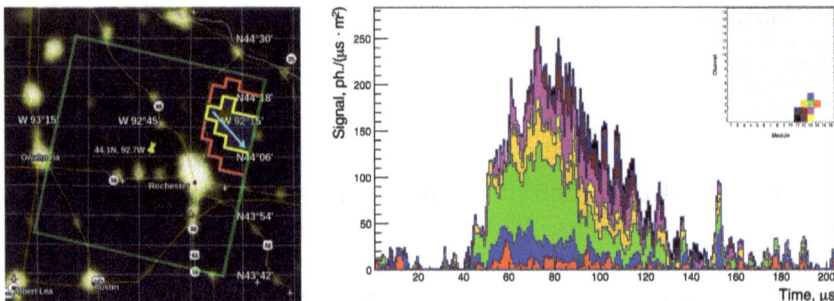

Fig. 7.   Left: FoV of TUS (green square) for the TUS161003 event overlaid on a Google Earth[d] map that includes NASA night Earth data. The yellow area shows the focal plane hit map while the red areas include the area with the effects of timing errors. Right: Light curve stacking the 10 hit channels for the TUS161003 event. The inset shows the location of the 10 hit pixels.[21]

aurora,[24] the last taking advantage of the Lomonosov satellite near-polar orbit.

The OWL study along with the ample work of the EUSO program, especially the MAPMT-based hardware for air-fluorescence telescopes, led to the development in 2018 of the Probe of Multi-Messenger Astrophysics (POEMMA), designed as a NASA Astrophysics Probe-class mission. POEMMA was selected as one of several different astrophysics missions to be developed under study program (NASA NNH16ZDA001N-APROBES) to help define the NASA Astrophysics Probe class, which is larger than an Explorer class but smaller than a Flagship mission. More importantly it provided resources to develop the POEMMA instruments, spacecraft, and mission designs at the Integrated Mission Design Center[11] at NASA/GSFC. Thus, the POEMMA study provided the opportunity for the UHECR community to develop as optimal a UHECR mission as possible, considering the relatively generous constraints defined for the probe class mission. The POEMMA UHECR exposure performance is shown in Fig. 8.

POEMMA is designed to observe cosmic neutrinos and to identify the sources of UHECRs with full-sky coverage. POEMMA consists of two spacecraft that co-view EASs while flying in a loose formation, separated by 300 km, at 525 km altitudes and at 28.5° inclination. Each spacecraft hosts a large Schmidt telescope with a novel focal plane, shown in Fig. 9, optimized to observe both the isotropic

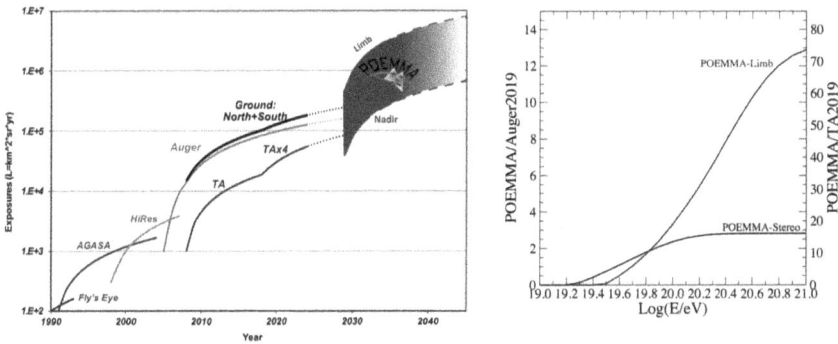

Fig. 8.   Left: The anticipated UHECR exposure growth curve for POEMMA compared to other UHECR experiments assuming nadir-pointing stereo-fluorescence measurements versus limb-pointing UHECR measurements but with less precision. Right: Comparison of five-year POEMMA exposure versus UHECR energy in terms of the PAO and TA exposures as of the 2019 ICRC.[25]

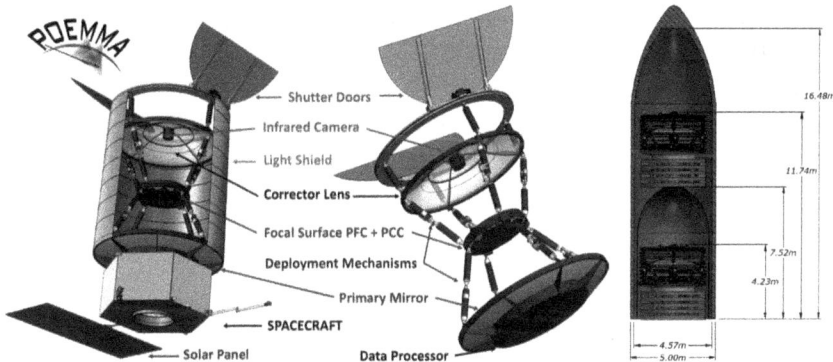

Fig. 9.   Schematics of a POEMMA satellite (left) and the Schmidt telescope (right) consisting of a 4 m diameter primary mirror, 3.3 m diameter corrector plate, and 1.6 m diameter focal surface composed of 126,720 pixels in the POEMMA Fluorescence Camera (PFC) and 15,360 pixels in the POEMMA Cherenkov Camera (PCC). Several components are detailed in the schematic, including infrared cameras which will measure cloud cover within the 45° full FoV of each telescope during science observations. Right: The stowed POEMMA spacecraft in a dual-launch fairing on the launch vehicle with length dimensions.[25]

near-UV fluorescence signal generated by EASs from UHECRs and UHE neutrinos and forward-beamed, optical Cherenkov signals from EAS. In neutrino limb-viewing Cherenkov mode, POEMMA will be sensitive to cosmic tau neutrinos above 20 PeV by observing the upward-moving EAS induced from tau neutrino interactions in the Earth. The POEMMA spacecraft are designed to quickly re-orient to a target-of-opportunity (ToO) neutrino mode to view and follow transient astrophysical sources with unique flux sensitivity over the full sky. Each POEMMA Schmidt telescope is composed of a 4 m diameter primary mirror and 3.3 m diameter corrector lens that yields nearly 6 m^2 effective on-axis light collecting ability to both perform precision UHECR measurements and search for UHE cosmic neutrinos. The large collection area provides for a UHECR detection threshold energy of 20 EeV, which is needed to have sufficient overlap with ground-based measurements performed by the Pierre Auger Observatory (PAO: in the Southern Hemisphere)[26,27] and the Telescope Array (TA: in the Northern Hemisphere).[28,29] Furthermore, UHECR measurements are required for $E_{CR} \approx 40$ EeV to be able to measure the anisotropic features measured by PAO (correlation to star-burst galaxies)[30] and TA (hot spot).[31] That is an experiment that has high UHECR sensitivity and will also have high sensitivity to UHE neutrino interactions that occur deep in the atmosphere that are well separated from UHECR EAS.[25] The high-statistics ($\geq$1,400 events for a five-year mission) full-sky UHECR measurements above 20 EeV using the stereo air-fluorescence technique would provide a major advance in discovering the sources of UHECRs.[25,32] POEMMA also provides unique sensitivity to UHE cosmic neutrino searches using stereo air-fluorescence measurements and an Earth limb-pointed mode to observe VHE Earth-interacting cosmic tau neutrinos using the beamed optical Cherenkov light generation from EAS for $E_\nu \gtrsim 20$ PeV.[33,34] Figure 10 illustrates the two science modes of POEMMA, while Table 1 summarizes the UHECR performance.

While the wavelength band for air fluorescence extends from below 200 nm to over 1000 nm,[35] the vast majority is in the band from 300 nm to 450 nm.[35,36] This allows for the use of UV filters to limit the effects of the dark-sky airglow background to be ~500 photons m^{-2} ns^{-1} sr^{-1}[37] and thus help reduce the UHECR EAS observational energy threshold. POEMMA, at 525 km altitude LEO, will view the air-fluorescence EAS signals using a large 45° full FoV. Thus, a vast amount of atmosphere will be monitored, i.e., ~10^{13} metric tons.[25,38] The EAS imaging requirements that correspond to resolving approximately 1 km spatial lengths on the Earth's surface

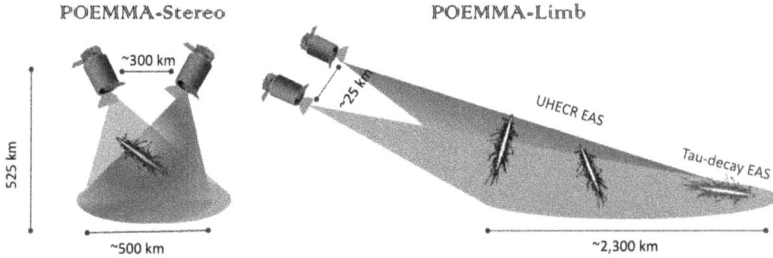

Fig. 10. The POEMMA science modes. Left: POEMMA-Stereo where the spacecraft are separated and viewing a common atmospheric volume to measure the fluorescence emission from EAS. Right: POEMMA-Limb where the instruments are tilted to view near and below the limb of the Earth for optical Cherenkov EAS induced by tau neutrino events in the Earth.[25]

Table 1. Overview of the simulated POEMMA UHECR measurement capabilities.[46]

Parameter	Performance
UHECR Stereo	260,000 km^2 sr (50 EeV)
Geometry Factor	400,000 km^2 sr ($\geq$100 EeV)
Obs Duty Cycle	13%
Physics Energy Thres	20 EeV
UHE Stereo Energy Res	< 19% (50 EeV)
UHE Angular Res	$\leq$ 1.2° (50 EeV)
UHE X_{max} Res	$\leq$ 30 g cm^{-2} (50 EeV)
CR rejection factor for UHE neutrinos	2 × 10^{-4}
Sky Coverage	±20% @ 50 EeV (1 Year)
Variability	±10% @ 50 EeV (5 Year)

Source: Adapted from Ref. [39].

leads to a iFoV= 0.084° for each pixel in the POEMMA focal plane shown in Fig. 14. This iFoV from 525 km altitude LEO sets the integration time of the imaged air fluorescence to be 1 μs to optimize the signal to dark-sky background ratio for EAS fluorescence detection.

The yield of fluorescence light (at STP) is of the order of 10 photons per MeV of ionization energy deposited by EAS particles.[40] For EAS viewed from space using a wide FoV telescope, the entire EAS development is within the FoV. Figure 11 shows the air-fluorescence yield as a function of altitude[39] and shows that the majority of the air-fluorescence signal is generated below an altitude of $\lesssim$30 km.

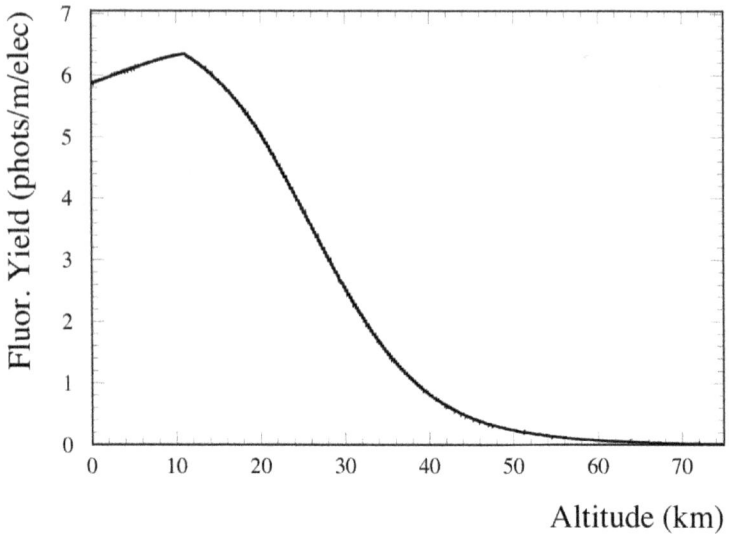

Fig. 11.   Air-fluorescence yield as a function of altitude.[39]

Viewed from a distance of more than 10 km, the EASs can be well approximated as a line along which the shower progresses, at first brightening then diminishing. A parametric model of the longitudinal development using the Gaisser–Hillas function[41] along the shower track combined with ray tracing of the generated isotropic light provides input for Monte Carlo simulations for the generation of EAS, optical (and radio) signals, propagation and attenuation through the atmosphere, and instrument response are used to model the detection of a space-based instrument.[42–44] The leftmost schematic in Fig. 10 illustrates stereo-fluorescence detection using two Schmidt telescopes for POEMMA.[25]

The EAS also creates an Cherenkov light signal due to the fact the charged particles are moving faster than the speed of light in the medium, i.e., the atmosphere. This Cherenkov light can scatter in the atmosphere and also be detected but tends to be a small fraction of the total EAS optical signal, ∼10% based on space-based UHECR simulation studies. The beamed Cherenkov light can also scatter off the ground and be used to determine the ground spot of the UHECR trajectory (see Fig. 5), which improves the accuracy in the reconstruction of the EAS trajectory.[44]

3.1. Optical detection instrumentation

The stereo-fluorescence technique, pioneered by the ground-based Fly's Eye experiment in Utah[45] provides a experimental methodology to precisely measure the development of EAS, thus achieving good angular, energy, and X_{max} resolutions for each EAS measurement. Figure 12 illustrates two different techniques. Monocular reconstruction uses a single detector location to measure the temporal and spatial evolution of the EAS in a segmented UV camera using the timing and geometry to determine the EAS trajectory and location of X_{max}. The second technique uses two (or more) EAS observations and stereo reconstruction to provide a more precise determination of the three-dimensional location of the EAS development and thus more precise angular, energy, and X_{max} measurements. When applied from a space-based experiment, these techniques allow for an immense amount of atmosphere to be used as a UHECR and UHE neutrino detector. In particular, the stereo reconstruction technique using an iFoV $\approx 0.1°$ from a LEO altitude of 525 km precisely defines the EAS three-dimensional trajectory. Figure 13 shows the simulation results for a 50 EeV UHECR proton event as seen in the focal plane

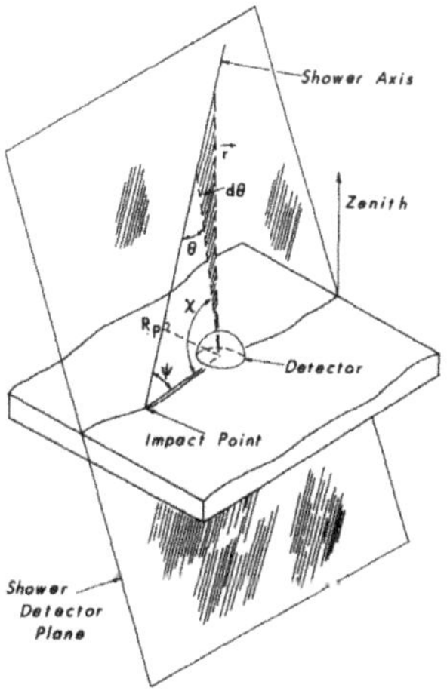

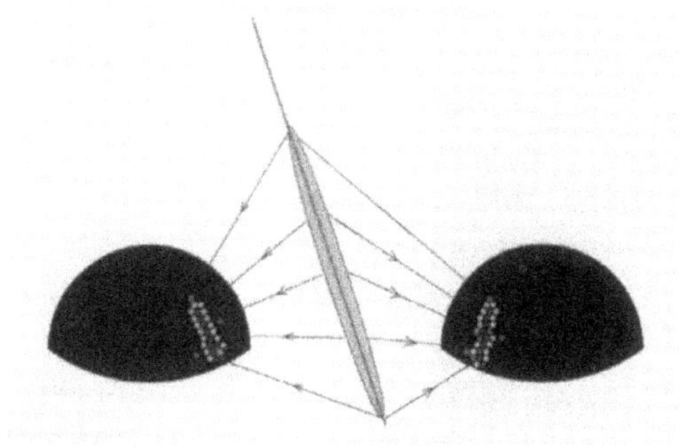

Fig. 12. Left: The UHECR reconstruction using a single site measurement of the fluorescence from an EAS. Right: Stereo observations of an EAS illustrating how the two EAS-detector planes intersect to form a line that defines the EAS trajectory.[45]

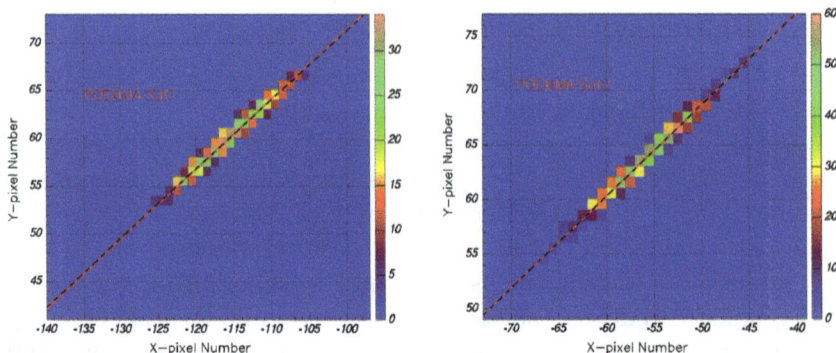

Fig. 13.   A co-viewed, stereo reconstructed 50 EeV simulated UHECR in the two POEMMA focal planes. The solid line shows the simulated trajectory, while the dashed line is defined by the reconstructed trajectory. The color map provided the photoelectron statistics simulated for each pixel.[46]

of the POEMMA telescopes, with the reconstructed and simulation MC-truth tracks superimposed on the pixel heat maps. The fact that these are virtually indistinguishable demonstrate the accuracy provided by using two well-separated EAS detector planes. This provides a straightforward geometrical method to determine the line where the two planes overlap, thus accurately determining the location of the EAS. Only in the case where the angle between the planes is small, e.g., $\lesssim 5°$, does the pure geometrical reconstruction have difficulty.[46] Timing information can be used to reconstruct this class of UHECR events. This leads to exceptional angular resolution ($\sigma \lesssim 1.5°$ above 20 EeV) of the EAS which is needed to obtain the required energy and X_{max} resolutions to accurately measure the UHECR spectrum and UHECR nuclear composition evolution.[46] Furthermore, the ability to precisely reconstruct the EAS trajectory, energy, and X_{max} provides a method to separate the more horizontal neutrino events from the much more copious UHECR events. For POEMMA, the selection of $X_{Start} \gtrsim 2000 \, \mathrm{g \, cm^{-2}}$ has been shown, based on the analysis of the X_{Start} distributions for UHECR protons for $E_{CR} \geq 40 \, \mathrm{EeV}$, to lead to a $\gtrsim 10^4$ rejection factor for UHECR background events.[46] These UHE neutrino measurements are performed whenever the POEMMA satellites are in stereo-fluorescence UHECR operational mode. It is noted that Earth-emergent EASs from neutrino interactions in the Earth also provide a fluorescence signal that can be observed from space if the instruments are pointed to the Earth's limb.[13]

Table 2. POEMMA Observatory Specifications: Observatory = Two Telescopes; Each Telescope = Instrument + Spacecraft.

Telescope	Instrument	
Optics	Schmidt	45° full FoV
	Primary Mirror	4 m diam.
	Corrector Lens	3.3 m diam.
	Focal Surface	1.6 m diam.
	Pixel Size	$3 \times 3 \,\text{mm}^2$
	Pixel FoV	0.084°
PFC	MAPMT (1 μs)	126,720 pixels
PCC	SiPM (20 ns)	15,360 pixels
Observatory	**Each Telescope**	
	Mass	1,550 kg
	Power (w/cont)	700 W
	Data	<1 GB/day
	Spacecraft	
	Slew rate	90° in 8 min
	Pointing Res.	0.1°
	Pointing Know.	0.01°
	Clock synch.	10 ns
	Data Storage	7 days
	Communication	S-band
	Wet Mass	3,450 kg
	Power (w/cont)	550 W
	Mission	**(2 Telescopes)**
	Lifetime	3 year (5 year goal)
	Orbit	525 km, 28.5° Inc
	Orbit Period	95 min
	Telescope Sep.	∼(25–1000) km

Source: Adapted from Ref. [25], % DOI: 0.1088/1475-7516/2021/06/007.

The specifications of the POEMMA Observatory, spacecraft, and telescopes are detailed in Table 2. Each POEMMA telescope employs a 1.6 m diameter hybrid focal plane, the POEMMA Fluorescence Camera (PFC) comprises the majority of the area, while the POEMMA Cherenkov Camera (PCC) takes up the rest, as shown

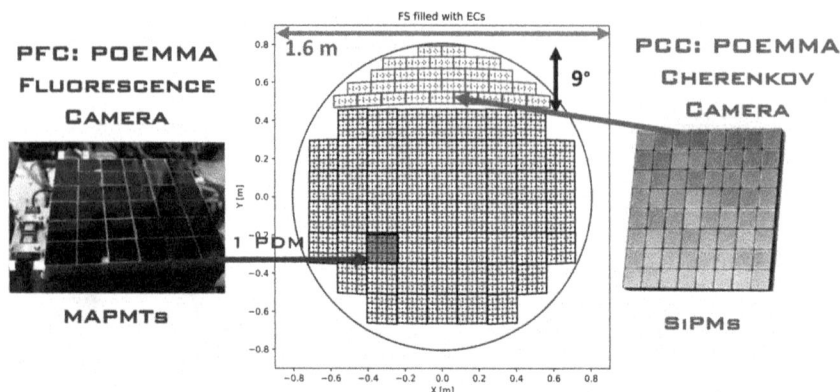

Fig. 14.   Layout of the hybrid focal plane of a POEMMA Schmidt telescope. The majority of the area is composed of PFC MAPMT modules with a UV filter to record the 300–500 nm air-fluorescence light in 1 μs snapshots. The PCC is composed of SiPM pixels whose 300–1000 nm wavelength response is well matched to that from the EAS optical Cherenkov signals and are recorded with 10 ns cadence.[25]

in Fig. 14. The PFC is optimized for the measurement of UHECR EAS longitudinal air-fluorescence evolution while the PCC is optimized for the measurement of the fast, Cherenkov signals generated by upward-moving and over-the-limb EAS observations. The PFC uses 55 photodetector modules (PDMs) based on the system developed for the JEM-EUSO instrument[47] and flown in a number of balloon flights.[15,17] Each PDM consists of 36 64-channel MAPMTs, with a BG3 filter located on each MAPMT to constrain the wavelength to that of the UV fluorescence band (300–500 nm) to minimize the effects of the atmospheric dark-sky airglow background. The PFC design consists of 26,720 $3 \times 3 \, mm^2$ pixels and will record signals using 1 μsec temporal sampling to measure the waxing and waning development of the UHECR-induced EASs over time frames of 10–100+μs. The smaller area PCC consists of 30 focal surface units (FSUs) with each composed of a 512-channel array of silicon photomultipliers (SiPMs), whose broader wavelength response, 300–1000 nm, is better matched to inherent spectral variability of the optical Cherenkov light measurement due to atmospheric attenuation effects. When the POEMMA telescopes are pointed toward the Earth's limb, a region corresponding to 9° in elevation angle (defined from the nadir direction) and approximately 30° in azimuth is used to observe for the Cherenkov signals, including searching for upward-moving EASs from tau neutrino interactions in the Earth. This angular mapping on the focal plane is illustrated by the SiPM (upper crescent) component in Fig. 14. The POEMMA PCC leads to a modest ~10% reduction in the UHECR instantaneous geometry factor for the PFC for fluorescence UHECR observations. In stereo-fluorescence mode, POEMMA will have remarkable sensitivity to UHE neutrinos above 20 EeV due to the precise angular, energy, and EAS profile measurements required for the UHECR spectrum and composition measurements. It should be noted that a comparison of the air-fluorescence response of SiPMs with significant PDE in the near UV demonstrates that, at least from a photodetection standpoint, currently available SiPMs have nearly identical response as MAPMTs, but each have specific UV filter requirements due to the much larger wavelength bandpass of SiPMs and that of the dark-sky airglow background.[39] Thus, the potential exists for space-based SiPM-based instruments tuned to the EAS air-fluorescence signal with less massive focal planes and without the need for high voltage.

4. POEMMA UHECR Measurement Performance and UHECR Science

Figure 8 illustrates the gains in exposure using space-based UHECR measurements in the context of POEMMA. Assuming five years of POEMMA-Stereo operation, the total exposure is expected to be ~ 8×10^5 km^2 ster years with precision measurements of UHECRs above 40 EeV: energy resolution of < 20%, an angular resolution of $\leq 1.5°$ above 40 EeV; and a X_{max} resolution of ≤ 30 g cm^{-2}, which allows for the identification of proton, helium, nitrogen, and iron in a mixed UHECR composition.[48] The left figure shows the exposure growth as a function of time and compares this to the growth anticipated from PAO and the TA×4 upgrade.[49] The right plot shows the anticipated five-year POEMMA exposure in context of that reported by PAO (right y-axis scale) and TA (left y-axis scale) at the 2019 ICRC. Above 40 EeV, the yearly UHECR exposure of POEMMA-Stereo is more than four times higher than that of the PAO ground array and 18 times higher than the TA ground array. Above 100 EeV, the POEMMA gain in exposure increases nearly twofold for each comparison. It should be noted that while ground arrays operate with nearly 100% duty cycle, the precision of the measurements is not yet at the level provided by the stereo-fluorescence technique, although improvements in electron/muon identification and radio measurements form the EAS could significantly improve the precision of ground-array EAS measurements.[50] Currently, the UHECR fluorescence measurements performed by PAO and TA have ~10% duty cycle, thus the ground-based fluorescence exposure is ~10% of that shown in left plot in Fig. 8. If POEMMA performs UHECR measurements in tilted, POEMMA-Limb mode, Fig. 8 shows the significant gain in exposure but at a cost of increased UHECR detection energy threshold and reduced precision on the EAS measurements. Conservatively assuming two monocular EAS measurements, simulation of the POEMMA-Limb UHECR response above 40 EeV yields an energy resolution of ~30%, an angular resolution of <10°, and an Xmax resolution of ~100 $g\,cm^{-2}$, which is sufficient to distinguish between proton and iron UHECR primaries.[46] While the POEMMA spacecraft are planned to be used in tandem, the ability to perform monocular measurements is needed for mission-risk mitigation.

As with ground-based UHECR experiments, space-based experiments rely on detailed Monte Carlo simulations to predict the accuracy and science return of the measurements. Here, those developed for POEMMA are summarized. Figure 15 shows the stereo-reconstructed angular resolution as a function of simulated UHECR energy, where the telescopes are separated by 300 km and tilted to view a common volume along the orbit path. The strength of the stereo reconstruction technique is evident as both the zenith and

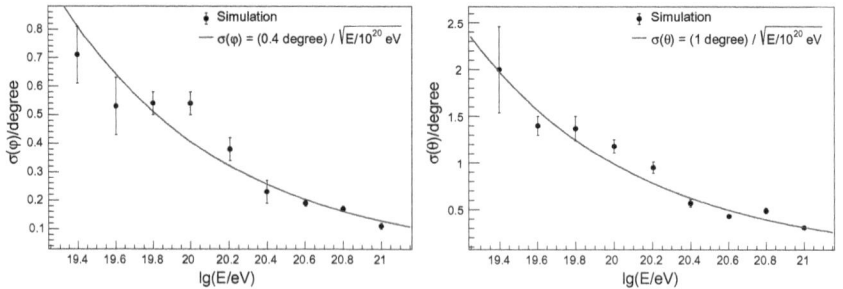

Fig. 15.   UHECR EASs simulated stereo-reconstructed angular resolution versus UHECR energy for POEMMA. Left: Azimuth. Right: Zenith.[46]

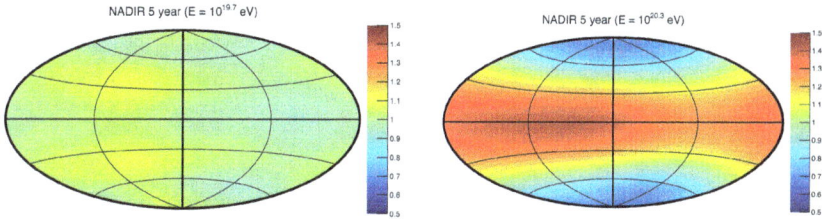

Fig. 16. Sky exposure for POEMMA-Stereo UHECR observations in declination versus right ascension. The color scale denotes the exposure variations accounting for the sun and moon positions during a five-year mission. Left: Sky exposure for UHECRs of 50 EeV. Right: Sky exposure for showers of 200 EeV.[21]

azimuth angular resolutions are $< 1.5°$ for $E_{CR} \geq 40$ EeV. This angular resolution with the stereo technique yields simulated UHECR energy resolution $< 20\%$ for $E_{CR} \geq 50$ EeV.[46] The UHECR sky coverage for these UHECR observations from a 525 km altitude and $28.5°$ inclination orbit are shown in Fig. 16 assuming five years of observations at 50 and 200 EeV. The results demonstrate full sky coverage and $\pm 5\%$ variation at 50 EeV and $\pm 50\%$ variation at 200 EeV.

The ground-based PAO and TA UHECR spectrum measurements demonstrate a significant difference, possibly indicating that the UHECR sources are different in the Northern and Southern Hemispheres. Assuming the PAO and TA UHECR spectral parameters reported at the 2019 ICRC and POEMMA's simulated response, Fig. 17 shows the predicted measurements assuming five years of both stereo and limb (tilted) observations, showing the possible extension well past 100 EeV. This capability allows for the search of spectral recovery at the highest energies[46] due to the 10^6 km^2 sr year scale exposures provided by five years of space-based UHECR observation.

The stereo reconstruction technique combined with fine sampling of the EAS development evolution provide a X_{max} resolution of $\lesssim 30$ g cm^{-2} that improves to < 20 g cm^{-2} above 100 EeV and thus provides accurate determination of the UHECR nuclear composition.[46] Figure 18 presents this capability in terms of extending to higher energy a model based on PAO lower energy measurements. The results show good determination above 100 EeV, well before statistical errors begin to dominate the results.

In five years of operation, POEMMA is anticipated to measure over 1,400 UHECRs above 40 EeV with high precision given the good measurement resolutions. Figure 19 shows the source sky maps

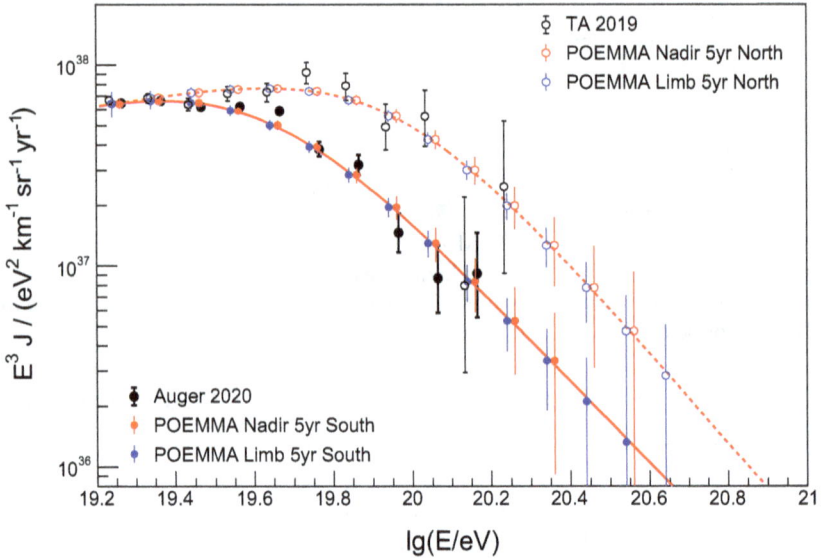

Fig. 17. Simulated POEMMA spectra extrapolated from and compared to the PAO 2020 spectrum (black dots and solid line)[51] and the extrapolation and comparison to the TA 2019 spectrum (black open circles and dotted line) from Ref. [52] for the POEMMA-Stereo (red) and POEMMA-Limb (blue) observational modes, for UHECRs above 16 EeV.[21]

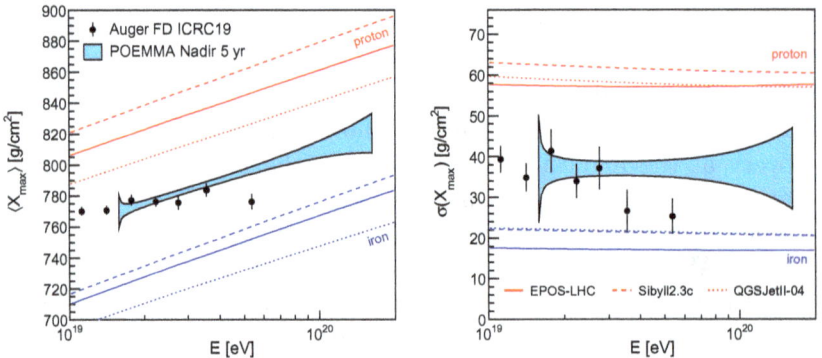

Fig. 18. Simulated capability of POEMMA to measure $\langle X_{max} \rangle$ and $\sigma(X_{max})$ for UHECR composition studies. A simple UHECR composition model based on lower energy PAO measurements was used to assess the performance. The blue band maps the statistical uncertainties in five years of POEMMA-Stereo measurements versus number of events per 0.1 in the logarithm of energy, the X_{max} resolution and efficiency for $\theta < 70°$ and an intrinsic EAS-to-EAS fluctuations of $40 \, g \, cm^{-2}$ was assumed. The black dots show the PAO fluorescence UHECR data presented at the 2019 ICRC.[53]

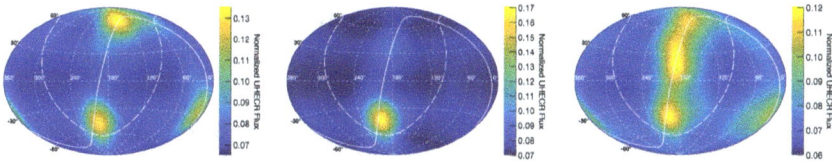

Fig. 19. Equatorial coordinate sky maps of simulated POEMMA UHECR measurements for different astrophysical catalogs using the best fit parameters reported by the PAO collaboration.[54] Left: Starburst galaxies with 11% anisotropy fraction. Middle: *Swift*-BAT AGNs with 8% anisotropy fraction. Right: 2MRS catalog with 19% anisotropy fraction. All assume 15% angular spread of the arrival directions.[21]

obtained using three different assumptions regarding the astrophysical distribution of UHECRs[21] based on the PAO results and correlation analysis with similar catalogs.[54] The different results in the upper and lower hemispheres predicted by these demonstrate why full sky coverage with precision UHECR measurements is important for definitively identifying the astrophysical sources of UHECRs.

It is noted that the UHECR measurement performance shown through modeling and simulations also expands the search for UHE neutrinos, UHE photons, and measurement of the proton–air cross section at $\sqrt{s} = 450\,\mathrm{TeV}$[46] while also providing unique sensitivity to the detection of super-heavy dark matter (SHDM) from decay or annihilation into UHE neutrinos[55] or UHE photons.[46]

4.1. *Effects due to the dark-sky airglow background and cloud cover*

Optical measurements of EASs are performed with a persistent atmospheric background even during astronomical night. The molecules and atoms in the upper atmosphere of the Earth are continuously ionized by solar and cosmic radiation leading to production of faint light that is commonly known as an airglow.[58,59] These chemiluminescent processes are concentrated in a ~10 km thick layer around 95 km in altitude over the entire planet and are different from the geomagnetic processes that cause aurora near the polar regions. When viewed from LEO, the dark-sky airglow light together with the reflected component of the starlight and zodiacal light act as a diffuse night atmospheric background that effectively sets the energy threshold for

EAS detection. For space-based UHECR EAS measurements, the directly viewed zodiacal light is not an issue, just the atmospheric scattered component. Airglow is a dynamic phenomenon that varies with geographical position, season, solar activity, geomagnetic activity, and changes in the Earth's atmosphere.[60–62] Additionally, the altitude of the observatory and its viewing direction in terms of angle away from nadir further yields to an increase in intensity, in part by viewing a longer path through the dark-sky airglow layer and is described by the van Rhijn formula.[63] Figure 20 presents the dark-sky airglow measurements by Hanuschik near maximal solar activity[56,57] for the wavelength band 300–1000 nm in units of photons m^{-2} ns^{-1} sr^{-1} as a cumulative sum as a function of wavelength. Given the strong wavelength dependence, the impact of dark-sky airglow on the measurements of EAS air fluorescence and optical Cherenkov needs to be considered and an analysis for both is detailed in Ref. [39]. Since the main contribution to the air fluorescence is within 300–450 nm, near-UV filters combined with the wavelength response of bi-alkali PMTs limit the airglow intensity. A value of ~500 photons m^{-2} ns^{-1} sr^{-1} for the dark-sky airglow in the 300–400 nm wavelength band is based on measurements from balloon altitudes by the NIGHTGLOW experiment nadir-viewing measurements,[37] slightly more but consistent with the measurements of Hanuschik.[56,57]

NIGHTGLOW also measured a van Rhijn–type enhancement as a function of viewing angle away from nadir. A straightforward calculation based on an instrument's effective optical collecting area, iFoV,

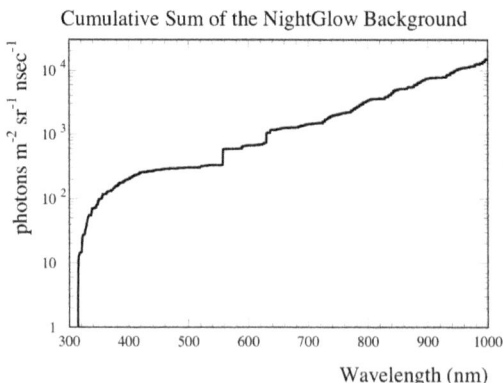

Fig. 20.   Cumulative sum of the dark-sky airglow background using the measurements of Hanuschik.[39,56,57]

and signal integration time provides a mean measurement of the air-glow background in a pixel of a telescope. Assuming $A_{Eff} = 6\,\mathrm{m}^2$, iFoV $= 0.084°$, and $1\,\mu s$ signal integration time, e.g., the POEMMA on-axis response, an airglow background of 500 photons $\mathrm{m}^{-2}\,\mathrm{ns}^{-1}\,sr^{-1}$ yields a per pixel mean photon background of ≈ 6.5 photons. Assuming an effective quantum efficiency of 30% for the PMT channel, this yields a background of ≈ 2 photoelectrons (PEs). Thus, any EAS fluorescence signal needs to be well above this to achieve a sufficient signal-to-background ratio to only trigger the instrument on EAS events versus the airglow background. In practice, any instrument will occasionally operate to trigger on the background to provide a calibration assessment.

The shape of the optical Cherenkov spectrum emitted from either upward-moving EAS sourced from tau neutrino interactions in the Earth or from viewing VHECRs above the Earth limb is highly variable over the band starting at 300 nm to above 1000 nm.[39,65] This is due to large differences in the wavelength-dependent atmospheric attenuation due to the different amounts of aerosols, ozone, and atmosphere itself that the Cherenkov light from different EASs must propagate through to reach the detector. This effect is quantified by examining the optical depth as a function of angle above the limb of the Earth[e] and is shown in Fig. 21. When the VHECR is viewed from a detector at 525 km altitude, for angles $\lesssim 1°$ above the limb, the attenuating effects of the atmosphere become quite small. However, for the 33 km altitude case, a significant column depth of ozone remains that significantly attenuates the Cherenkov signal at $\sim 350\,\mathrm{nm}$,[42] even out to the detector horizontal angle (90° from nadir). Given the extreme variability of the wavelength dependence of the optical depth, the Cherenkov spectrum at the instrument can peak throughout the range of $<300–1000$ nm for VHECRs and upward-moving Earth-emergent τ-lepton EASs.[33,34,64,65] This motivates the use of SiPMs as the photodetectors since they have a response matched to the optical Cherenkov wavelength band, including a SiPM variety with peak response around 450 nm.[66] Over the entire wavelength band shown in Fig. 20, the total count rate is $\sim 15,000$ photons $\mathrm{m}^{-2}\,\mathrm{ns}^{-1}\,sr^{-1}$, but the fast, ~ 10 ns time spread

---

[e]presented as angle from nadir, where angle from the nadir for the limb is defined by the altitude of observation and the radius of the Earth.

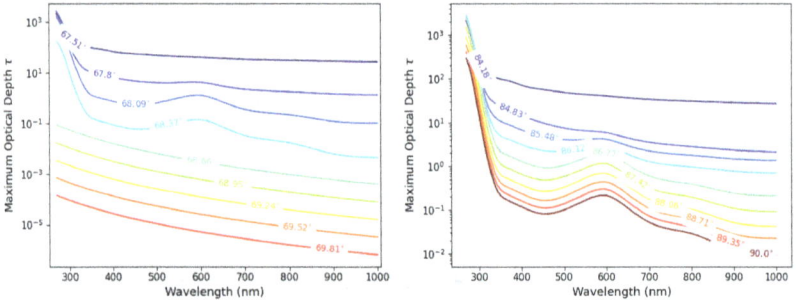

Fig. 21.   Optical depth as a function of wavelength for viewing the EAS optical Cherenkov from over the limb of the Earth VHECRs. The different curves are the optical depths for different angles away from nadir. Left: The optical depth for viewing from 525 km altitude orbit where the limb of the Earth is at an angle 67.51° away from nadir. Right: The optical depth for viewing from 33 km altitude orbit where the limb of the Earth is at an angle 84.18° away from nadir.[64]

of the optical Cherenkov signals when viewed a few degrees from the EAS trajectory[33,65,67] helps to reduce the effects of the airglow background. SiPMs with a blue-peaked response also have photodetection efficiencies (PDE) $\lesssim 20\%$ above 700 nm, which also helps to reduce the airglow background. For an $A_{Eff} = 1$ m^2, iFoV= 0.1°, $\Delta t = 10$ ns, and an effective PDE = 30% over the 200–1000 nm band, the mean background would be ≈ 0.14 PEs.

Since space-based instruments view different atmospheric conditions as the instruments orbit the Earth, the distribution of clouds are inherently variable, and data from Earth-observing satellites together with UHECR simulations provide a mechanism to assess the impact.[14,14,68,69] The left panel in Fig. 22 shows the distribution of observed 100 EeV UHECR EASs simulated for the OWL mission overlaid on a particular frame of MODIS cloud measurements.[68] When cloud height measurements are included and using long-duration cloud measurement datasets, they predict 50–60% of the observable portion of the simulated EASs are above the clouds. Furthermore, if the location of EAS signal maximum occurs above the cloud, a high fraction of events that enter clouds can be accurately reconstructed, especially if the precise spatial location of the EAS is provided by stereo-fluorescence measurements.[70]

The level of scattered moonlight also affects the duty cycle of UHECR observations, and the level in terms of the units of dark-sky background versus moon phase is shown in the right panel of Fig. 22.

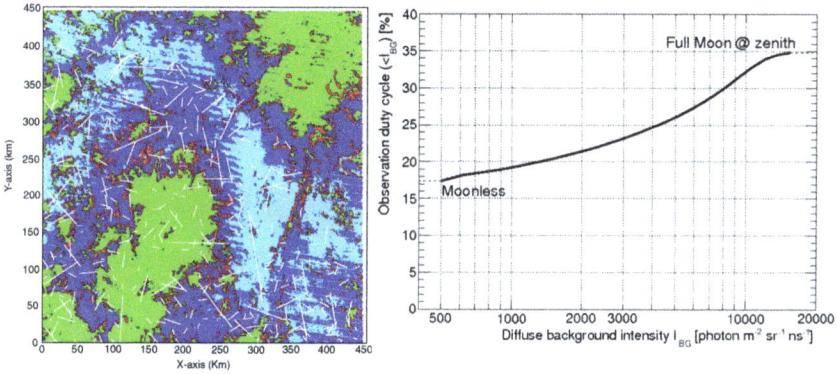

Fig. 22. Left: The projected 100 EeV simulated observed UHECR EAS tracks (white lines), which trigger the OWL instrument, superimposed over measured MODIS cloud cover, in approximately one quadrant of the square defined by the embedded quasi-circular OWL-viewed ground area. Key: (light blue) high-confidence cloud free; (dark blue) cloud free; (red) probably cloudy; (green) cloudy. For this particular cloud frame, 20% of the simulated track sample have a "clear" aperture as defined by not cloudy or probably cloudy MODIS pixels within 3 km of a track. From the MODIS database used in this study, ∼50% of the cloud heights were ≤ 4 km and below the visible portion of the EAS.[68,69] Right: The night-sky observational duty cycle (η) as a function of astronomical dark-sky and moonlight-induced background as a function of moon phase.[14]

Practically, accepting more moonlight for UHECR observations effectively raises the detection energy threshold due to the higher moonlight (plus airglow) background. Flight dynamic studies performed for the JEM-EUSO ISS implementation (400 km altitude, 51.6° inclination)[69] and POEMMA (525 km altitude, 28.5° inclination)[46] calculate that ∼20% of the time, the spacecraft orbit will be in astronomical night with minimal effects from moonlight. Assuming that 60% of UHECR EASs are observable, a 12% total duty cycle is calculated for UHECR experiments in LEO.

5. UHECR EAS Optical Measurements Using Balloon-borne Instruments

While orbiting neutrino-detection missions using the EAS optical signals can be performed in multi-year mission time frames, the availability of ULDB flights, potentially of 100-day duration allows for meaningful UHECR measurements in a relatively cost-effective

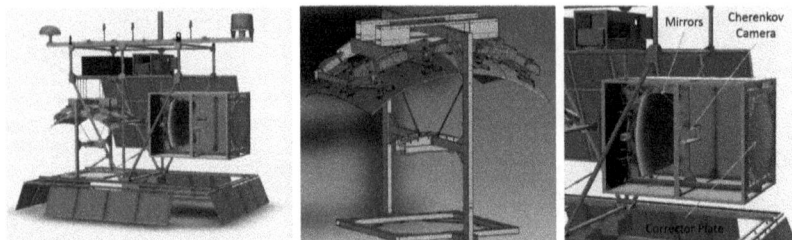

Fig. 23. The EUSO-SPB2 payload configuration. Left: Schematic of the payload.[72] Middle: Schematic of the fluorescence telescope.[73] Right: Schematic of the Cherenkov telescopes.[77] Not shown is the infrared cloud camera.[76]

suborbital space flight platform. These ∼30 km altitude flights, which are launched from latitudes, e.g., Wanaka, New Zealand, that allow for astronomical night operation, provide an environment to view the air-fluorescence signal of downward-moving EASs from UHECRs as well as the optical Cherenkov signal above-the-limb VHECRs while searching for neutrino-induced, Earth-emergent upward-moving EAS. The EUSO-SB2 ULDB experiment (see Fig. 23), scheduled to launch in 2023, consists of a downward-looking fluorescence telescope (FT) to record the development of the fluorescence EAS signals from UHECRs.[71–73] The FT[72] consists of a ∼1 m diameter optical aperture Schmidt telescope with a $37.4° \times 11.4°$ FoV. The focal plane of the 6912 pixels in the FT are contained in three photodetection modules (PMDs) developed for the EUSO mission,[74] which are planned for use in the POEMMA mission. Each PDM consisting of 36 multi-anode photomultipler tubes (MAPMTs) and SPACIROC3 front-end electronic (FEE) readouts that employ $1\,\mu s$ sampling of the UHECR EAS development and have a channel to record fast, $\gtrsim 10\,ns$ optical signals such as that from Cherenkov reflection off the ground. This PDM design has been flown in the ULDB EUSO-SB1 mission in 2017,[75] the EUSO-Balloon test flight mission[15] flown in 2014, and the Mini-EUSO instrument[16] operating on the ISS starting in 2019. This PDM development path illustrates one of the benefits of using relatively inexpensive balloon-borne instruments to develop technologies for eventual space-based, orbiting missions. Simulation studies have predicted the UHECR observation rate for EUSO-SPB2 for UHECR fluorescence events is about 1 event in 8 hours. This rate is for energies slightly above 1 EeV, assuming clear observation during dark moonless nights,[73] which reflects the inherent difficulty of performing a UHECR air-fluorescence experiment even from a 100-day ULDB balloon mission.

The EUSO-SPB2 payload includes a Cherenkov telescope (CT)[71,78] and a Schmidt telescope with a 12.8° × 6.4° horizontal by vertical (above or below the viewed Earth limb) FoV. The focal plane is composed of a 512 pixel SiPM camera using the MUSIC ASIC[79] for FEE readout using a 100 MegaSample/second digitization. The choice of the SiPM is such that it has relatively high sensitivity in the 200–1000 nm wavelength band needed due to the inherent variability of the Cherenkov signal.[39] The 10 ns digitization span is optimized for the fast Cherenkov signal that is produced within ~1.5° from the EAS trajectory. The CT optics have a relatively large pixel, or instantaneous, iFoV = 0.4°, implying the count rate from the dark-sky background will be significant and dominate over the background due to SiPM dark-count rates.[78] One other novel feature of the CT is that a bifocal mirror splits the 0.785 m² effective area into two separate images on the focal plane. This provides a coincidence trigger using spatially separated events but at a cost of modestly increasing the effects of the dark-sky background. The goal of the bifocal design is to determine whether event backgrounds due to ionization in the SiPM pixels, i.e., due to traversing cosmic rays or other radiation backgrounds, require the need of such a spatially separated trigger.

Figure 24 presents the simulated integral count rates above a VHECR threshold energy for the EUSO-SPB2 and POEMMA

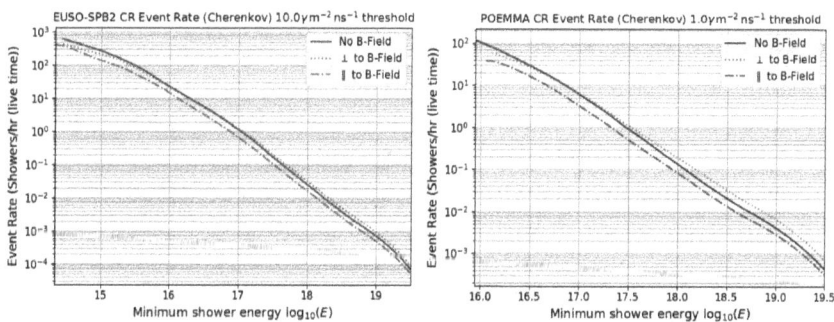

Fig. 24. Simulated integral cosmic-ray event rates (above a threshold E) for observing over-the-limb events via the optical Cherenkov signal for EUSO-SPB2 (left) and POEMMA (right). The effects of the geomagnetic field are provided by assuming the acceptance for three different conditions, no $\vec{B}$-field, $\vec{B}$-field parallel to the EAS trajectory, and $\vec{B}$-field perpendicular to the trajectory.[64]

missions. The results show that the EUSO-SPB2 ULDB experiment should have a VHECR threshold energy well below 1 PeV with a rate of several hundred observed events per hour of live time above threshold. The rates are based on three different models of the geomagnetic field to estimate its possible effect. In practice, setting different analysis thresholds on the Cherenkov signal strength should map out the VHECR rate since the cosmic-ray spectrum in this energy range is experimentally fairly well-measured. The simulated rates for POEMMA are also shown in Fig. 24. These over-the-limb VHECR events do form a background for upward-moving EAS neutrino searches due to the $\lesssim 1°$ atmospheric refraction of the VHECR optical Cherenkov signal when viewed near the limb.[34] Thus, this measurement provides a measure of this potential background to upward, Earth-emerging events from tau neutrino interactions while also providing a coarse calibration signal for the Cherenkov telescope.

The results presented in Fig. 24 are based on simulating the electromagnetic properties of the EAS, thus ignoring the $\sim 10\%$ muon contribution. Since muons are relatively long lived in terms of EAS age, they could provide an additional Cherenkov signature.[80] If there is a way to use the differences in muon versus electromagnetic EAS Cherenkov signatures, which are strongly dependent on atmospheric attenuation, i.e., viewing angle from the instrument, this would allow a measurement of the hadronic content in the detected EAS.

However, a single measurement within the Cherenkov light pool does not precisely constrain the actual geometry of the observed EAS. This translates into a more coarse resolution for both the UHECR energy and location of X_{max}. However, the concept of using SmallSat constellations for a variety of space-based measurements is currently being implemented.[81] In principle, a constellation of modest, meter-diameter Cherenkov telescopes could provide enough sampling of the Cherenkov light and temporal pool[67,82,83] to more precisely measure the properties of the VHECRs, as is done in ground-based experiments such as the non-imaging Cherenkov array at the Bolivian Air Shower Joint Experiment (BASJE) site at Mount Chacaltaya[84] and the Non-Imaging CHErenkov (NICHE) array at the TA Middle Drum site in Utah.[85] Both of these ground-based experiments use arrays of optical Cherenkov detectors to sufficiently sample the EAS Cherenkov light spatial and temporal profiles to yield good measurements on the UHECR direction, energy, and particle type.

6. UHECR EAS Radio Measurements Using Balloon-borne Instruments

To date, the most successful balloon-borne cosmic neutrino and UHECR measurements have been made using the EAS radio measurement technique pioneered by the Antarctic Impulsive Transient Antenna (ANITA) experiment. The two pictures on the left of Fig. 25 show the launch preparation and instrument from the ANITA-I flight, showing the upper 8 horn and 16 horn lower arrays used to measure the EAS-induced and in-ice Askaryan radio signals in the 200–1200 MHz band. During the four LDB flights of the ANITA instrument,[86–89] the experiment has searched for neutrino-induced particle cascades, e.g., showers, in the Antarctic ice, using beamed Askaryan radio emission[90] but has also detected direct geomagnetic radio emission from UHECRs viewed above the Earth limb as well as the reflected UHECR EAS radio signal off of the ice.[91,92] The ANITA-IV radio measurements[89,93] also included a class of events consistent with Earth-emerging τ-leptons observed near the Earth's limb from $E_\nu \gtrsim 1$ EeV. The ANITA-IV results also include two events consistent with over-the-limb directly observed UHECRs events. The analysis included modeling the effects of atmospheric refraction for viewing just above the Earth limb events. Atmospheric refraction of optical and radio signals is a potential VHECR/UHECR background to events from tau neutrinos interacting in the Earth that lead to upward-moving EASs near the limb of the Earth.[34,89,93]

Fig. 25. Leftmost: The ANITA-I long-duration balloon payload preparing for launch in Antarctica. Middle: The ANITA-I instrument showing the upper eight-horn radio antennae array and lower 16-horn array before its 35-day LDB flight in December 2006 through January 2007.[86] Rightmost: An artist rendering of the PUEO long-duration balloon payload to be launched in Antarctica, showing the various antenna arrays including a low-frequency drop down antenna.[94]

The payload for ultra-high-energy observations (PUEO)[94] is a NASA Pioneer-class mission for radio detection of neutrino events in the Antarctic ice, τ-lepton -induced, upward-moving EASs from neutrino interactions in the Earth, and direct and reflected UHECR signals. The PUEO is shown in the rightmost panel in Fig. 25 and is built on the ANITA development, flight, and analysis experience. PUEO employs a larger radio array in the 300–1200 MHz band, including a drop-down low-frequency, 50–300 MHz radio antenna. PUEO also includes an interferometric phased-array trigger that coherently sums radio waveforms from each antennae with the appropriate propagation time delays to provide an increase in the signal-to-noise of $\approx \sqrt{N_{\text{antennae}}}$ compared to using a single antenna. This phased-array trigger technique has been demonstrated in the Askaryan Radio Array (ARA) experiment.[95]

At frequencies <10 GHZ, well above the EAS radio detection frequency band, the atmosphere and clouds are radio transparent and do not affect the ability to observe radio emission of EAS. However, knowledge of the atmospheric properties is important for radio emission since the local index of refraction and the relative orientation of the geomagnetic field determines the EAS radio beam pattern.[96] Similar issues are inherent with single measurements of the optical Cherenkov radiation, e.g., a signal measurement does not precisely constrain the actual geometry of the observed EAS that is needed to measure the UHECR energy and location of X_{max} with high resolution.

Several ideas for future space-based radio UHECR cascade detection, including the Earth-orbiting observatories, such as the synoptic wideband orbiting radio detector (SWORD), to detect over-the-limb and reflected EAS signals,[97] and lunar-orbiting missions, including the Zettavolt Askaryan Polarimeter (ZAP) that detects cascades in the lunar regolith via the Askaryan effect[98] have been studied. One issue for Earth-orbiting experiments is dealing with ionospheric dispersion of the EAS radio signal.[97]

7. Summary

Space-based UHECR missions provide extremely high UHECR exposure, $\sim 10^6$ km^2 ster years in a five-year mission. Using the stereo-fluorescence technique with at least two spacecraft also provides precision measurements to accurately determine the incident direction, energy, and nuclear composition of UHECRs. Combined with full-sky sensitivity, space-based measurements provide a mechanism to study the astrophysics of UHECR sources. While the majority of the development effort is to build and launch the spacecraft, once in orbit the operation can be relatively seamless and potentially longer than the designed lifetime, just dependent on expendables such as propulsion requirements for orbit maintenance, which is important for large, multi-meter-sized spacecraft in LEO. Furthermore, the avionics (reaction wheels and magnetic torquer bars) required for satellite operation also allow for the spacecraft to be slewed and pointed to any direction along the orbit with a minimal, if any, propulsion requirement. The POEMMA spacecraft contain extra propulsion to allow the satellite separations to be changed multiple times per the mission.[34] Combined, the slewing and adjustment of the spacecraft separation allows tuning of the UHECR aperture. Thus, if a recovery in the UHECR spectrum is observed above 100 EeV, the POEMMA telescope configuration could be optimized to increase the geometry factor at the highest energies similar to that when POEMMA is in Earth-limb observing mode. The next generation of large ground-based UHECR experiments, such as the Global Cosmic-Ray Observatory (GCOS) currently under study[99] and the Giant Radio Array for Neutrino Detection (GRAND)[100] if designed to cover an area of $\sim 4 \times 10^4$ km^2 would have slightly larger exposure to the space-based POEMMA experiment but must be composed of >10,000 individual surface detectors. However, ground-detectors with the ability to discern electrons from muons as the EAS hits the Earth, especially combined with radio and air-fluorescence measurements, would provide an unique measure of the hadronic interactions at the UHECR energy scale, which is well above that currently achievable at terrestrial colliders, e.g., $\sqrt{s} = 450$ TeV for a 100 EeV proton. It should be noted that precision UHECR measurements can infer the proton–air cross section at these energies since the UHECR composition, including proton content, is measured, e.g., see Ref. [46].

With the expected flight of EUSO-SPB2 in 2023, the technique of measuring VHECRs via the optical Cherenkov radiation will be demonstrated, while the flight of PUEO will make further EAS radio measurements on UHECRs. This is a potential new technique for space-based VHECR and UHECR measurements and could blossom with the use of a small constellation of SmallSats each with a modest-sized Cherenkov telescope. Combined with NASA Probe-class missions such as POEMMA, the future of space-based UHECR measurements looks very promising, and these high-statistics high-resolution measurements could finally provide an astrophysics accounting of the UHECR sources.

References

1. F. Schroeder, News from cosmic ray air showers (Cosmic Ray Indirect — CRI Rapporteur), *PoS.* **ICRC2019**, 030 (2019). doi: 10.22323/1.358.0030.
2. R. Gelaro, W. McCarty, M. J. Suarez, R. Todling, A. Molod, L. Takacs, C. A. Randles *et al.*, The modern-era retrospective analysis for research and applications, version 2 (merra-2), *J. Clim.* **30**(14), 5419–5454 (2017). doi: 10.1175/JCLI-D-16-0758.1. https://journals.ametso c.org/view/journals/clim/30/14/jcli-d-16-0758.1.xml.
3. S. Hoover, J. Nam, P. W. Gorham, E. Grashorn, P. Allison, S. W. Barwick, J. J. Beatty *et al.*, Observation of ultrahigh-energy cosmic rays with the ANITA balloon-borne radio interferometer, *Phys. Rev. Lett.* **105**(15), 151101 (October, 2010). doi: 10.1103/PhysRevLett.105.151101.
4. R. Benson and J. Linsley, Satellite observation of cosmic-ray air showers. *Bull. Am. Astron. Soc.* **12**, 818 (September, 1980).
5. R. Benson and J. Linsley, Satellite observation of cosmic-ray air showers. In *International Cosmic Ray Conference*, Vol. 8, *International Cosmic Ray Conference*, p. 145 (January, 1981).
6. Y. Takahashi, Maximum-energy Auger Air Shower Satellite (MASS) for observing cosmic rays in the energy region 10^{19-22} eV. In *International Cosmic Ray Conference*, Vol. 3, *International Cosmic Ray Conference*, p. 595 (January, 1995).
7. Y. Takahashi, R. A. Chipman, J. O. Dimmock, L. W. Hillmann, D. J. Lamb, T. M. Leslie, J. J. Weimer *et al.*, Maximum-energy Auger (AIR)-shower satellite (MASS/AIRWATCH), In B. D. Ramsey and T. A. Parnell (eds.), *Proceedings of the SPIE*, Vol. 2806, *Society of Photo-Optical Instrumentation Engineers (SPIE) Conference Series*, pp. 102–112 (1996). doi: 10.1117/12.253970.

8. Y. Takahasi, Cosmic ray measurements from space. Technical Report NASA-CR-204041, University of Alabama, Hunstville (1996). MASS/AIRWATCH Huntsville Workshop Report.
9. J. F. Krizmanic, J. W. Mitchell, R. E. Streitmatter, and Owl Collaboration. Optimization of the Orbiting Wide-angle Light collectors (OWL) mission for charged-particle and neutrino astronomy. In *International Cosmic Ray Conference*, Vol. 33, *International Cosmic Ray Conference*, p. 2334 (January, 2013).
10. J. Krizmanic and J. Mitchell. The potential of spaced-based high-energy neutrino measurements via the airshower Cherenkov signal. In *Proceedings of the 32nd ICRC (Beijing)*, Vol. 4, p. 316 (2011). doi: 10.7529/ICRC2011/V04/1331.
11. GSFC Integrated Design Center. https://idc.nasa.gov/idc/.
12. OWL IDC Study report (2002).
13. G. Vankova, S. Mladenov, M. Bogomilov, R. Tsenov, M. Bertaina, and A. Santangelo, Detection of Earth-skimming UHE tau neutrino with the JEM-EUSO detector, *arXiv e-prints*. art. arXiv:1509.05995 (September, 2015).
14. J. H. Adams, S. Ahmad, J. N. Albert, D. Allard, M. Ambrosio, L. Anchordoqui, A. Anzalone *et al.*, An evaluation of the exposure in nadir observation of the JEM-EUSO mission, *Astropart. Phys.* **44**, 76–90 (April, 2013). doi: 10.1016/j.astropartphys.2013.01.008.
15. M. E. Bertaina, P. von Ballmoos, and JEM-EUSO Collaboration. Results of the EUSO-Balloon flight. In *35th International Cosmic Ray Conference (ICRC2017)*, Vol. 301, *International Cosmic Ray Conference*, p. 445 (January, 2017).
16. S. Bacholle, P. Barrillon, M. Battisti, A. Belov, M. Bertaina, F. Bisconti, C. Blaksley *et al.*, Mini-EUSO mission to study Earth UV emissions on board the ISS, *Astrophys. J. Suppl. Ser.* **253**(2), 36 (April, 2021). doi: 10.3847/1538-4365/abd93d.
17. G. Osteria, V. Scotti, and JEM-EUSO Collaboration, EUSO-SPB: In-flight performance, *Nucl. Instrum. Methods Phys. Res. A.* **936**, 237–238 (August, 2019). doi: 10.1016/j.nima.2018.10.213.
18. V. Scotti, G. Osteria, and JEM-EUSO Collaboration, The EUSO-SPB2 mission, *Nucl. Instrum. Methods Phys. Res. A.* **958**, 162164 (April, 2020). doi: 10.1016/j.nima.2019.05.005.
19. P. A. Klimov, M. I. Panasyuk, B. A. Khrenov, G. K. Garipov, N. N. Kalmykov, V. L. Petrov, S. A. Sharakin *et al.*, The TUS detector of extreme energy cosmic rays on board the Lomonosov satellite, *Space Sci. Rev.* **212**(3-4), 1687–1703 (November, 2017). doi: 10.1007/s11214-017-0403-3.

20. P. Klimov, Ultra-high energy cosmic ray detector TUS: Preliminary results of the first year of measurements, *PoS.* **ICRC2017**, 1098 (2018). doi: 10.22323/1.301.1098.

21. B. A. Khrenov, G. K. Garipov, M. A. Kaznacheeva, P. A. Klimov, M. I. Panasyuk, V. L. Petrov, S. A. Sharakin *et al.*, An extensive-air-shower-like event registered with the TUS orbital detector, *J. Cosmol. Astropart. Phys.* **2020**(3), 033 (March, 2020). doi: 10.1088/1475-7516/2020/03/033.

22. P. Klimov, S. Sharakin, M. Zotov, F. Bertaina, and M. Fenu, Lomonosov-UHECR/TLE Collaboration. Main results of the TUS experiment on board the Lomonosov satellite. In *37th International Cosmic Ray Conference (ICRC2021)*, Vol. 37, *International Cosmic Ray Conference*, p. 316 (July, 2021).

23. P. Klimov, B. Khrenov, M. Kaznacheeva, G. Garipov, M. Panasyuk, V. Petrov, S. Sharakin *et al.*, Remote sensing of the atmosphere by the ultraviolet detector TUS onboard the Lomonosov satellite, *Remote Sens.* **11**(20), 2449 (October, 2019). doi: 10.3390/rs11202449.

24. P. A. Klimov and K. F. Sigaeva, Fast near-UV radiation pulsations measured by the space telescope TUS in the auroral region, *J. Atmos. Sol. Terr. Phys.* **220**, 105672 (September, 2021). doi: 10.1016/j.jastp.2021.105672.

25. POEMMA Collaboration, A. V. Olinto, J. Krizmanic, J. H. Adams, R. Aloisio, L. A. Anchordoqui, A. Anzalone *et al.*, The POEMMA (Probe of extreme multi-messenger astrophysics) observatory, *J. Cosmol. Astropart. Phys.* **2021**(6), 007 (June, 2021). doi: 10.1088/1475-7516/2021/06/007.

26. J. Abraham, P. Abreu, M. Aglietta, E. Ahn, D. Allard, I. Allekotte, J. Allen *et al.*, Trigger and aperture of the surface detector array of the pierre auger observatory, *Nucl. Instrum. Methods Phys. Res. Sec. A.* **613**(1), 29–39 (2010). ISSN 0168-9002. doi: 10.1016/j.nima.2009.11.018. https://www.sciencedirect.com/science/article/pii/S0168900209021688.

27. J. Abraham, P. Abreu, M. Aglietta, C. Aguirre, E. Ahn, D. Allard, I. Allekotte *et al.*, The fluorescence detector of the pierre auger observatory, *Nucl. Instrum. Methods Phys. Res. Sec. A.* **620**(2), 227–251 (2010). ISSN 0168-9002. doi: 10.1016/j.nima.2010.04.023. https://www.sciencedirect.com/science/article/pii/S0168900210008727.

28. T. Abu-Zayyad, R. Aida, M. Allen, R. Anderson, R. Azuma, E. Barcikowski, J. Belz *et al.*, The surface detector array of the telescope array experiment, *Nucl. Instrum. Methods Phys. Res. Sec.* **689**, 87–97 (2012). ISSN 0168-9002. doi: 10.1016/j.nima.2012.05.079. https://www.sciencedirect.com/science/article/pii/S0168900212005931.

29. H. Tokuno, Y. Tameda, M. Takeda, K. Kadota, D. Ikeda, M. Chikawa, T. Fujii *et al.*, New air fluorescence detectors employed in the telescope array experiment, *Nucl. Instrum. Methods Phys. Res. Sec. A.* **676**, 54–65(2012). ISSN 0168-9002. doi: 10.1016/j.nima.2012.02. 044. https://www.sciencedirect.com/science/article/pii/S016890021 2002422.

30. P. Abreu, M. Aglietta, J. M. Albury, I. Allekotte, A. Almela, J. Alvarez-Muniz, R. Alves Batista *et al.*, The ultra-high-energy cosmic-ray sky above 32 EeV viewed from the Pierre Auger Observatory, *PoS.* **ICRC2021**, 307 (2021). doi: 10.22323/1.395.0307.

31. J. Kim, D. Ivanov, K. Kawata, H. Sagawa, and G. Thomson, Hotspot update, and a new excess of events on the sky seen by the telescope array experiment, *PoS.* **ICRC2021**, 328 (2021). doi: 10.22323/1.395. 0328.

32. J. Alvarez-Muñiz, A. Romero-Wolf, and E. Zas, Čerenkov radio pulses from electromagnetic showers in the time domain, *Phys. Rev. D.* **81**(12), 123009 (June, 2010). doi: 10.1103/PhysRevD.81.123009.

33. A. Romero-Wolf, S. A. Wissel, H. Schoorlemmer, W. R. Carvalho, J. Alvarez-Muñiz, E. Zas, P. Allison *et al.*, Comprehensive analysis of anomalous ANITA events disfavors a diffuse tau-neutrino flux origin, *Phys. Rev. D.* **99**(6), 063011 (March, 2019). doi: 10.1103/PhysRevD. 99.063011.

34. T. M. Venters, M. H. Reno, J. F. Krizmanic, L. A. Anchordoqui, C. Guépin, and A. V. Olinto, POEMMA's target-of-opportunity sensitivity to cosmic neutrino transient sources, *Phys. Rev. D.* **102**(12), 123013 (December, 2020). doi: 10.1103/PhysRevD.102. 123013.

35. G. Davidson and R. O'Neil, Optical radiation from nitrogen and air at high pressure excited by energetic electrons, *J. Chem. Phys.* **41**(12), 3946–3955 (December, 1964). doi: 10.1063/1.1725841.

36. A. N. Bunner. Cosmic ray detection by atmospheric fluorescence. PhD thesis, Cornell University (January, 1967).

37. L. M. Barbier, R. Smith, S. Murphy, E. R. Christian, R. Farley, J. F. Krizmanic, J. W. Mitchell *et al.*, NIGHTGLOW: An instrument to measure the Earth's nighttime ultraviolet glow — results from the first engineering flight, *Astropart. Phys.* **22**(5-6), 439–449 (January, 2005). doi: 10.1016/j.astropartphys.2004.10.002.

38. F. W. Stecker, J. F. Krizmanic, L. M. Barbier, E. Loh, J. W. Mitchell, P. Sokolsky, and R. E. Streitmatter, Observing the ultrahigh energy universe with OWL eyes, *Nucl. Phys. B Proc. Suppl.* **136**, 433–438 (November, 2004). doi: 10.1016/j.nuclphysbps.2004.10.027.

39. J. F. Krizmanic and Poemma Collaboration, Space-based extensive air shower optical Cherenkov and fluorescence measurements using SiPM

detectors in context of POEMMA, *Nucl. Instrum. Methods Phys. Res. A.* **985**, 164614 (January, 2021). doi: 10.1016/j.nima.2020.164614.

40. See Particle Data Group, "Particle Detectors for Non-Accelerator Physics", https://pdg.lbl.gov.

41. T. K. Gaisser and A. M. Hillas. Reliability of the method of constant intensity cuts for reconstructing the average development of vertical showers. In *Proceedings of the 15th ICRC (Plovdiv)*, Vol. 8, pp. 353–357 (1977).

42. J. Krizmanic, Performance of the Orbiting Wide-angle Light collector (OWL/AirWatch) Experiment via Monte Carlo Simulation. In *26th International Cosmic Ray Conference (ICRC26)*, Vol. 2, *International Cosmic Ray Conference*, p. 388 (August, 1999).

43. P. Mikulski, Correlating primary type with the longitudinal profile. In *26th International Cosmic Ray Conference (ICRC26)*, Vol. 1, *International Cosmic Ray Conference*, p. 445 (January, 1999).

44. C. Berat, S. Bottai, D. De Marco, S. Moreggia, D. Naumov, M. Pallavicini, R. Pesce *et al.*, Full simulation of space-based extensive air showers detectors with ESAF, *Astropart. Phys.* **33**(4), 221–247 (May, 2010). doi: 10.1016/j.astropartphys.2010.02.005.

45. R. M. Baltrusaitis, R. Cady, G. L. Cassiday, R. Cooperv, J. W. Elbert, P. R. Gerhardy, S. Ko *et al.*, The Utah Fly's Eye detector, *Nucl. Instrum. Methods Phys. Res. A.* **240**(2), 410–428 (October, 1985). doi: 10.1016/0168-9002(85)90658-8.

46. L. A. Anchordoqui, D. R. Bergman, M. E. Bertaina, F. Fenu, J. F. Krizmanic, A. Liberatore, A. V. Olinto *et al.*, Performance and science reach of the Probe of Extreme Multimessenger Astrophysics for ultrahigh-energy particles, *Phys. Rev. D.* **101**(2), 023012 (January, 2020). doi: 10.1103/PhysRevD.101.023012.

47. J. H. Adams, S. Ahmad, J. N. Albert, D. Allard, L. Anchordoqui, V. Andreev, A. Anzalone *et al.*, The JEM-EUSO instrument, *Exp. Astron.* **40**(1), 19–44 (November, 2015). doi: 10.1007/s10686-014-9418-x.

48. J. Krizmanic, D. Bergman, and P. Sokolsky. The modeling of the nuclear composition measurement performance of the non-imaging cherenkov array (niche) (2013).

49. E. Kido, Status and prospects of the TAx4 experiment, *EPJ Web Conf.* **210**, 06001 (2019). doi: 10.1051/epjconf/201921006001.

50. T. P. A. Collaboration, A. Aab, P. Abreu, M. Aglietta, E. J. Ahn, I. A. Samarai, I. F. M. Albuquerque *et al.*, The pierre auger observatory upgrade — preliminary design report (2016).

51. A. Aab, P. Abreu, M. Aglietta, J. M. Albury, I. Allekotte, A. Almela, J. Alvarez Castillo *et al.*, Features of the energy spectrum of cosmic rays above 2.5×10^{18} eV using the pierre auger observatory, *Phys.*

Rev. Lett. **125**, 121106 (September, 2020). doi: 10.1103/PhysRevLett. 125.121106.

52. D. Ivanov, Energy spectrum measured by the telescope array, *PoS.* **ICRC2019**, 298 (2019). doi: 10.22323/1.358.0298.

53. A. Yushkov, Mass composition of cosmic rays with energies above $10^{17.2}$ eV from the hybrid data of the Pierre Auger Observatory, *PoS.* **ICRC2019**, 482 (2020). doi: 10.22323/1.358.0482.

54. *The Pierre Auger Observatory: Contributions to the 36th International Cosmic Ray Conference (ICRC 2019)*, Madison, Wisconsin, USA, July 24–August 1, 2019 (September, 2019).

55. C. Guépin, R. Aloisio, A. Cummings, L. A. Anchordoqui, J. F. Krizmanic, A. V. Olinto, M. H. Reno, and T. M. Venters, Indirect dark matter searches at ultrahigh energy neutrino detectors, *Phys. Rev. D.* **104**(8), 083002 (October, 2021). doi: 10.1103/PhysRevD.104. 083002.

56. R. W. Hanuschik, A flux-calibrated, high-resolution atlas of optical sky emission from UVES, *Astron. Astrophys.* **407**, 1157–1164 (September, 2003). doi: 10.1051/0004-6361:20030885.

57. P. C. Cosby, B. D. Sharpee, T. G. Slanger, D. L. Huestis, and R. W. Hanuschik, High-resolution terrestrial nightglow emission line atlas from UVES/VLT: Positions, intensities, and identifications for 2808 lines at 314–1043 nm, *J. Geophys. Res. (Space Physics)* **111**(A12), A12307 (December, 2006). doi: 10.1029/2006JA012023.

58. R. R. Meier, Ultraviolet spectroscopy and remote sensing of the upper atmosphere, *Space Sci. Rev.* **58**, 1–185 (1991). doi: 10.1007/ BF01206000.

59. C. Leinert, S. Bowyer, L. K. Haikala, M. S. Hanner, M. G. Hauser, A. C. Levasseur-Regourd, I. Mann *et al.*, The 1997 reference of diffuse night sky brightness, *Astron. Astrophys. Suppl. Ser.* **127**, 1–99 (January, 1998). doi: 10.1051/aas:1998105.

60. G. G. Shepherd, Y.-M. Cho, G. Liu, M. G. Shepherd, and R. G. Roble, Airglow variability in the context of the global mesospheric circulation, *J. Atmos. Sol. Terr. Phys.* **68**, 2000–2011 (December, 2006). doi: 10. 1016/j.jastp.2006.06.006.

61. K. A. Deutsch and G. Hernandez, Long-term behavior of the OI 558 nm emission in the night sky and its aeronomical implications, *J. Geophys. Res. (Space Physics)* **108**, 1430 (December, 2003). doi: 10. 1029/2002JA009611.

62. R. F. Pfaf, The near-Earth plasma environment, *Space Sci. Rev.* **168**, 23–112 (2002).

63. F. E. Roach and A. B. Meinel, The height of the nightglow by the Van Rhijn method, *ApJ.* **122**, 530 (November, 1955). doi: 10.1086/146115.

64. A. L. Cummings, R. Aloisio, J. Eser, and J. F. Krizmanic, Modeling the optical cherenkov signals by cosmic ray extensive air showers directly observed from suborbital and orbital altitudes, *Phys. Rev. D.* **104**, 063029 (September, 2021). doi: 10.1103/PhysRevD.104.063029.

65. A. L. Cummings, R. Aloisio, and J. F. Krizmanic, Modeling of the tau and muon neutrino-induced optical Cherenkov signals from upward-moving extensive air showers, *Phys. Rev. D.* **103**(4), 043017 (February, 2021). doi: 10.1103/PhysRevD.103.043017.

66. A. N. Otte, D. Garcia, T. Nguyen, and D. Purushotham, Characterization of three high efficiency and blue sensitive silicon photomultipliers, *Nucl. Instrum. Methods Phys. Res. Sec. A.* **846**, 106–125 (2017). ISSN 0168-9002. doi: 10.1016/j.nima.2016.09.053. http://www.sciencedirect.com/science/article/pii/S0168900216309901.

67. J. R. Patterson and A. M. Hillas, The relation of Cerenkov time profile widths to the distance to maximum of air showers, *J. Phys. G Nucl. Phys.* **9**, 323–337 (March, 1983). doi: 10.1088/0305-4616/9/3/013.

68. P. Sokolsky and J. Krizmanic, Effect of clouds on apertures of space-based air fluorescence detectors, *Astropart. Phys.* **20**(4), 391–403 (January, 2004). doi: 10.1016/S0927-6505(03)00196-8.

69. J. F. Krizmanic, P. Sokolsky, and R. Streitmatter. Limitations on space-based air fluorescence detector apertures obtained from IR cloud measurements. In *International Cosmic Ray Conference*, Vol. 2, *International Cosmic Ray Conference*, p. 639 (July, 2003).

70. T. Abu-Zayyad, C. C. H. Jui, and E. C. Loh, The effect of clouds on air showers observation from space, *Astropart. Phys.* **21**, 163–182 (2004). doi: 10.1016/j.astropartphys.2004.01.001.

71. J. Eser, A. V. Olinto, and L. Wiencke, Science and mission status of EUSO-SPB2, *PoS.* **ICRC2021**, 404 (2021). doi: 10.22323/1.395.0404.

72. G. Osteria, J. H. Adams, M. Battisti, A. S. Belov, M. E. Bertaina, F. Bisconti, F. Saverio Cafagna *et al.*, The fluorescence telescope on board EUSO-SPB2 for the detection of ultra high energy cosmic rays, *PoS.* **ICRC2021**, 206 (2021). doi: 10.22323/1.395.0206.

73. G. Filippatos, M. Battisti, M. E. Bertaina, F. Bisconti, J. Eser, G. Osteria, F. Sarazin, and L. Wiencke, Expected performance of the EUSO-SPB2 fluorescence telescope, *PoS.* **ICRC2021**, 405 (2021). doi: 10.22323/1.395.0405.

74. G. Abdellaoui, S. Abe, A. Acheli, J. H. Adams, S. Ahmad, A. Ahriche, J. N. Albert *et al.*, Cosmic ray oriented performance studies for the JEM-EUSO first level trigger, *Nucl. Instrum. Methods Phys. Res. A.* **866**, 150–163 (September, 2017). doi: 10.1016/j.nima.2017.05.043.

75. G. Osteria, V. Scotti, and JEM-EUSO Collaboration, EUSO-SPB: In-flight performance, *Nucl. Instrum. Methods Phys. Res. A.* **936**, 237–238 (August, 2019). doi: 10.1016/j.nima.2018.10.213.

76. R. Diesing, A. Bukowski, N. Friedlander, A. Miller, S. Meyer, and A. V. Olinto, UCIRC2: EUSO-SPB2's infrared cloud monitor, *PoS*. **ICRC2021**, 214 (2021). doi: 10.22323/1.395.0214.

77. Š. Mackovjak, P. Bobík, J. Baláž, I. Strhárský, M. Putiš, and P. Gorodetzky, Airglow monitoring by one-pixel detector, *Nucl. Instrum. Methods Phys. Res. A.* **922**, 150–156 (April, 2019). doi: 10.1016/j.nima.2018.12.073.

78. M. Bagheri, P. Bertone, I. Fontane, E. Gazda, E. G. Judd, J. F. Krizmanic, E. N. Kuznetsov *et al.*, Overview of Cherenkov telescope on-board EUSO-SPB2 for the detection of very-high-energy neutrinos, *PoS*. **ICRC2021**, 1191 (2021). doi: 10.22323/1.395.1191.

79. S. Gómez, D. Gascón, G. Fernández, A. Sanuy, J. Mauricio, R. Graciani, and D. Sanchez, MUSIC: An 8 channel readout ASIC for SiPM arrays. In F. Berghmans and A. G. Mignani (eds.), *Optical Sensing and Detection IV*, Vol. 9899, pp. 85–94, SPIE (2016). doi: 10.1117/12.2231095.

80. A. Neronov, D. V. Semikoz, I. Vovk, and R. Mirzoyan, Cosmic-ray composition measurements and cosmic ray background-free γ-ray observations with Cherenkov telescopes, *Phys. Rev. D.* **94**(12), 123018 (December, 2016). doi: 10.1103/PhysRevD.94.123018.

81. G. Curzi, D. Modenini, and P. Tortora, Large constellations of small satellites: A survey of near future challenges and missions, *Aerospace.* **7**, 133 (September, 2020). doi: 10.3390/aerospace7090133.

82. A. M. Hillas, The sensitivity of Cerenkov radiation pulses to the longitudinal development of cosmic-ray showers, *J. Phys. G.* **8**, 1475–1492 (1982).

83. A. M. Hillas, Angular and energy distributions of charged particles in electron-photon cascades in air, *J. Phys. G.* **8**, 1461–1473 (1982).

84. Y. Tsunesada, R. Katsuya, Y. Mitsumori, K. Nakayama, F. Kakimoto, H. Tokuno, N. Tajima *et al.*, New air Cherenkov light detectors to study mass composition of cosmic rays with energies above knee region, *Nucl. Instrum. Methods Phys. Res. A.* **763**, 320–328 (November, 2014). doi: 10.1016/j.nima.2014.06.054.

85. D. R. Bergman, Y. Tsunesada, J. F. Krizmanic, and Y. Omura. NICHE: Air-Cherenkov observation at the TA site. In *European Physical Journal Web of Conferences*, Vol. 210, p. 05001 (October, 2019). doi: 10.1051/epjconf/201921005001.

86. Anita Collaboration, P. W. Gorham, P. Allison, S. W. Barwick, J. J. Beatty, D. Z. Besson, W. R. Binns *et al.*, The Antarctic Impulsive Transient Antenna ultra-high energy neutrino detector: Design, performance, and sensitivity for the 2006–2007 balloon flight, *Astropart. Phys.* **32**(1), 10–41 (August, 2009). doi: 10.1016/j.astropartphys.2009.05.003.

87. P. W. Gorham, P. Allison, B. M. Baughman, J. J. Beatty, K. Belov, D. Z. Besson, S. Bevan *et al.*, Observational constraints on the ultrahigh energy cosmic neutrino flux from the second flight of the ANITA experiment, *Phys. Rev. D.* **82**(2), 022004 (July, 2010). doi: 10.1103/PhysRevD.82.022004.

88. A. Vieregg. Results from the third flight of ANITA. In *European Physical Journal Web of Conferences*, Vol. 216, p. 01009 (June, 2019). doi: 10.1051/epjconf/201921601009.

89. P. W. Gorham, P. Allison, O. Banerjee, L. Batten, J. J. Beatty, K. Belov, D. Z. Besson *et al.*, Constraints on the ultrahigh-energy cosmic neutrino flux from the fourth flight of ANITA, *Phys. Rev. D.* **99**(12), 122001 (June, 2019). doi: 10.1103/PhysRevD.99.122001.

90. G. A. Askar'yan, Excess negative charge of an electron-photon shower and its coherent radio emission, *Zh. Eksp. Teor. Fiz.* **41**, 616–618 (1961).

91. S. Hoover, J. Nam, P. W. Gorham, E. Grashorn, P. Allison, S. W. Barwick, J. J. Beatty *et al.*, Observation of ultrahigh-energy cosmic rays with the ANITA balloon-borne radio interferometer, *Phys. Rev. Lett.* **105**(15), 151101 (October, 2010). doi: 10.1103/PhysRevLett.105.151101.

92. P. W. Gorham, A. Ludwig, C. Deaconu, P. Cao, P. Allison, O. Banerjee, L. Batten *et al.*, Unusual near-horizon cosmic-ray-like events observed by ANITA-IV, *Phys. Rev. Lett.* **126**(7), 071103 (February, 2021). doi: 10.1103/PhysRevLett.126.071103.

93. ANITA Collaboration, P. W. Gorham, A. Ludwig, C. Deaconu, P. Cao, P. Allison, O. Banerjee *et al.*, Unusual near-horizon cosmic-ray-like events observed by ANITA-IV, *arXiv e-prints*. art. arXiv:2008.05690 (August, 2020).

94. Q. Abarr, P. Allison, J. Ammerman Yebra, J. Alvarez-Muñiz, J. J. Beatty, D. Z. Besson, P. Chen *et al.*, The Payload for Ultrahigh Energy Observations (PUEO): A white paper, *arXiv e-prints*. art. arXiv:2010.02892 (October, 2020).

95. P. Allison, S. Archambault, R. Bard, J. Beatty, M. Beheler-Amass, D. Besson, M. Beydler *et al.*, Design and performance of an interferometric trigger array for radio detection of high-energy neutrinos, *Nucl. Instrum. Methods Phys. Res. Sec. A.* **930**, 112–125 (2019). ISSN 0168-9002. doi: 10.1016/j.nima.2019.01.067. https://www.sciencedirect.com/science/article/pii/S016890021930124X.

96. J. Alvarez-Muñiz, W. R. Carvalho, and E. Zas, Monte Carlo simulations of radio pulses in atmospheric showers using ZHAireS, *Astropart. Phys.* **35**, 325–341 (January, 2012). doi: 10.1016/j.astropartphys.2011.10.005.

97. A. Romero-Wolf, P. Gorham, K. Liewer, J. Booth, and R. Duren, Concept and analysis of a satellite for space-based radio detection of ultra-high energy cosmic rays, *ArXiv e-prints* (February, 2013).

98. A. Romero-Wolf, J. Alvarez-Muñiz, L. A. Anchordoqui, D. Bergman, J. Carvalho, Washington, A. L. Cummings, P. Gorham *et al.*, Radio detection of ultra-high energy cosmic rays with low lunar orbiting smallSats, *arXiv e-prints*. art. arXiv:2008.11232 (August, 2020).

99. J. R. Hörandel, The global cosmic ray observatory — GCOS, *PoS*. **ICRC2021**, 027 (2021). doi: 10.22323/1.395.0027.

100. K. Fang, J. Alvarez-Muniz, R. Alves Batista, M. Bustamante, W. Carvalho, D. Charrier, I. Cognard *et al.*, The Giant Radio Array for Neutrino Detection (GRAND): Present and perspectives, *ArXiv e-prints* (August, 2017).

Chapter 5

Neutrinos: Scientific Motivation

Francis Halzen

Department of Physics, Wisconsin IceCube
Particle Astrophysics Center, University of Wisconsin–Madison,
Madison, WI 53706, USA
halzen@icecube.wisc.edu

The IceCube project transformed a cubic kilometer of transparent natural Antarctic ice into a Cherenkov detector. It discovered PeV-energy neutrinos originating beyond our galaxy with an energy flux that is comparable to that of GeV-energy gamma rays and EeV-energy cosmic rays. These neutrinos provide the only unobstructed view of the cosmic accelerators that power the highest-energy radiation reaching us from the universe. We review the results from IceCube's first decade of operations, which exploited multiple techniques for the measurement of the diffuse neutrino flux from the universe. Reaching beyond 10 PeV, this flux also provides a beam to study neutrinos themselves. We subsequently review the multimessenger data that identified the supermassive black hole TXS 0506+056 as a source of cosmic neutrinos. We draw attention to accumulating indications that cosmic neutrinos are associated with gamma-ray-obscured active galaxies, i.e., the energy of the photons that inevitably accompanies cosmic neutrinos emerges at MeV level or below.

1. Neutrino Astronomy: A Brief Introduction

The shortest-wavelength radiation reaching us from the universe is not radiation at all; it consists of cosmic rays — high-energy nuclei, mostly protons. Some reach us with extreme energies exceeding 10^8 TeV from a universe beyond our galaxy that is obscured by gamma rays and from which only neutrinos can reach us as astronomical messengers.[1] Their origin is still unknown, but the identification of a supermassive black hole powering a cosmic-ray accelerator[2,3] represents a breakthrough toward a promising path for resolving the century-old puzzle of the origin of cosmic rays: multimessenger astronomy.

The rationale for searching for cosmic-ray sources by observing neutrinos is straightforward: In relativistic particle flows near neutron stars or black holes, some of the gravitational energy released in the accretion of matter is transformed into the acceleration of protons or heavier nuclei, which subsequently interact with ambient radiation or dust, hydrogen, and dense molecular clouds to produce pions and other secondary particles that decay into neutrinos. Isospin dictates that both neutral and charged secondary pions are produced; while charged pions decay into neutrinos, neutral pions decay into gamma rays. The fact that cosmic neutrinos are inevitably accompanied by high-energy photons transforms neutrino astronomy into multimessenger astronomy.

A main challenge of multimessenger astronomy is to separate these photons, which we will refer to as *pionic* photons, from photons radiated by electrons that may be accelerated along with the cosmic-ray protons. Another challenge is to identify the electromagnetic energy associated with the pionic photons because they do not reach our telescopes with their initial energy but do so after losses suffered in the extragalactic background light (EBL), mostly microwave photons, and possibly, in the source. As is the case for constructing a neutrino beam in a particle physics laboratory, neutrinos are produced in a so-called beam dump with a target transforming the energy of the proton beam into neutrinos. Powerful neutrino sources within reach of IceCube's sensitivity require a dense target that is likely to be obscured by pionic gamma rays; they are likely to be gamma-ray-obscured sources. We review the accumulating evidence that this is indeed the case.

With 10 years of data, the emergence of active galaxies as sources of cosmic rays was not unexpected.[4,5] The detailed blueprint for a cosmic-ray accelerator must meet two challenges: The highest-energy particles in the beam must reach energies beyond 10^8 TeV for extragalactic sources and their luminosity must accommodate the observed flux. Both requirements represent severe constraints that have guided theoretical speculations toward active galaxies; for a recent review, see Ref. [6]. Acceleration of protons (and nuclei) to TeV energy and above requires massive bulk flows of relativistic charged particles. While accelerating the protons, active galaxies also boost large masses to relativistic velocities. The radio emission reveals that the plasma in the jets of active galaxies flows with velocities of $0.99\,c$. A fraction of a solar mass per year can be accelerated to relativistic Lorentz factors of order 10, leading to a luminosity of $10^{46}\,\mathrm{erg\,s^{-1}}$, which is close to the Eddington limit.

In contrast to our own galaxy hosting a black hole at its center that is mostly dormant, in an active galaxy, the rotating supermassive black hole absorbs the matter in the host galaxy at a very high rate. Fast-spinning matter falling onto the active galactic nucleus (AGN) swirls around the black hole in an accretion disk, like the water approaching the drain of your bath tub. When the accretion disk comes in contact with the rotating black hole, its space–time drags the magnetic field, winding it into a tight cone around the rotation axis into a jet of particles; see Fig. 1. Not just particles but huge "blobs" of plasma from the accretion disk are flung out along these field lines. It is not clear whether it is the rotational energy of the black hole or the magnetic energy in the rotating plasma that powers the accelerator. If the jet runs into a target material, for instance the ubiquitous $10\,\mathrm{eV}$ ultraviolet photons in some galaxies, neutrinos can be produced. Production of high-energy neutrinos in the cores of the AGN may also result from the acceleration of cosmic rays in the high-field regions associated with the accretion disk or the corona surrounding it.[7]

Two general scenarios can accommodate the diffuse cosmic neutrinos observed by IceCube: protons accelerated at the core interacting with high-density targets or strong fields near the black hole,[8,9] or alternatively, diffusion through the galaxy to produce neutrinos in collisions with interstellar matter.[10] We will return to these speculations after a review of the neutrino observations.

2. Discovery of High-Energy Cosmic Neutrinos

Near the National Science Foundation's research station located at the geographic South Pole, the IceCube project[11] transformed one

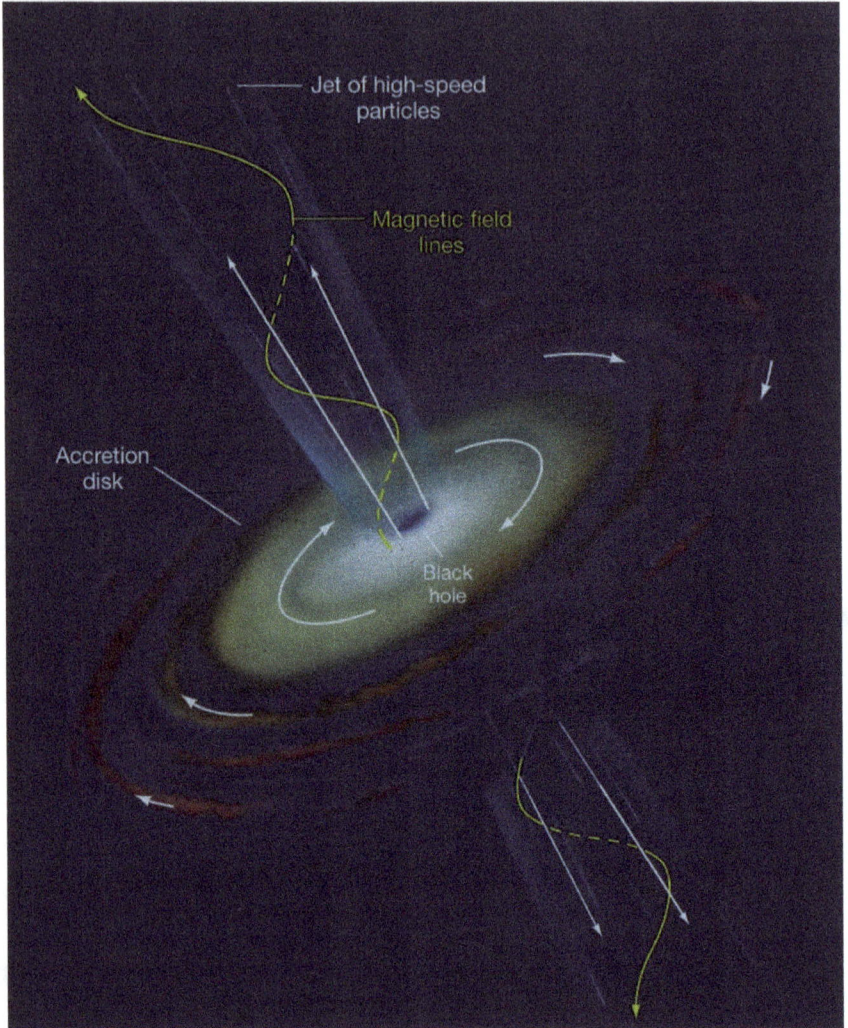

Fig. 1. Accretion disk meets the spinning black hole that winds up the disk's magnetic field lines. Credit: Pearson Education, Inc., Upper Saddle River, New Jersey.

cubic kilometer of natural Antarctic ice into a Cherenkov detector. The deep ice of the Antarctic glacier constitutes the detector, forming both a support structure and Cherenkov medium. Below a depth of 1,450 m, a cubic kilometer of glacial ice is instrumented with 86 cables called "strings", each of which is equipped with 60 optical

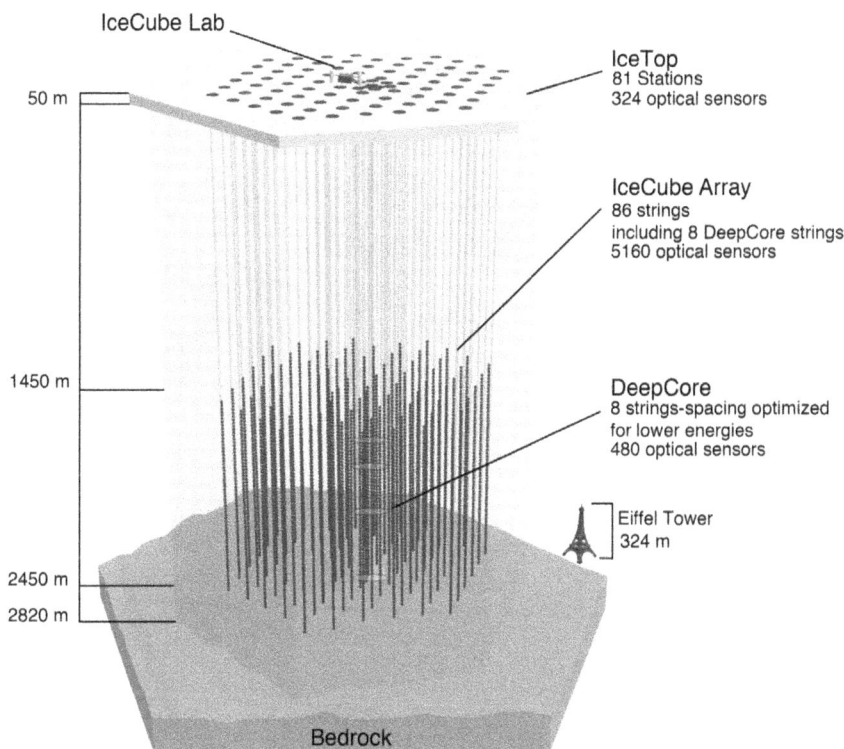

Fig. 2.   Architecture of the IceCube observatory.

sensors; see Fig. 2. Each digital optical module (DOM) consists of a glass sphere containing the photomultiplier and the electronics board that captures and digitizes the signals locally using an onboard computer; see Fig. 3. The digitized signals are given a global time stamp with residuals accurate to 2 ns and are subsequently transmitted to the surface. Processors at the surface continuously collect the time-stamped signals from the optical modules, each of which functions independently. The digital messages are sent to a string processor and a global event trigger. They are subsequently sorted into the Cherenkov patterns emitted by secondary muon tracks produced by muon neutrinos interacting in the ice or by particle showers for the case of electron and tau neutrinos. These reveal the flavor, energy, and direction of the incident neutrino.[12] Constructed between 2004 and 2010, IceCube has now collected 10 years of data with the completed detector.

Fig. 3.   Digital optical module showing the down-facing 10-inch photomultiplier and the associated electronics that digitize the light signals.

In neutrino astronomy, the first challenge is to select a pure sample of neutrinos, more than 100,000 per year above a threshold of 0.1 TeV, from a background of 10 billion atmospheric cosmic-ray muons. The second is to identify the small fraction of these neutrinos that are astrophysical in origin. Atmospheric neutrinos are a background for cosmic neutrinos, at least at neutrino energies below ~100 TeV. Above this energy, the atmospheric neutrino flux reduces to a few events per year, even in a kilometer-scale detector, and thus neutrinos well above that energy are cosmic in origin.

The arrival directions of a secondary muon track and an electromagnetic shower initiated by an electron or tau neutrino are determined by the arrival times of the Cherenkov photons at the optical sensors, while the number of photons is a proxy for the energy deposited by secondary particles in the detector. Although the detector only records the energy of the secondary muon inside the detector, from Standard Model physics, we can infer the energy spectrum of the parent neutrino.

Tracks resulting from muon neutrino interactions can be traced back to their sources with a $\leq 0.4°$ angular resolution for the highest-energy events. By contrast, the reconstruction of cascade directions, in principle possible to within a few degrees, is still in the development stage at IceCube, achieving $8°$ resolution.[13,14] On the other hand, determining their energy from the observed light pool is straightforward, and a resolution better than 15% can be achieved. For example, we contrast in Fig. 4, the Cherenkov patterns initiated by an electron (or tau) neutrino of 1 PeV energy (top) and a neutrino-induced muon losing 2.6 PeV energy while traversing the detector (bottom).

Searching for high-energy neutrinos of cosmic origin, IceCube continuously monitors the whole sky, collecting very-high-statistics datasets of atmospheric neutrinos. Neutrino energies cover more than six orders of magnitude, from $\sim$5 GeV in the highly instrumented inner core, labeled DeepCore in Fig. 2, to extreme energies beyond 10 PeV. Soon after the completion of the detector, with two years of data, IceCube discovered an extragalactic flux of cosmic neutrinos[1] with an energy flux, $E^2 dN/dE$, in the local universe that is, surprisingly, similar to that of gamma rays.[15,16]

Two principal methods are used to identify cosmic neutrinos. The first method reconstructs upgoing muon tracks initiated by muon neutrinos, and the second identifies neutrinos of all flavors that interact inside the instrumented volume of the detector. We describe these methods in turn.

With the first method, muon neutrinos can be detected even when interacting outside the detector because of the kilometer range of the secondary muons. IceCube has collected large samples of muon neutrinos with high purity, often above 99%, and measured the atmospheric neutrino flux over more than five orders of magnitude in energy with a result that is consistent with theoretical calculations. The tracks can be well reconstructed and separated from the background of atmospheric muons using the Earth as a filter. More importantly, IceCube also observes an excess of neutrino events at energies beyond 100 TeV[17–19] that cannot be accounted for by the

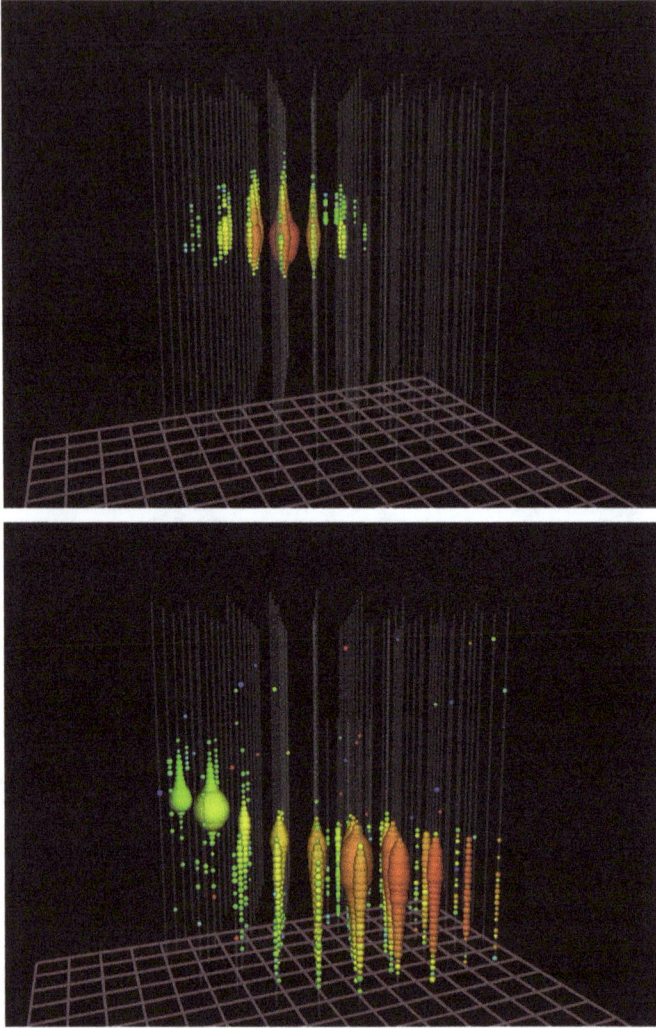

Fig. 4. Top Panel: Light pool produced in IceCube by a shower initiated by an electron or tau neutrino of 1.14 PeV, which represents a lower limit of the energy of the neutrino that initiated the shower. White dots represent sensors with no signal. For the colored dots, the color indicates arrival time, from red (early) to purple (late) following the rainbow, and size reflects the number of photons detected. Bottom Panel: A muon track coming up through the Earth traverses the detector at an angle of 11° below the horizon. The deposited energy, i.e., the energy equivalent of the total Cherenkov light of all charged secondary particles inside the detector, is 2.6 PeV.

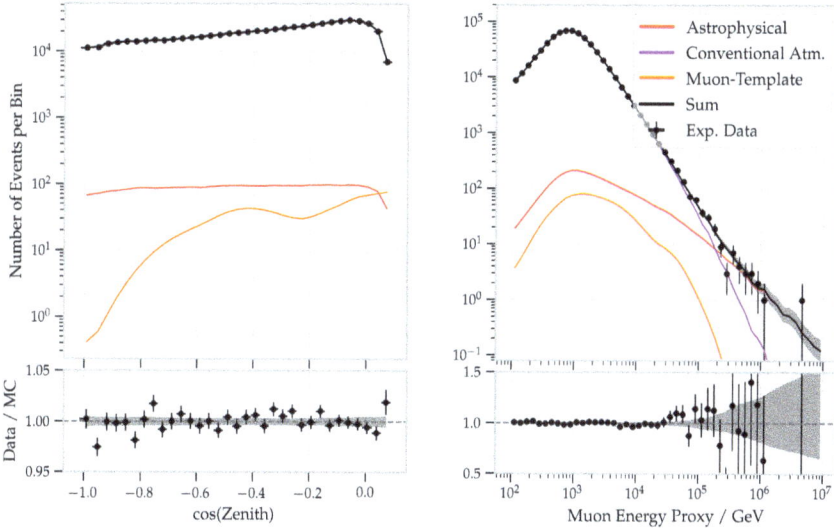

Fig. 5. Distributions of muon tracks arriving from the Northern Hemisphere, i.e., with declination greater than $-5°$, for the period 2010–2018, are shown as a function of reconstructed zenith (left)[20] and muon energy (right). The full dataset consists of about 650,000 neutrino events with a purity of 99.7%. The best-fit expectation for astrophysical and atmospheric neutrinos are superimposed. Statistical errors are shown as crosses, the gray bands in the ratio plots show an estimate of the systematics obtained by varying all fit parameters within their uncertainties.

atmospheric flux. The recent energy flux measurement covering 9.5 years of data representing a sample of 650,000 neutrinos with 99.7% purity is shown in Fig. 5. The excess cosmic neutrino flux (red) over the atmospheric background (blue) is well described by a power law with a spectral index of -2.37 ± 0.09 and a normalization at 100 TeV neutrino energy of $(1.36^{+0.24}_{-0.25}) \times 10^{-18}\,\mathrm{GeV}^{-1}\mathrm{cm}^{-2}\mathrm{sr}^{-1}\mathrm{s}^{-1}$.[20] The residual atmospheric muon background is small (yellow). For more details, see Ref. [21].

The arrival directions of the astrophysical muon tracks are isotropically distributed over the sky. Surprisingly, there is no evidence for a correlation to nearby sources in the galactic plane; Ice-Cube is recording a diffuse flux of extragalactic sources. Only after analyzing 10 years of data[22] has evidence emerged at the 3σ level that the neutrino sky is not completely isotropic. The anisotropy results from four sources — TXS 0506+056 among them (more about that source later) — that emerge as point sources above the $\sim 4\sigma$ level (pretrial); see Fig. 6. The strongest of these sources is the nearby active galaxy NGC 1068, also known as Messier 77.

The second method for separating cosmic from atmospheric neutrinos exclusively identifies high-energy neutrinos interacting inside the instrumented detector volume, so-called starting events. It divides the instrumented volume of ice into an outer veto shield and a ~ 500-million ton inner fiducial volume. The advantage of focusing on neutrinos interacting inside the instrumented volume of ice is that the detector functions as a total absorption calorimeter,[13] allowing for a good energy measurement that separates cosmic from lower-energy atmospheric neutrinos. With this method, neutrinos from all directions in the sky and of all flavors can be identified, including both muon tracks and secondary showers produced by charged-current interactions of electron and tau neutrinos and neutral current interactions of neutrinos of all flavors. A sample event with a light pool of roughly one hundred thousand photoelectrons extending over more than 500 m is shown in the top panel of Fig. 4.

The starting event sample revealed the first evidence for neutrinos of cosmic origin.[1,23] Events with PeV energies and with no trace of accompanying muons from an atmospheric shower are highly unlikely to be of atmospheric origin. The present seven-year dataset contains a total of 60 neutrino events with deposited energies ranging from 60 TeV to 10 PeV that are likely to be of cosmic origin. The deposited energy and zenith dependence of the high-energy starting events[19,24] are compared to those of the atmospheric background in Fig. 7. A purely atmospheric explanation of the observation is excluded at 8σ.

The flux of cosmic neutrinos has by now been characterized with a range of methods. Figure 8 shows the results of a search exclusively identifying showers that have been isolated from the atmospheric background down to energies below 10 TeV.[26] The energy spectrum of $E^{-2.5}$ agrees with the measurement for upgoing muons with a spectral index of $E^{-2.4}$ above an energy of ~ 100 TeV.[19] In general, analyses reaching lower energies exhibit larger spectral indices with the updated 7.5 years starting-event sample,[24] yielding a spectral index value of -2.87 ± 0.2 for the 68.3% confidence interval.

In summary, IceCube has observed cosmic neutrinos using both methods for rejecting background. Based on different methods for

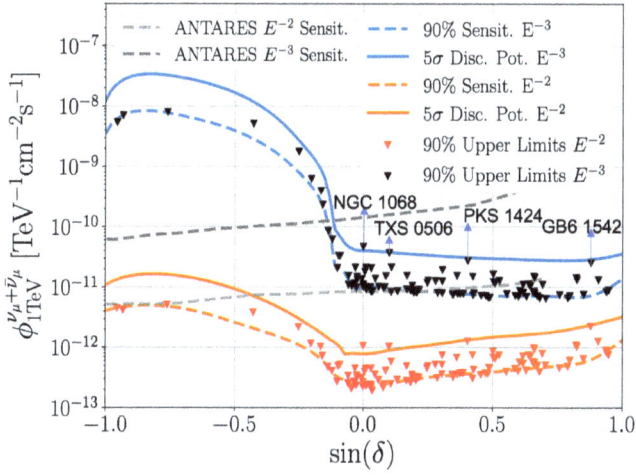

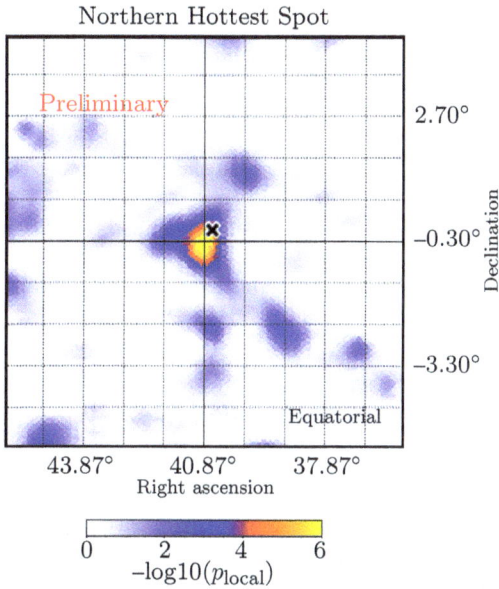

Fig. 6. Top Panel: Upper limits on the flux from candidate point sources of neutrinos in 10 years of IceCube data assuming two spectral indices of the flux. Four sources exceed the 4σ level (pretrial) and collectively result in a 3σ anisotropy of the sky map. Bottom Panel: Association of the hottest source in the sky map as well as the strongest source in the list of candidate sources with the active galaxy NGC 1068.

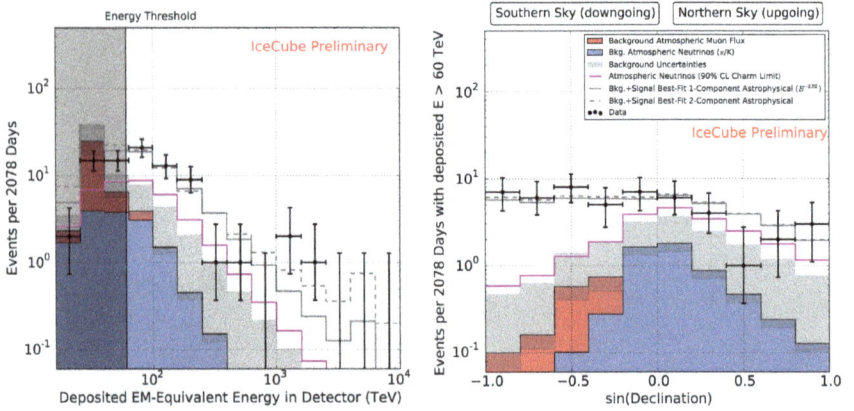

Fig. 7.   Left Panel: Deposited energies, by neutrinos interacting inside IceCube, observed in six years of data.[19] The gray region shows uncertainties in the sum of all backgrounds. The atmospheric muon flux (blue) and its uncertainty is computed from simulation to overcome statistical limitations in our background measurement and scaled to match the total measured background rate. The atmospheric neutrino flux is derived from previous measurements of both the π, K, and charm components of the atmospheric spectrum.[25] Also shown are two fits to the spectrum, assuming a simple power law (solid gray) and a broken power law (dashed gray). Right Panel: The same data and models, but now showing the distribution of events with deposited energy above 60 TeV in declination. At the South Pole, the declination angle δ is equivalent to the distribution in zenith angle θ related by the identity, $\delta = \theta - \pi/2$. It is clearly visible that the data is flat in the Southern Hemisphere, as expected from the contribution of an isotropic astrophysical flux.

reconstruction and energy measurement, their results agree, pointing at extragalactic sources whose flux has equilibrated in the three flavors after propagation over cosmic distances,[27] with $\nu_e : \nu_\mu : \nu_\tau \sim 1 : 1 : 1$.

We should comment at this point that there is yet another method to conclusively identify cosmic neutrinos: the observation of very-high-energy tau neutrinos. Tau neutrinos are produced in the atmosphere by the oscillations of muon neutrinos into tau neutrinos but only for neutrino energies well below 100 GeV. Above this energy, they must be of cosmic origin, produced in cosmic accelerators whose neutrino flux has approximately equilibrated between the three flavors after propagating over cosmic distances. Tau neutrinos produce two spatially separated showers in the detector: one from the interaction of the tau neutrino and the second from the tau decay. The mean

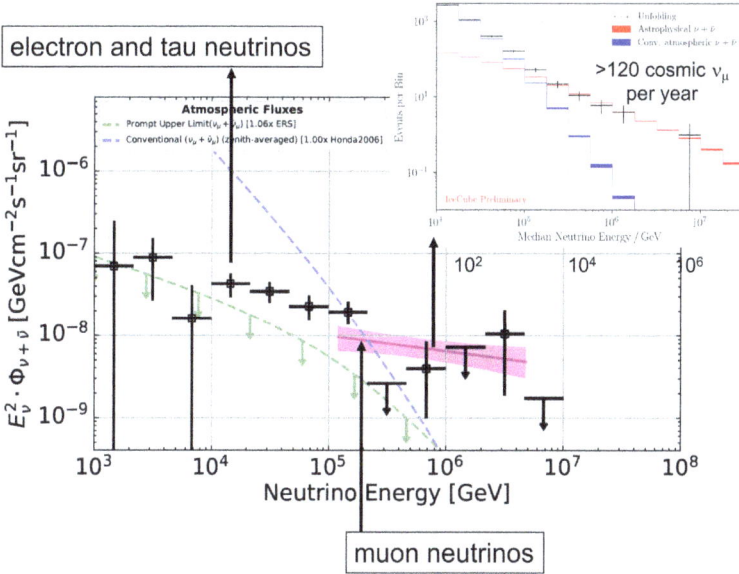

Fig. 8. Flux of cosmic muon neutrinos[19] inferred from the eight-year upgoing-muon track analysis (red solid line) with 1σ uncertainty range (shaded range; from fit shown in upper-right inset) is compared with the flux of showers initiated by electron and tau neutrinos.[26] The measurements are consistent assuming that each flavor contributes an identical flux to the diffuse spectrum.

tau lepton decay length is $\lambda_\tau = (E_\tau/m)\,c\tau \approx 50$ m $\times (E_\tau/\mathrm{PeV})$, where m, τ, and E_τ are the mass, lifetime, and energy of the tau, respectively. Two such candidate events have been identified.[28] An event with a decay length of $17\,\mathrm{m}$ and a probability of 98% of being produced by a tau neutrino is shown in Fig. 9.

Yet another independent confirmation of the observation of neutrinos of cosmic origin appeared in the form of the Glashow resonance event shown in Fig. 10. The event was identified in a dedicated search for partially contained events: an antielectron neutrino interacting with an atomic electron produced an event compatible with an incident neutrino energy of 6.3 PeV, characteristic of the resonant production of a weak intermediate W^-.[29]

Given its energy and direction, the event is classified as an astrophysical neutrino at the 5σ level. Furthermore, the presence of muons in the shower as well as the measured energy are consistent with the hadronic decay of a W^- produced at the Glashow resonance. In the

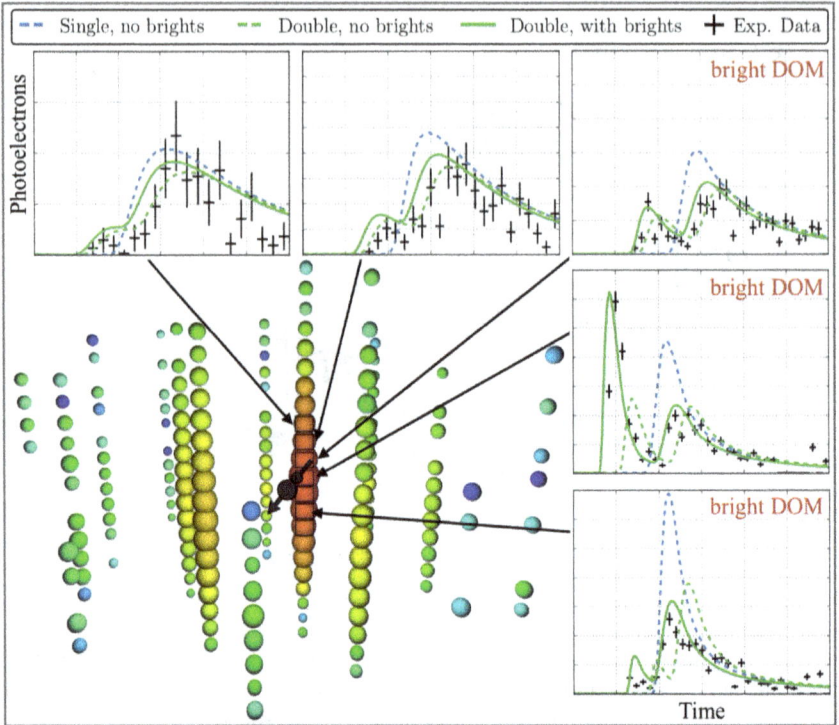

Fig. 9.   Event view of a tau neutrino. The Cherenkov photons associated with the production and subsequent decay of the tau neutrino are identified by the double-peaked photon count as a function of time for the bright DOMs, for instance, the one shown in the top-right corner. The best fit (solid line) corresponds to a 17 m decay length and is far superior to fits assuming a single electromagnetic or hadronic shower (dashed lines).

observer frame, where the electron mass ($m_e = 0.511\,\text{MeV}$) is at rest, the resonance energy is given by $E_R = M_W^2/2m_e = 6.32\,\text{PeV}$ for $M_W = 80.38\,\text{GeV}$. The energy deposited in the detector of $6.05 \pm 0.72\,\text{PeV}$ translates into a neutrino energy of $6.3\,\text{PeV}$ after correcting the visible energy produced by the hadronic decay of the W for shower particles that do not radiate. Taking into account the detector's energy resolution, the probability that the event is produced off resonance by deep inelastic scattering is only 0.01, assuming a spectrum with a spectral index of $\gamma = -2.5$. Given the Standard Model resonant cross section, we expect 1.55 events in the data sample searched assuming an antineutrino:neutrino ratio of 1:1,

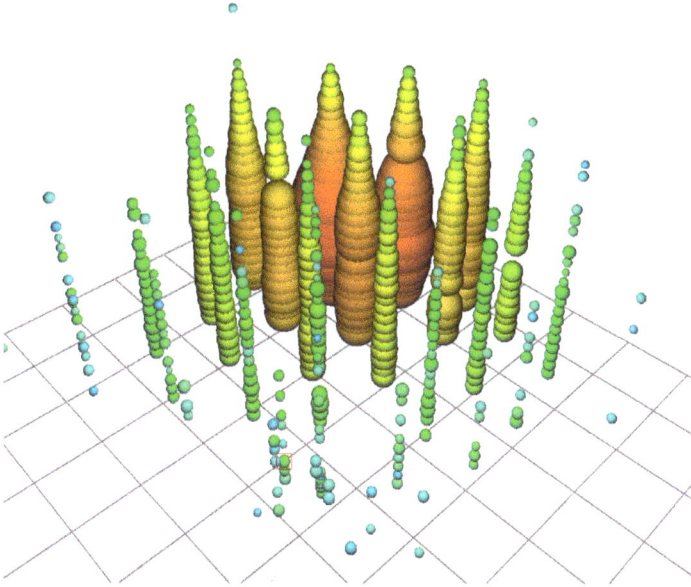

Fig. 10. Particle shower created at the Glashow resonance. The energy, measured via Cherenkov radiation in the Antarctic ice sheet, is reconstructed at the resonant energy for the production of a weak intermediate boson W^- in the interaction of an antielectron neutrino with an atomic electron in the ice. The properties of the secondary muons produced in the particle shower are consistent with the hadronic decay of a resonant W^- boson.

characteristic of a cosmic source producing an equal number of pions of all three electric charges.

The observation of a Glashow resonant event heralds the presence of electron antineutrinos in the cosmic neutrino flux. Its unique signature illustrates a method to disentangle neutrinos from antineutrinos, thus opening a path toward distinguishing astronomical accelerators that produce neutrinos via hadronuclear or photohadronic interactions with or without strong magnetic fields.[29] As such, knowledge of both the flavor and charge of the incident neutrino will add a new tool for doing neutrino astronomy.

3. Multimessenger Astronomy

The most important message emerging from the IceCube measurements may not be apparent yet: the prominent and surprisingly important role of neutrinos relative to photons in the nonthermal universe. To illustrate this point, we have shown the energy fluxes in Fig. 11, $E^2 dN/dE$, of neutrinos and gamma rays in the universe, with the energy density of cosmic neutrinos comparable to that of gamma rays observed with the NASA Fermi satellite.[15] This may indicate a common origin and, in any case, creates an excellent opportunity for multimessenger studies.

Photons are inevitably produced in association with neutrinos when accelerated cosmic rays produce both neutral and charged pions in interactions with target material in the vicinity of the accelerator. While neutral pions decay into two gamma rays, $\pi^0 \rightarrow \gamma + \gamma$, the charged pions decay into three high-energy neutrinos (ν) and antineutrinos ($\bar{\nu}$) via the decay chain $\pi^+ \rightarrow \mu^+ + \nu_\mu$ followed by $\mu^+ \rightarrow e^+ + \bar{\nu}_\mu + \nu_e$ and the charged-conjugate process. On average, the four final-state leptons equally share the energy of the charged pion. With these approximations, gamma rays and neutrinos carry on average one-half and one-fourth of the energy of the parent pion.

The neutrino production rate Q_{ν_α} (with typical units $\text{GeV}^{-1}\text{s}^{-1}$ and α labeling the neutrino flavor) can be related to the one for charged pions Q_π as

$$\sum_\alpha E_\nu Q_{\nu_\alpha}(E_\nu) \simeq 3 \left[E_\pi Q_{\pi^\pm}(E_\pi) \right]_{E_\pi \simeq 4E_\nu}, \tag{1}$$

while similarly, the production rate of pionic gamma rays is related to the one for neutral pions as

$$E_\gamma Q_\gamma(E_\gamma) \simeq 2 \left[E_\pi Q_{\pi^0}(E_\pi) \right]_{E_\pi \simeq 2E_\gamma}. \tag{2}$$

Pion production in the interactions of cosmic rays with strong photon fields proceeds resonantly via the processes $p + \gamma \rightarrow \Delta^+ \rightarrow \pi^0 + p$ and $p + \gamma \rightarrow \Delta^+ \rightarrow \pi^+ + n$. These channels produce charged and neutral pions with probabilities $2/3$ and $1/3$, respectively. However, the additional contribution of nonresonant pion production changes these ratios to approximately $1/2$ and $1/2$, respectively. In contrast, cosmic rays interacting with matter produce equal numbers of pions of all three charges: $p + p \rightarrow n_\pi \left[\pi^0 + \pi^+ + \pi^- \right] + X$, where n_π is the pion multiplicity. We thus obtain a charge ratio $K_\pi = n_{\pi^\pm}/n_{\pi^0} \simeq 2, 1$ for pp and $p\gamma$ interactions, respectively.

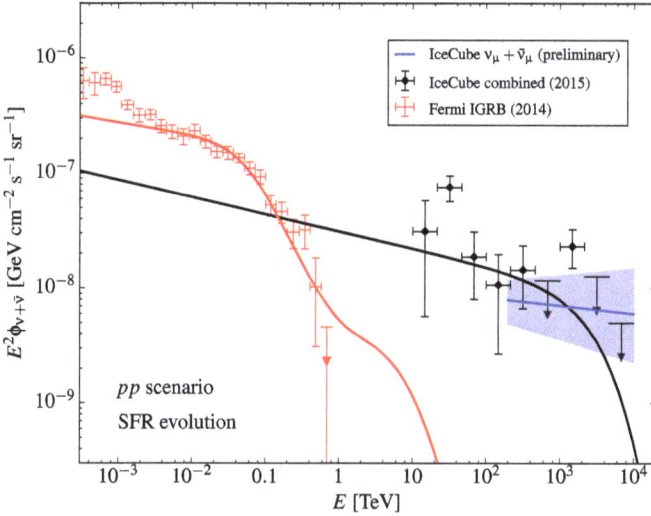

Fig. 11. Calculation illustrating that the astrophysical neutrino flux (black line) observed by IceCube qualitatively matches the corresponding cascaded gamma-ray flux (red line) observed by Fermi. We assume transparent sources. The result suggests that the decay products of neutral and charged pions from pp interactions may be responsible for the nonthermal emission in the universe.[32] The black data points are early IceCube results.[33,34] Also shown as a blue band is the best fit to the flux of high-energy muon neutrinos.[17-19] Introducing the cutoff on the high-energy flux, shown in the figure, does not affect the result.

Equations (1) and (2) can now be combined to obtain a direct relation between the gamma-ray and neutrino production rates:

$$\frac{1}{3}\sum_\alpha E_\nu^2 Q_{\nu_\alpha}(E_\nu) \simeq \frac{K_\pi}{4}\left[E_\gamma^2 Q_\gamma(E_\gamma)\right]_{E_\gamma=2E_\nu}, \tag{3}$$

where the factor $1/4$ accounts for the fact that two gamma rays are produced in the neutral pion decay with twice the energy of the accompanying neutrino, $\langle E_\nu\rangle/\langle E_\gamma\rangle \simeq 1/2$. Note that the relative production rate of gamma rays and neutrinos only depends on the ratio of charged-to-neutral pions produced without any reference to the cosmic-ray beam that initiates their production in the target. This powerful relation follows from the fact that pion production conserves isospin and nothing else.

Before applying this relation to data, one must recall that the universe is not transparent to PeV gamma rays. These will interact with microwave photons and other components of the EBL to initiate an electromagnetic cascade that reaches Earth in the form of multiple photons of lower energy. The electromagnetic shower subdivides the initial PeV photon energy, leading to multiple photons in the range of GeV–TeV energies by the time the gamma rays arrive at Earth.[30,31] If the source itself is opaque to gamma rays, the high-energy gamma rays will lose energy even before reaching the EBL to possibly emerge at Earth below the threshold of Fermi with MeV or lower energies.

As discussed in the introduction, production of neutrinos requires a beam and a target, and a powerful neutrino source requires a dense target that will render the source opaque to high-energy gamma rays. While the beam loses energy in the target producing high-energy neutrinos, the energy of the accompanying photons may be spread over a wide range of the electromagnetic spectrum before reaching our telescopes. This is very likely to be the case for the most powerful neutrino sources, which exceed the sensitivity of IceCube at this time. The energy of the photons associated with cosmic neutrinos may thus emerge below the threshold of our gamma-ray instruments, for instance, the NASA Fermi gamma-ray satellite; we refer to these as (gamma) dark sources.

For example, we calculate the gamma-ray flux accompanying the diffuse cosmic neutrino flux observed by IceCube, which is described by a simple power law with spectral index of -2.15, consistent with the neutrino data above an energy of 100 TeV. The result is shown in Fig. 11 assuming transparent sources and equal multiplicities; $K_\pi = 2$. The matching energy densities of the extragalactic gamma-ray flux detected by Fermi and the high-energy neutrino flux measured by IceCube suggest that, rather than detecting some exotic sources, it is more likely that IceCube to a large extent observes the same universe astronomers do. The finding implies that a significant fraction of the energy in the nonthermal universe originates in hadronic processes, indicating a larger role than previously thought.

Clearly, in this exercise, the slope and overall normalization of the neutrino spectrum have been adjusted to not exceed the isotropic gamma-ray background observed by the Fermi satellite. We conclude that the high-energy cosmic neutrino flux above 100 TeV, shown in Fig. 11, saturates this limit; higher normalization and larger spectral indices will result in a gamma-ray result that exceeds the Fermi data, as illustrated in Fig. 12. A larger spectral index results in a larger neutrino energy flux at energies below 100 TeV for both neutrinos

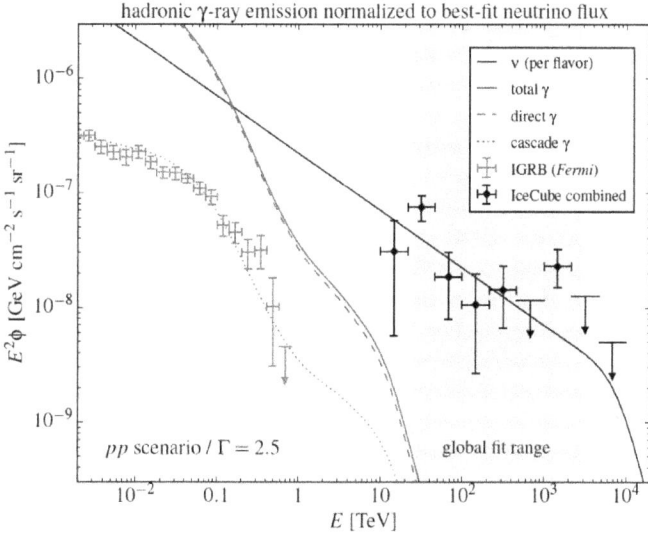

Fig. 12. Same calculation as in Fig. 11 with the spectral index of -2.15 replaced by -2.5, closer to what is suggested by the present IceCube measurements. The predicted gamma-ray flux exceeds the Fermi observations, implying that the assumption of transparent sources is not tenable. The excess flux is shifted below the Fermi threshold, to MeV energies or below, by cascading of the pionic gamma rays in the source before reaching the EBL. Introducing the cutoff on the high-energy flux, shown in the figure, does not affect the result.

and accompanying photons. The latter will, after cascading in the EBL, exceed the Fermi observations. There is no conflict here; the resolution is that the sources themselves must be opaque to photons that lose energy in the source even before entering the EBL and as a result reach Earth with energies that are below the threshold of the Fermi satellite, at MeV energy or below.

Alternatively, the target for producing the neutrinos may be photons. This changes the value of K_π and, more importantly, the shape of the energy spectrum; for a detailed discussion see Ref. [35]. Yielding an energy spectrum that peaks near PeV energies, as shown in Fig. 13, the contribution to the Fermi flux is suppressed at lower energies relative to the power law assumed in Fig. 11. However, as was the case for pp interactions, fits that do not exceed the Fermi data tend not to accommodate the cosmic neutrino spectrum below 100 TeV. If sources of the TeV–PeV neutrinos are transparent to gamma rays with respect to two-photon annihilation, tensions with

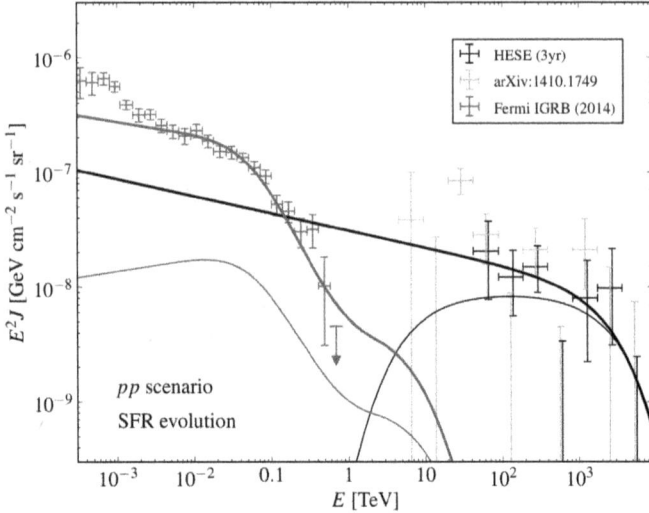

Fig. 13. The calculation in Fig. 11 is compared to an identical calculation adopting a spectral shape characteristic for the production of neutrinos on a gamma-ray target in the source. While the pionic gamma-ray energy flux is now suppressed relative to the Fermi observations, the conclusion that the sources are likely obscured is recovered after correct normalization to the most up-to-date IceCube measurements.[32]

the isotropic diffuse gamma-ray background measured by Fermi seem unavoidable, independent of the production mechanism.[32]

The conclusion is inescapable that the energy densities of neutrinos and gamma rays in the extreme universe are qualitatively the same. To the extent that the IceCube observations favor steeper energy spectra than $E^{-2.15}$, we anticipate contributions to the diffuse flux from hidden sources. Interestingly, the common energy density of photons and neutrinos is also comparable to that of the ultra-high-energy extragalactic cosmic rays.[36] We anticipate that multimessenger studies of gamma-ray and neutrino data will be powerful tools to study the neutrino production mechanism and to constrain neutrino source models. Accordingly, IceCube developed methods, most promisingly real-time multiwavelength observations with astronomical telescopes, to identify the sources and build on the discovery of cosmic neutrinos to launch a new era in astronomy.[37,38]

4. Identifying Neutrino Sources: The Supermassive Black Hole TXS 0506+056

Phenomenological studies[39,40] and recent data analyses[41–43] have converged on the fact that Fermi's extragalactic gamma-ray flux, shown in Fig. 11, is dominated by blazars, AGN with jets pointing at Earth. The qualitative matching of the energy densities of photons and neutrinos, discussed in the previous section, may suggest that the unidentified neutrino sources contributing to the diffuse flux might have already been observed as strong gamma-ray emitters. However, a dedicated IceCube study[44] of the directions of cosmic neutrinos in the direction of the blazars observed by Fermi shows no evidence of neutrino emission from these sources. The inferred limit on their contribution to IceCube's diffuse neutrino flux leaves room for a contribution below the 10% level. Nevertheless, the multimessenger campaign[2] launched by the neutrino alert IC-170922A identified the first source of cosmic neutrinos as a Fermi "blazar"; the multiwavelength data will shed light on the apparent contradiction.

Since 2016, the IceCube multimessenger program has grown from galactic supernova alerts and attempts to match neutrino observations with early LIGO/Virgo gravitational wave candidates to a steadily expanding set of automatic filters that selects in real time rare, very-high-energy neutrino events that are likely to be cosmic in origin.[45] Within less than a minute of their detection in the deep Antarctic ice, the arrival directions of the neutrinos are reconstructed and automatically sent to the Gamma-ray Coordinate Network for potential follow-up by astronomical telescopes.

4.1. *Observation of a cosmic neutrino source: TXS 0506+056*

On September 22, 2017, the tenth alert, IceCube-170922A,[46] reported a well-reconstructed muon that deposited 180 TeV inside the detector, corresponding to an energy of the parent neutrino of 290 TeV. Its arrival direction was aligned with the coordinates of a known Fermi blazar, TXS 0506+056, to within 0.06°. The source was "flaring" with a gamma-ray flux that had increased by a factor of seven in recent months. A variety of estimates converged on a probability on the order of 10^{-3} that the coincidence was accidental. The identification of the neutrino with the source reached the level of evidence but not more than that. What clinched the association was a series of subsequent observations, culminating with the optical observation of the source switching from an "off" to an "on" state two hours after the emission of IC-170922A, conclusively associating the neutrino with TXS 0506+056[47]:

- The redshift of the host galaxy was measured to be $z \simeq 0.34$.[48] It is important to realize that nearby blazars, like the Markarian sources, are at a distance that is ten times closer, and therefore TXS 0506+056, with a similar flux despite the greater distance, is one of the most luminous sources in the universe. This suggests that it belongs to a special class of sources that accelerates proton beams in dense environments, revealed by the neutrino. That the source is special eliminates any conflict between its observation and the lack of correlation between the arrival directions of Ice-Cube neutrinos and the bulk of the blazars observed by Fermi.[44] Such limits implicitly assume that all sources in an astronomical category are identical, and this is a strong, unstated assumption as underscored by this observation.

- Originally detected by NASA's Fermi[49] and Swift[50] satellites, the alert was followed up by ground-based air Cherenkov telescopes.[51] MAGIC detected the emission of gamma rays with energies exceeding 100 GeV starting several days after the observation of the neutrino.[52] Given its distance, this establishes the source as a relatively rare TeV blazar.

- Given where to look, IceCube searched its archival neutrino data up to and including October 2017 for evidence of neutrino emission at the location of TXS 0506+056.[3] When searching the sky for point sources of neutrinos, two analyses have been routinely performed: one that searches for steady emission of neutrinos and one that searches for flares over a variety of timescales. Evidence was found for 19 high-energy neutrino events on a background of fewer than six in a burst lasting 110 days. This burst dominates the integrated flux from the source over the last 9.5 years of archival IceCube data, leaving the 2017 flare as a second subdominant feature. We note that this analysis applied a published prescription to data; the chance that this observation was a fluctuation is small.

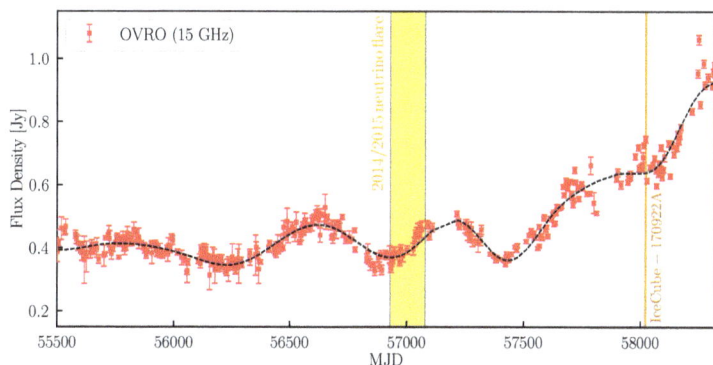

Fig. 14. TXS 0506+056 radio light curve from Owen Valley Radio Observatory (OVRO) at 15 GHz (red). The dashed line illustrates the pattern of the radio flux density. The 2014–2015 110-day neutrino flare (yellow band) and the IceCube-170922A episodes are shown. Radio data suggest that the neutrinos arrive during periods of enhanced radio emission.

- Radio interferometric images[53,54] of the source revealed a jet that loses its tight collimation beyond 5 milliarcseconds running into material that is likely the target for producing the neutrinos. The nature of this target material is still a matter of debate. Speculations include the merger with another galaxy that may supply plenty of material to interact with the jet of the dominant galaxy. Alternatively, the jet may interact with the dense molecular clouds of a star-forming region or simply with supermassive stars in the central region of the host galaxy.[53,54] Also, in a so-called structured jet, the accelerated protons may collide with a slower-moving and denser region of jetted photons. Additionally, the VLBA data reveal that the neutrino burst occurs at the peak of enhanced radio emission at 15 GHz, which started five years ago; see Fig. 14. The radio flare may be a signature of a galaxy merger; correlations of radio bursts with the process of merging supermassive black holes have been anticipated.[55]
- The MASTER robotic optical telescope network has been monitoring the source since 2005 and found the strongest time variation of the source in the last 15 years to occur two hours after the emission of IC170922, with a second variation following the 2014–2015 burst.[47] The blazar switches from the "off" to the "on" state two hours after the emission of the neutrino. They argue that the observation conclusively associates the source with the neutrino.[47]

Additionally, it is important to note the striking fact that all high-energy spectra, for both photons and neutrinos and for both the 2014 and 2017 bursts, are consistent with a hard E^{-2} spectrum, which is expected for a cosmic accelerator. In fact, the gamma-ray spectrum hints at flattening beyond that during the 110-day period of the 2014 burst.[56,57]

In summary, both the multiwavelength campaign[2] and the observation of the source in archival neutrino data[3] provide statistically independent evidence for TXS 0506+056 as a source of high-energy neutrinos. When combined, they reach a level of 4.4σ. This does not take into account the significance contributed by the optical and TeV associations on timescales of hours and days discussed above. It is challenging to evaluate the final combined significance because of the *a posteriori* nature of these considerations but conclude that the association of neutrinos with the source summarized above is totally compelling.

Other IceCube alerts have triggered intriguing observations. Following up on a July 31, 2016 neutrino alert, the AGILE collaboration, which operates an orbiting X-ray and gamma-ray telescope, reported a day-long blazar flare in the direction of the neutrino one day before the neutrino detection.[58] A tentative but very intriguing association of an IceCube alert[59] has been made with a tidal disruption event, an anticipated source of high-energy neutrino emission. Even before IceCube issued automatic alerts, in April 2016, the TANAMI collaboration argued for the association of the highest-energy IceCube event at the time, dubbed "Big Bird", with the flaring blazar PKS B1424-418.[60] Interestingly, the event was produced at a minimum of the Fermi flux,[61] as expected for a neutrino source and as was also the case for PKS 1502+106, which we will discuss further on. AMANDA, IceCube's predecessor, observed three neutrinos in coincidence with a rare flare of the blazar 1ES 1959+650, detected by the Whipple telescope in 2002.[62] However, none of these identifications reach the significance of the observations triggered by IC-170922A.

4.2.   *Blueprint of the TXS 0506+056 beam dump?*

The gamma-ray community has developed a routine procedure for modeling the spectrum of blazars with two contributions: a lower-energy component produced by synchrotron radiation by the electron beam and a high-energy component resulting from the inverse Compton scattering of (possibly the same) photons by accelerated electrons; for a recent discussion, see Ref. [63]. This model cannot accommodate the TXS observations for two reasons: An electron beam does not produce pions that decay into neutrinos and, even in the presence of a proton beam, a target is required to produce the parent pions. A neutrino source requires an accelerator and a target — it must be a "beam dump". It is evident that a source that is transparent to high-energy gamma rays is unlikely to host the target material to produce neutrinos with the opacity for $\gamma\gamma$ interactions absorbing photons typically two orders of magnitude larger than the one for $p\gamma$ interactions producing neutrinos.[64] We will work through these arguments quantitatively in the following, but it is clear that TXS 0506+056 is not a high-energy photon blazar at the times of neutrino emission. That TXS 0506+056 belongs to a different class of sources is reinforced by the fact that attempts at conventional blazar modeling of the multiwavelength spectrum have been unsuccessful despite the opportunity of tuning 14 free parameters; see e.g., Refs. [63,65].

We will scrutinize the multiwavelength information gathered on TXS 0506+056 rather than try to match it to, or model it as, a known class of astronomical objects. Our conclusion will be that the source is special, with properties unlikely to match any astronomical classification, which should not be surprising[64] given that it emits neutrinos.

First, we would like to draw attention to a more recent alert, IC-190730A, sent by IceCube on July 30, 2019. A well-reconstructed 300-TeV muon neutrino was observed in spatial coincidence with the blazar PKS 1502+106.[66] With a reconstructed energy just exceeding that of IC-170922A, it is the highest-energy neutrino alert so far. OVRO radio observations[67] show that the neutrino is coincident with the peak flux density of a flare at 15 GHz that started five years prior,[68] matching the similar long-term radio outburst observed from TXS 0506+056 at the time of IC-170922A; see Fig. 14. Even more intriguing is the fact that the gamma-ray flux observed by Fermi shows a clear minimum at the time that the neutrino is emitted; see Fig. 15. We infer that at this time the jet meets the target that produces the neutrino. Inevitably, the accompanying high-energy gamma rays will be absorbed and their electromagnetic

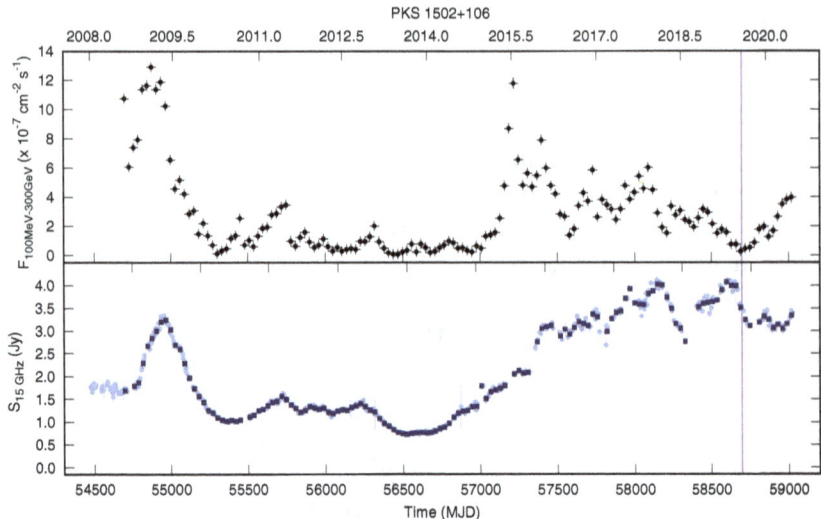

Fig. 15.   Temporal variation of the γ-ray and radio brightness of PKS 1502+106. Top Panel: Fermi-LAT likelihood light curve integrated between 100 MeV and 300 GeV (marked by black dots with error bars). Bottom Panel: OVRO flux density curve of PKS 1502+106 plotted with light-blue dots, which is superimposed by the radio flux density curve binned to the Fermi-LAT light curve (marked with dark-blue squares). The detection time of the neutrino IC-190730A is labeled by a vertical purple line.

energy cascade down to energies below the Fermi threshold, i.e., MeV or X-rays. For a discussion, see Ref. [61], where we argue that cosmic neutrinos are produced by temporarily gamma-suppressed blazars or, more likely, any other category of AGN.

We have already pointed out that the large distance of TXS 0506+056 suggests that the source is special. So does the large neutrino flux observed: we illustrate this by showing that a subset of 1–10 % of sources with a density similar to blazars, bursting at the level of TXS 0506+056 in the period of 10 years covered by the IceCube data, is sufficient to accommodate the total diffuse flux observed by IceCube.

The diffuse neutrino flux corresponding to a population of sources with source density ρ and neutrino luminosity L_ν is given by[64,69]

$$E^2 \frac{dN}{dE} = \frac{1}{4\pi} \int d^3 r \frac{L_\nu}{4\pi r^2} \rho, \qquad (4)$$

which can be simplified to

$$E^2 \frac{dN}{dE} = \frac{c}{4\pi} t_H \xi L_\nu \rho, \tag{5}$$

where ξ is the result of the integration over the redshift history of the sources and t_H is the Hubble time corresponding to a Hubble distance $R_H \simeq 4.3\,\text{Gpc}$. For instance, for a spectral index of $\gamma \simeq 2$ and no source evolution in the local universe, $\xi_z \simeq 0.5$, while for sources following the star-formation rate, $\xi_z \simeq 2.6$.[69] The above relation can be adapted for the case of sources that flare for a duration Δt in a total time of observation T_{obs}:

$$E^2 \frac{dN}{dE} = \frac{c}{4\pi} t_H \xi L_\nu \rho \frac{\Delta t}{T_{obs}} \mathcal{F}. \tag{6}$$

Assuming that blazars are the sources of the diffuse flux, we would conclude that

$$3 \times 10^{-11}\,\text{TeVcm}^{-2}\text{s}^{-1}\text{sr}^{-1} = \frac{\mathcal{F}}{4\pi} \left(\frac{R_H}{3\,\text{Gpc}}\right)\left(\frac{\xi}{0.7}\right)\left(\frac{L_\nu}{1.2 \times 10^{47}\,\text{erg/s}}\right)$$
$$\times \left(\frac{\rho}{10^{-8}\,\text{Mpc}^{-3}}\right)\left(\frac{\Delta t}{110\,\text{d}} \frac{10\,\text{yr}}{T_{obs}}\right), \tag{7}$$

which results in $\mathcal{F} = 0.05$, where $\mathcal{F}$ is the fraction of some astronomical category of source that is neutrino beam dumps. We here normalized $\mathcal{F}$ to the density of blazars, but we could have alternatively normalized to a smaller fraction of all AGNs. We conclude in any case that if TXS 0506+056 were a source typical of all blazars, it would overproduce the diffuse flux observed by IceCube. If neutrino sources only represent a few percent of those labeled by astronomers as blazars, attempts by IceCube and others to find correlations between the directions of high-energy neutrinos and *all* Fermi blazars must inevitably be unsuccessful, as was the case in Ref. [44].

The energy in neutrinos is consistent with their production by the sources of the highest-energy cosmic rays[64]:

$$E^2 \frac{dN}{dE} \simeq \frac{c}{4\pi} \left(\frac{1}{2}(1 - e^{-\tau_{p\gamma}})\xi t_H \frac{dE}{dt}\right). \tag{8}$$

From the total cosmic ray injection rate $dE/dt = (1\sim2) \times 10^{44}\,\text{erg}\,\text{Mpc}^{-3}\text{yr}^{-1}$ of extragalactic cosmic rays into the universe, we can determine the average opacity of the sources to protons $\tau_{p\gamma}$ in the target that produces the observed diffuse neutrino flux and achieve the match of Eq. (7):

$$\left(\frac{L_\nu}{1.2 \times 10^{47}\,\mathrm{erg/s}}\right)\left(\frac{\rho}{10^{-8}\,\mathrm{Mpc}^{-3}}\right)\left(\frac{\Delta t}{110\,\mathrm{d}}\frac{10\,\mathrm{yr}}{T_{\mathrm{obs}}}\right)\left(\frac{\mathcal{F}}{0.05}\right)$$

$$\simeq \frac{1}{2}(1 - e^{-\tau_{p\gamma}})\frac{dE/dt}{(1-2)\times 10^{44}\,\mathrm{erg\,Mpc^{-3}yr^{-1}}}. \tag{9}$$

We find that $\tau_{p\gamma} > 0.8$. This can be achieved with a proton beam with low boost factor interacting with the intense photon fields near the core of an active galaxy. With the opacity to photons about two orders of magnitude larger, the target producing the neutrinos is not transparent to TeV photons, only to photons with tens of GeV, and that is indeed what is observed by Fermi at the time of the 2014 flare.

There is in fact evidence that, temporarily, TXS 0506+056 was a gamma-ray-obscured source at the time the IC-170922A neutrino was emitted. Recall that the optical observations show a dramatic transition of the blazar from the "off" to the "on" state two hours after the emission of the neutrino, resulting additionally in the doubling of its total luminosity.[47] Also, the atmospheric gamma-ray telescopes observe rapid variations in the flux around the time of the neutrino emission,[2] with the gamma-ray emission observed by MAGIC only emerging after several days.[52]

The observation that the energy flux in neutrinos and very-high-energy cosmic rays are similar supports the fact that the cosmic rays must be highly efficient at producing neutrinos, requiring a large target density that renders them opaque to high-energy gamma rays. A consistent picture emerges when the source opacity $\tau_{p\gamma}$ exceeds a value of 0.8,[64] resulting in a gamma-ray cascade where photons lose energy in the source before cascading to yet lower energies in the EBL. Some of their energy emerges below the Fermi threshold by the time they reach Earth. This is reinforced by the discussion in the previous section that the diffuse spectrum indicates a contribution of obscured sources.

The nature of this special class of sources has not been settled. One straightforward explanation could be a subclass of blazars or, more likely, a smaller fraction of all AGNs, selected by redshift evolution: powerful proton accelerators producing neutrinos may have been active in the past but are no longer today. This accommodates the large redshift of TXS 0506+056, which would be the closest among a set of sources that only accelerated cosmic rays at early redshifts.[70]

Alternatively, in merging galaxies, there is plenty of material for accelerated cosmic rays to interact with the jet of the dominant galaxy. Merger activity in active galaxies in not uncommon. The fresh material provides optically thick environments and allows for rapid variation of the Lorentz factors. A cursory review of the literature on the production of neutrinos in galaxy mergers is sufficient to conclude that they can indeed accommodate the observations of both the individual sources discussed above and the total flux of cosmic neutrinos.[71-73] Besides mergers, some form of structured jet where the accelerated protons collide with a slower-moving and denser region of jetted photons is a possibility. The jet could interact with dense molecular clouds of a star-forming region or simply with supermassive stars near the central region of the host galaxy.[53,54]

We previously mentioned the evidence emerging from 10 years of IceCube data that the arrival directions of cosmic neutrinos are no longer isotropic.[74] The anisotropy results from four sources — TXS 0506+056 among them — that show evidence for clustering above the 4σ level. The strongest of these sources is the nearby active galaxy NGC 1068. There is evidence for shocks near the core and for molecular clouds with column density reaching $\sim 10^{25}$ cm^{-2}.[75] Similar to TSX 0506+056, a merger onto the black hole is observed — either with a satellite galaxy or, more likely, with a star-forming region.[76] This major accretion event may be the origin of the increased neutrino emission.

Although it is obviously challenging to provide a final conclusion on the origin of neutrinos and cosmic rays, we should not lose sight of the fact that high-energy neutrino astronomy exists and that IceCube has demonstrated it has the tools to reveal the extreme universe with more data or, more realistically, with a larger detector.

5. From Discovery to Astronomy: Larger Telescopes with Better Resolution

Following the pioneering work of DUMAND,[77] several neutrino telescope projects were initiated in the Mediterranean Sea during the

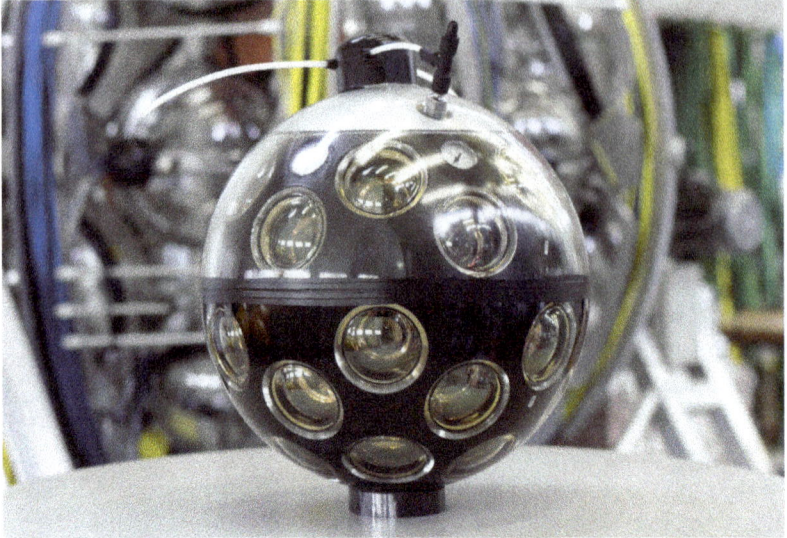

Fig. 16. KM3NeT optical module.[81] The optical module consists of a glass sphere with a diameter of 42 cm, housing 31 photosensors (yellowish disks). The glass sphere can withstand the pressure of the water and is transparent to the faint light that must be detected to see neutrinos.

1990s.[78–80] In 2008, the construction of the ANTARES detector off the coast of France was completed. It demonstrated the feasibility of neutrino detection in the deep sea and has provided a wealth of technical experience and design solutions for deep-sea components. More recently, an international collaboration has started construction of a multi-cubic-kilometer neutrino telescope in the Mediterranean Sea, KM3NeT.[81] They developed a digital optical module that incorporates 31 three-inch photomultipliers instead of one large photomultiplier tube; see Fig. 16. The advantages are a tripling of the photocathode area per optical module, a segmentation of the photocathode allowing for a clean identification of coincident Cherenkov photons, some directional sensitivity, and a reduction of the overall number of penetrators and connectors, which are expensive and prone to failure. KM3NeT in its second phase[81] will consist of two units for astrophysical neutrino observations, each consisting of 115 strings carrying more than 2,000 optical modules.

A parallel effort is underway in Lake Baikal with the construction of the deep-underwater neutrino telescope Baikal-GVD (Gigaton Volume Detector).[82] The first GVD cluster was upgraded in the spring of 2016 to its final size: 288 optical modules, a geometry of 120 m in diameter and 525 m in height, and an instrumented volume of 6 Mt. Each of the eight strings consists of three sections with 12 optical modules. At present, 7 of the 14 clusters have been deployed, reaching a sensitivity close to the diffuse cosmic neutrino flux observed by IceCube.

IceCube itself is deploying seven new strings at the bottom of the detector array that have been designed as an incremental extension of the DeepCore detector and as a test bed for the technologies of the next-generation detector. The new instrumentation will dramatically boost IceCube's performance at the lowest energies, increasing the samples of atmospheric neutrinos by a factor of 10. New calibration devices will advance our understanding of the response of the light sensors in both current and new strings, resulting in improved reconstructions of cascade events, better identification of tau neutrinos, and an enhanced pointing resolution of muon neutrinos that could approach the 0.1 degree level for the high-energy events. The improved calibration of the existing sensors will also enable a reanalysis of more than 10 years of archival data and significantly increase the discovery potential for neutrino sources before the construction of a second-generation instrument.

Further progress requires a larger instrument. Therefore, as a next step, IceCube proposes to instrument $10 \, \text{km}^3$ of glacial ice at the South Pole, capitalizing on the large absorption length of light in ice to thereby increase IceCube's sensitive volume by an order of magnitude.[83–85] This large gain is made possible by the unique optical properties of the Antarctic glacier revealed by the construction of IceCube. Exploiting the extremely long photon absorption lengths in the deep Antarctic ice, the spacing between strings of light sensors will be increased from 125 m to close to 250 m without significant loss of performance of the instrument. The instrumented volume can therefore grow by one order of magnitude while keeping the instrumentation and its budget at the level of the current IceCube detector.

The new facility will increase the rates of cosmic events from hundreds to thousands over several years. The superior angular resolution of the longer muon tracks will allow for the discovery of cosmic neutrino sources, currently seen at the 3σ level in the 10-year sky map.

Acknowledgments

Discussion with collaborators inside and outside the IceCube Collaboration, too many to be listed, have greatly shaped this presentation. Thanks. This research was supported in part by the U.S. National Science Foundation under grants PLR-1600823 and PHY-1913607 and by the University of Wisconsin Research Committee with funds granted by the Wisconsin Alumni Research Foundation.

References

1. M. G. Aartsen *et al.*, Evidence for high-energy extraterrestrial neutrinos at the IceCube detector, *Science*, **342**, 1242856 (2013). doi: 10.1126/science.1242856.
2. M. Aartsen *et al.*, Multimessenger observations of a flaring blazar coincident with high-energy neutrino IceCube-170922A, *Science*. **361**(6398), eaat1378 (2018). doi: 10.1126/science.aat1378.
3. M. Aartsen *et al.*, Neutrino emission from the direction of the blazar TXS 0506+056 prior to the IceCube-170922A alert, *Science*. **361**(6398), 147–151 (2018). doi: 10.1126/science.aat2890.
4. T. K. Gaisser, F. Halzen, and T. Stanev, Particle astrophysics with high-energy neutrinos, *Phys. Rept.* **258**, 173–236 (1995). doi: 10.1016/0370-1573(95)00003-Y.
5. J. Learned and K. Mannheim, High-energy neutrino astrophysics, *Ann. Rev. Nucl. Part. Sci.* **50**, 679–749 (2000). doi: 10.1146/annurev.nucl.50.1.679.
6. R. Blandford, D. Meier, and A. Readhead, Relativistic jets from active galactic nuclei, *Annu. Rev. Astron. Astrophys.* **57**(1), 467–509 (Aug, 2019). doi: 10.1146/annurev-astro-081817-051948.
7. K. Murase, S. S. Kimura, and P. Meszaros, Hidden cores of active galactic nuclei as the origin of medium-energy neutrinos: Critical tests with the MeV gamma-ray connection, *Phys. Rev. Lett.* **125**(1), 011101 (2020). doi: 10.1103/PhysRevLett.125.011101.

8. F. Halzen and E. Zas, Neutrino fluxes from active galaxies: A model independent estimate, *Astrophys. J.* **488**, 669–674 (1997). doi: 10.1086/304741.

9. J. Becker Tjus, B. Eichmann, F. Halzen, A. Kheirandish, and S. Saba, High-energy neutrinos from radio galaxies, *Phys. Rev.* **D89**(12), 123005 (2014). doi: 10.1103/PhysRevD.89.123005.

10. D. Hooper, A case for radio galaxies as the sources of IceCube's astrophysical neutrino flux, *JCAP.* **1609**(09), 002 (2016). doi: 10.1088/1475-7516/2016/09/002.

11. M. G. Aartsen *et al.*, The IceCube neutrino observatory: Instrumentation and online systems, *JINST.* **12**(03), P03012 (2017). doi: 10.1088/1748-0221/12/03/P03012.

12. F. Halzen, Astroparticle physics with high energy neutrinos: From amanda to IceCube, *Eur. Phys. J.* **C46**, 669–687 (2006). doi: 10.1140/epjc/s2006-02536-4.

13. M. Aartsen *et al.*, Energy reconstruction methods in the IceCube neutrino telescope, *JINST.* **9**, P03009 (2014). doi: 10.1088/1748-0221/9/03/P03009.

14. T. Yuan, Improving the angular resolution in IceCube cascade reconstruction, *TeV Particle Astrophysics 2017 (TeVPA2017), Columbus, Ohio, USA* (2017).

15. M. Ackermann *et al.*, The spectrum of isotropic diffuse gamma-ray emission between 100 MeV and 820 GeV, *Astrophys. J.* **799**, 86 (2015). doi: 10.1088/0004-637X/799/1/86.

16. K. Fang and K. Murase, Linking high-energy cosmic particles by black hole jets embedded in large-scale structures, *Nat. Phys.* **14**(4), 396–398 (2018). doi: 10.1038/s41567-017-0025-4.

17. M. G. Aartsen *et al.*, Evidence for astrophysical muon neutrinos from the Northern sky with IceCube, *Phys. Rev. Lett.* **115**(8), 081102 (2015). doi: 10.1103/PhysRevLett.115.081102.

18. M. G. Aartsen *et al.*, Observation and characterization of a cosmic muon neutrino flux from the Northern hemisphere using six years of IceCube data, *Astrophys. J.* **833**(1), 3 (2016). doi: 10.3847/0004-637X/833/1/3.

19. M. G. Aartsen *et al.*, The IceCube neutrino observatory – contributions to ICRC 2017 Part II: Properties of the atmospheric and astrophysical neutrino flux, *PoS.* **ICRC2017** (2017).

20. J. Stettner, Measurement of the diffuse astrophysical muon-neutrino spectrum with ten years of IceCube data, *PoS.* **ICRC2019**, 1017 (2020). doi: 10.22323/1.358.1017.

21. J. B. Stettner. *Measurement of the energy spectrum of astrophysical muon-neutrinos with the IceCube observatory.* Dissertation, RWTH

Aachen University, Aachen (2021). https://publications.rwth-aachen.d e/record/811376. Veröffentlicht auf dem Publikationsserver der RWTH Aachen University; Dissertation, RWTH Aachen University, 2021.

22. M. G. Aartsen *et al.*, Time-integrated neutrino source searches with 10 years of IceCube data, *Phys. Rev. Lett.* **124**(5), 051103 (2020). doi: 10.1103/PhysRevLett.124.051103.

23. M. Aartsen *et al.*, First observation of PeV-energy neutrinos with Ice-Cube, *Phys. Rev. Lett.* **111**, 021103 (2013). doi: 10.1103/PhysRevLett. 111.021103.

24. R. Abbasi *et al.*, The IceCube high-energy starting event sample: Description and flux characterization with 7.5 years of data (11, 2020).

25. M. Aartsen *et al.*, Search for neutrino-induced particle showers with IceCube-40, *Phys. Rev.* **D89**, 102001 (2014). doi: 10.1103/PhysRevD. 89.102001.

26. M. Aartsen *et al.*, Characteristics of the diffuse astrophysical electron and tau neutrino flux with six years of IceCube high energy cascade data (1, 2020).

27. M. G. Aartsen *et al.*, Flavor ratio of astrophysical neutrinos above 35 TeV in IceCube, *Phys. Rev. Lett.* **114**(17), 171102 (2015). doi: 10.1103/ PhysRevLett.114.171102.

28. R. Abbasi *et al.*, Measurement of astrophysical tau neutrinos in Ice-Cube's high-energy starting events, *to be published in PLR* (11, 2020).

29. M. G. Aartsen *et al.*, Detection of a particle shower at the Glashow resonance with IceCube, *Nature.* **591**(7849), 220–224 (2021). doi: 10. 1038/s41586-021-03256-1.

30. R. J. Protheroe and T. Stanev, Electron and photon cascading of very high-energy gamma-rays in the infrared background, *Mon. Not. R. Astron. Soc.* **264**(1), 191–200 (1993). doi: 10.1093/mnras/264.1.191. http://dx.doi.org/10.1093/mnras/264.1.191.

31. M. Ahlers, L. Anchordoqui, M. Gonzalez-Garcia, F. Halzen, and S. Sarkar, GZK neutrinos after the Fermi-LAT diffuse photon flux measurement, *Astropart. Phys.* **34**, 106–115 (2010). doi: 10.1016/j. astropartphys.2010.06.003.

32. K. Murase, M. Ahlers, and B. C. Lacki, Testing the Hadronuclear origin of PeV neutrinos observed with IceCube, *Phys. Rev.* **D88**(12), 121301 (2013). doi: 10.1103/PhysRevD.88.121301.

33. M. Aartsen *et al.*, Observation of high-energy astrophysical neutrinos in three years of IceCube data, *Phys. Rev. Lett.* **113**, 101101 (2014). doi: 10.1103/PhysRevLett.113.101101.

34. M. G. Aartsen *et al.*, Atmospheric and astrophysical neutrinos above 1 TeV interacting in IceCube, *Phys. Rev.* **D91**(2), 022001 (2015). doi: 10.1103/PhysRevD.91.022001.

35. S. Yoshida and K. Murase, Constraining photohadronic scenarios for the unified origin of IceCube neutrinos and ultrahigh-energy cosmic rays (7, 2020).

36. A. Aab *et al.*, The Pierre Auger observatory: Contributions to the 34th International Cosmic Ray Conference (ICRC 2015). In *Proceedings, 34th International Cosmic Ray Conference (ICRC 2015): The Hague, The Netherlands, July 30–August 6, 2015* (2015). http://inspirehep. net/record/1393211/files/arXiv:1509.03732.pdf.

37. M. G. Aartsen *et al.*, Very high-energy gamma-ray follow-up program using neutrino triggers from IceCube, *JINST.* **11**(11), P11009 (2016). doi: 10.1088/1748-0221/11/11/P11009.

38. M. G. Aartsen *et al.*, The IceCube realtime alert system, *Astropart. Phys.* **92**, 30–41 (2017). doi: 10.1016/j.astropartphys.2017.05.002.

39. M. Ajello *et al.*, The luminosity function of fermi-detected flat-spectrum radio Quasars, *Astrophys. J.* **751**, 108 (2012). doi: 10.1088/0004-637X/ 751/2/108.

40. M. Di Mauro, F. Donato, G. Lamanna, D. A. Sanchez, and P. D. Serpico, Diffuse γ-ray emission from unresolved BL Lac objects, *Astrophys. J.* **786**, 129 (2014). doi: 10.1088/0004-637X/786/2/129.

41. M. Ackermann *et al.*, Resolving the extragalactic γ-ray background above 50 GeV with the fermi large area telescope, *Phys. Rev. Lett.* **116**(15), 151105 (2016). doi: 10.1103/PhysRevLett.116.151105.

42. H.-S. Zechlin, A. Cuoco, F. Donato, N. Fornengo, and A. Vittino, Unveiling the gamma-ray source count distribution below the fermi detection limit with photon statistics, *Astrophys. J. Suppl.* **225**(2), 18 (2016). doi: 10.3847/0067-0049/225/2/18.

43. M. Lisanti, S. Mishra-Sharma, L. Necib, and B. R. Safdi, Deciphering contributions to the extragalactic gamma-ray background from 2 GeV to 2 TeV, *Astrophys. J.* **832**(2), 117 (2016). doi: 10.3847/0004-637X/ 832/2/117.

44. M. G. Aartsen *et al.*, The contribution of Fermi-2LAC blazars to the diffuse TeV-PeV neutrino flux, *Astrophys. J.* **835**(1), 45 (2017). doi: 10.3847/1538-4357/835/1/45.

45. R. Abbasi *et al.*, Follow-up of astrophysical transients in real time with the IceCube neutrino observatory, *Astrophys. J.* **910**(1), 4 (2021). doi: 10.3847/1538-4357/abe123.

46. C. Kopper and E. Blaufuss, IceCube-170922A — IceCube observation of a high-energy neutrino candidate event, *GRB Coordinates Network, Circular Service, No. 21916, #1 (2017).* **21916** (2017).

47. V. M. Lipunov, V. G. Kornilov, K. Zhirkov, E. Gorbovskoy, N. M. Budnev, D. A. H. Buckley, R. Rebolo, M. Serra-Ricart, R. Podesta, N. Tyurina, O. Gress, Y. Sergienko, V. Yurkov, A. Gabovich, P. Balanutsa, I. Gorbunov, D. Vlasenko, F. Balakin, V. Topolev,

A. Pozdnyakov, A. Kuznetsov, V. Vladimirov, A. Chasovnikov, D. Kuvshinov, V. Grinshpun, E. Minkina, V. B. Petkov, S. I. Svertilov, C. Lopez, F. Podesta, H. Levato, A. Tlatov, B. Van Soelen, S. Razzaque, and M. Böttcher, Optical observations reveal strong evidence for high-energy neutrino progenitor, *ApJ. Lett.* **896**(2), L19 (June, 2020). doi: 10.3847/2041-8213/ab96ba.

48. S. Paiano, R. Falomo, A. Treves, and R. Scarpa, The redshift of the BL Lac object TXS 0506+056, *Astrophys. J.* **854**(2), L32 (2018). doi: 10.3847/2041-8213/aaad5e.

49. Y. T. Tanaka, S. Buson, and D. Kocevski, Fermi-LAT detection of increased gamma-ray activity of TXS 0506+056, located inside the IceCube-170922A error region, *The Astronomer's Telegram.* **10791** (September, 2017).

50. P. A. E. A. Keivani, J. A. Kennea, D. B. Fox, D. F. Cowen, J. P. Osborne, F. E. Marshall, and Swift-IceCube Collaboration, Further Swift-XRT observations of IceCube 170922A, *The Astronomer's Telegram.* **10792** (September, 2017).

51. R. Mirzoyan, First-time detection of VHE gamma rays by MAGIC from a direction consistent with the recent EHE neutrino event IceCube-170922A, *The Astronomer's Telegram.* **10817** (October, 2017).

52. S. Ansoldi *et al.*, The blazar TXS 0506+056 associated with a high-energy neutrino: Insights into extragalactic jets and cosmic ray acceleration, *Astrophys. J. Lett.* **863**, L10 (2018). doi: 10.3847/2041-8213/aad083.

53. S. Britzen, C. Fendt, M. Böttcher, M. Zajacek, F. Jaron, I. Pashchenko, A. Araudo, V. Karas, and O. Kurtanidze, Cosmic collider: IceCube neutrino generated in a precessing jet-jet interaction in TXS 0506+056? *To be published in Astron & Astrophys.*

54. E. Kun, P. L. Biermann, and L. Gergely, VLBI radio structure and core-brightening of the high-energy neutrino emitter TXS 0506+056 (2018).

55. L. Á. Gergely and P. L. Biermann, The Spin-flip phenomenon in supermassive black hole binary mergers, *Astrophys. J.* **697**(2), 1621–1633 (May, 2009). doi: 10.1088/0004-637x/697/2/1621. https://10.1088%2F 0004-637x%2F697%2F2%2F1621.

56. P. Padovani, P. Giommi, E. Resconi, T. Glauch, B. Arsioli, N. Sahakyan, and M. Huber, Dissecting the region around IceCube-170922A: The blazar TXS 0506+056 as the first cosmic neutrino source, *Mon. Not. Roy. Astron. Soc.* **480**, 192 (2018). doi: 10.1093/mnras/sty1852.

57. S. Garrappa *et al.*, Gamma ray counterparts to IceCube high-energy tracks, *8th International Fermi Symposium* (2018).

58. F. Lucarelli *et al.*, AGILE detection of a candidate gamma-ray precursor to the ICECUBE-160731 neutrino event, *Astrophys. J.* **846**(2), 121 (2017). doi: 10.3847/1538-4357/aa81c8.

59. R. Stein *et al.*, A high-energy neutrino coincident with a tidal disruption event (5, 2020). doi: 10.1038/s41550-020-01295-8.

60. M. Kadler *et al.*, Coincidence of a high-fluence blazar outburst with a PeV-energy neutrino event (2016).

61. E. Kun, I. Bartos, J. Becker Tjus, P. L. Biermann, F. Halzen, and G. Mező, Neutrino emission during the γ-suppressed state of blazars (9, 2020).

62. M. K. Daniel *et al.*, Spectrum of very high energy gamma-rays from the blazar 1ES1959+650 during flaring activity in 2002, *Astrophys. J.* **621**, 181 (2005). doi: 10.1086/427406.

63. J. Biteau, E. Prandini, L. Costamante, M. Lemoine, P. Padovani, E. Pueschel, E. Resconi, F. Tavecchio, A. Taylor, and A. Zech, Progress in unveiling extreme particle acceleration in persistent astrophysical jets, *Nat. Astron.* **4**(2), 124–131 (Jan., 2020). ISSN 2397-3366. doi: 10.1038/s41550-019-0988-4. http://dx.doi.org/10.1038/s41550-019-0988-4.

64. F. Halzen, A. Kheirandish, T. Weisgarber, and S. P. Wakely, On the neutrino flares from the direction of TXS 0506+056, *Astrophys. J.* **874**(1), L9 (2019). doi: 10.3847/2041-8213/ab0d27.

65. A. Reimer, M. Boettcher, and S. Buson, Cascading constraints from neutrino-emitting blazars: The case of TXS 0506+056, *Astrophys. J.* **881**(1), 46 (2019). doi: 10.3847/1538-4357/ab2bff. (Erratum: *Astrophys. J.* 899, 168 (2020)).

66. V. Lipunov, E. Gorbovskoy, V. Kornilov, N. Tyurina, F. Balakin, V. Vladimirov, P. Balanutsa, A. Kuznetsov, D. Vlasenko, I. Gorbunov, A. Pozdnyakov, D. Zimnukhov, V. Senik, A. Chasovnikov, V. Grinshpun, T. Pogrosheva, R. Rebolo, M. Serra, R. Podesta, C. Lopez, F. Podesta, C. Francile, H. Levato, D. Buckley, O. Gress, N. M. Budnev, O. Ershova, A. Tlatov, D. Dormidontov, V. Yurkov, A. Gabovich, and Y. Sergienko, IceCube-190730A: MASTER alert observations and analysis, *The Astronomer's Telegram.* **12971**, 1 (July, 2019).

67. S. Kiehlmann, T. Hovatta, M. Kadler, W. Max-Moerbeck, and A. C. S. Readhead, Neutrino candidate source FSRQ PKS 1502+106 at highest flux density at 15 GHz, *The Astronomer's Telegram.* **12996**, 1 (August, 2019).

68. V. Karamanavis, L. Fuhrmann, T. P. Krichbaum, E. Angelakis, J. Hodgson, I. Nestoras, I. Myserlis, J. A. Zensus, A. Sievers, and S. Ciprini, PKS 1502+106: A high-redshift fermi blazar at extreme angular resolution. Structural dynamics with VLBI imaging up to 86

GHz, *Astron. Astrophys.* **586**, A60 (February, 2016). doi: 10.1051/
0004-6361/201527225.

69. M. Ahlers and F. Halzen, Pinpointing extragalactic neutrino sources in
light of recent IceCube observations, *Phys. Rev.* **D90**(4), 043005 (2014).
doi: 10.1103/PhysRevD.90.043005.

70. A. Neronov and D. V. Semikoz, Self-consistent model of extragalactic
neutrino flux from evolving blazar population (2018).

71. K. Kashiyama and P. Mészáros, Galaxy mergers as a source of cosmic
rays, neutrinos, and gamma rays, *Astrophys. J.* **790**, L14 (2014). doi:
10.1088/2041-8205/790/1/L14.

72. C. Yuan, P. Mészáros, K. Murase, and D. Jeong, Cumulative neutrino
and gamma-ray backgrounds from halo and galaxy mergers, *Astrophys.
J.* **857**(1), 50 (2018). doi: 10.3847/1538-4357/aab774.

73. C. Yuan, K. Murase, and P. Mészáros, Secondary radio and X-ray
emissions from galaxy mergers, *Astrophys. J.* **878**(2), 76 (2019). doi:
10.3847/1538-4357/ab1f06.

74. T. Carver, Ten years of all-sky neutrino point-source searches. In *36th
International Cosmic Ray Conference (ICRC 2019) Madison, Wisconsin, USA, July 24–August 1, 2019* (2019).

75. A. Marinucci *et al.*, NuSTAR catches the unveiling nucleus of NGC
1068, *Mon. Not. Roy. Astron. Soc.* **456**(1), L94–L98 (2016). doi: 10.
1093/mnrasl/slv178.

76. S. García-Burillo, F. Combes, A. Usero, S. Aalto, M. Krips, S. Viti,
A. Alonso-Herrero, L. K. Hunt, E. Schinnerer, A. J. Baker, F. Boone,
V. Casasola, L. Colina, F. Costagliola, A. Eckart, A. Fuente, C. Henkel,
A. Labiano, S. Martín, I. Márquez, S. Muller, P. Planesas, C. Ramos
Almeida, M. Spaans, L. J. Tacconi, and P. P. van der Werf, Molecular
line emission in NGC 1068 imaged with ALMA. I. An AGN-driven outflow in the dense molecular gas, *Astron. Astrophys.,* **567**, A125 (July,
2014). doi: 10.1051/0004-6361/201423843.

77. J. Babson *et al.*, Cosmic ray muons in the deep ocean, *Phys. Rev.* **D42**,
3613–3620 (1990). doi: 10.1103/PhysRevD.42.3613.

78. G. Aggouras *et al.*, A measurement of the cosmic-ray muon flux with a
module of the NESTOR neutrino telescope, *Astropart. Phys.* **23**, 377–
392 (2005). doi: 10.1016/j.astropartphys.2005.02.001.

79. J. Aguilar *et al.*, First results of the instrumentation line for the deep-
sea ANTARES neutrino telescope, *Astropart. Phys.* **26**, 314–324 (2006).
doi: 10.1016/j.astropartphys.2006.07.004.

80. E. Migneco, Progress and latest results from Baikal, Nestor, NEMO
and KM3NeT, *J. Phys. Conf. Ser.* **136**, 022048 (2008). doi: 10.1088/
1742-6596/136/2/022048.

81. S. Adrian-Martinez *et al.*, Letter of intent for KM3NeT 2.0, *J. Phys.* **G43**(8), 084001 (2016). doi: 10.1088/0954-3899/43/8/084001.

82. A. D. Avrorin *et al.*, Status and recent results of the BAIKAL-GVD project, *Phys. Part. Nucl.* **46**(2), 211–221 (2015). doi: 10.1134/S1063779615020033.

83. M. G. Aartsen *et al.*, IceCube-Gen2 — The next generation neutrino observatory at the South Pole: Contributions to ICRC 2015 (2015).

84. M. Aartsen *et al.*, Neutrino astronomy with the next generation Ice-Cube neutrino observatory (November, 2019).

85. M. Aartsen *et al.*, IceCube-Gen2: The window to the extreme universe (August, 2020).

Chapter 6

New Techniques for Neutrino Detection

Amy Connolly

The Ohio State University &
Center for Cosmology and AstroParticle Physics,
191 W. Woodruff Ave., Columbus, OH 43210, USA
connolly@physics.osu.edu

In the past two decades, new techniques have emerged to target astro-physical neutrinos above 10 PeV (ultra-high-energy neutrinos), where the flux is expected to be so low that the detection volumes necessary for astrophysics measurements are of order 100 km^3. This chapter first reviews the most common of these relatively new neutrino detection strategies which utilize radio emission from the Askaryan effect in dense media and geomagnetic emission from air showers. The focus will be on established experiments and their major results before reviewing plans for the next stage of the field. This chapter also discusses newer, poten-tially transformational strategies that are being developed. The field is currently reaching discovery-level sensitivity while planning the next-generation phase that will transition to an era of precision measure-ments that are of importance to high-energy astrophysics and funda-mental physics.

1. Motivation

The discovery of a high energy astrophysical neutrino flux by the IceCube Observatory up to $\mathcal{O}(10\,\mathrm{PeV})$[1] is an exciting development in the rapidly expanding field of multi-messenger astronomy, which in the last decade has also seen the discovery of gravitational waves from catastrophic, distant astrophysical events.[2] Neutrinos at energies $> 10\,\mathrm{PeV}$, ultra-high-energy (UHE) neutrinos, have not yet been detected and are an important missing piece of multi-messenger astrophysics as they are the only particles that we can use to probe cosmic distances at the highest energies.[3] UHE neutrinos will also allow us to search for physics beyond the Standard Model at center-of-mass energies unreachable by particle physics accelerators on Earth.[4]

While detecting neutrinos via optical Cherenkov emission has been tremendously successful from $\sim 1\,\mathrm{MeV}$ to $10\,\mathrm{PeV}$, at higher energies new strategies are needed. This is because, above $10\,\mathrm{PeV}$, the flux of neutrinos falls below $\sim 1000\,/\mathrm{km}^2/\mathrm{year}$, and they interact over distances of order thousands of kilometers in ice, so we can expect less than one interaction per year in a cubic-kilometer detector. This means that we need to reach detection volumes much larger than the $\sim 1\,\mathrm{km}^3$ occupied by the IceCube Observatory to detect the order of tens of UHE neutrinos that would be needed for a rich science program with these unique particles in this regime.

In the commonly used detection media of polar ice and seawater, optical light is scattered and absorbed in $\sim 100\,\mathrm{m}$,[5,6] and this sets the required spacing of detector elements for that technique. Since it is not practical to instrument tens of km^3 with such a densely spaced array, new strategies are needed where the relevant absorption lengths are longer and/or the sensors are not embedded in the detection medium so that this problem can be evaded.

New detection strategies targeting the highest energy neutrinos take advantage of one or both of the following. First, attenuation lengths in pure ice are much longer at radio wavelengths than optical, of order 1 km. Second, when viewing a medium where neutrinos have the opportunity to interact from a high altitude or a far distance, an observer can monitor immense volumes with less instrumentation than would be needed to view it from within.

In this chapter, I will first motivate the reasons we expect the astrophysics neutrino flux to extend to the UHE regime and then review some of the fundamentals of UHE neutrino interactions. Next, I will describe the methodology behind recently developed techniques for neutrino detection, followed by the status of the major experiments utilizing each technique and major results. I will start with the radio technique in dense media, followed by others based on the detection of neutrino-induced air showers. Finally, I will give a broad survey of planned next-generation experiments using both established techniques and new techniques under development.

2. UHE Neutrino Flux

That there is a detectable flux of neutrinos extending to the UHE regime is strongly motivated by existing particle astrophysics measurements coupled with well-understood Standard Model physics. I briefly summarize these motivations here, followed by the impact that measurements of UHE neutrinos would have on open questions in astrophysics and particle physics.

2.1. *Motivation*

That cosmic-ray nuclei have been observed up to above 10^{20} eV gives us strong motivation to expect an observable flux of UHE neutrinos. Called ultra-high-energy cosmic rays (UHECRs), their origin remains an open question, with major candidates including active galactic nuclei, gamma-ray bursts, and radio galaxies. Despite questions surrounding their origin, the measured UHECRs give us reason to expect that UHE neutrinos will be produced both in astrophysical sources and via interactions undergone by UHECRs after leaving their sources.

If the mechanisms that are accelerating UHECRs in the astrophysical sources are carried out through hadronic processes,[7,8] then they would also be expected to produce neutrinos through photohadronic interactions,[9] such as:

$$p + \gamma \to n + \pi^+, \tag{1}$$

where the charged pion subsequently decays via $\pi^+ \to \mu^+ \nu_\mu$ and the muon to $\mu^+ \to e^+ \nu_\mu \nu_e$. Similar interactions to the one in Eq. (1) occur if the initial particle is a heavy nucleus instead of a proton. The neutron will produce an electron neutrino via $n \to p + e^- + \nu_e$, although the neutrino end product from the neutron decay will be

lower in energy by about a factor of 100 than the neutrinos from
pion decay due to the small $n - p$ mass difference.[10] Thus, the flavor
ratio at the source is expected to be $\nu_e : \nu_\mu : \nu_\tau = 1 : 2 : 0$ before
oscillating to a more even flavor distribution before detection.

This picture of neutrino production through Eq. (1) can become
more complex if we look deeper. The interaction can occur via the
Δ^* resonance, which is an excited proton state of mass 1232 MeV,
via higher resonances, or can instead result in multi-pion production
instead of just the two-body decay. Also, we should keep in mind that
although hadronic models are the most common models for UHECR
acceleration,[7,8,11] leptonic models offer an alternative that would not
induce a neutrino flux from the source.[12,13]

The interaction in Eq. (1) can also lead to UHE neutrinos when a
proton encounters a photon after having escaped its source. Within
about 50 Mpc, a UHE nucleus will interact with a cosmic microwave
background photon and this is called the Greisen–Zatsepin–Kuzmin
(GZK) mechanism.[14,15] The distance over which the interaction
occurs is called the GZK horizon. Using basic relativistic kinematics,
and accounting for the Maxwell–Boltzmann distribution of 3 K CMB
photons, the protons observed at Earth will begin to be suppressed
above about 5×10^{19} eV. That this mechanism would produce an
observable flux of neutrinos[10] was first pointed out by Berensinsky
and Zatsepin,[16] and these are often called cosmogenic neutrinos.

2.2. *Science impact*

If there is such a strong expectation for a UHE neutrino flux, why
go to the trouble to seek them out? What would we learn? UHE
neutrinos will answer important questions in both astrophysics and
particle physics that cannot be answered through other means.

An important outstanding question in astrophysics is the nature
of the accelerators that are producing cosmic rays up to above
10^{20} eV, and UHE neutrinos have a unique role to play in unrav-
eling the answers. They will provide a first-ever glimpse of the most
powerful particle accelerators at distances beyond our local universe
($\gtrsim 50$ Mpc), aside from the spectacular events revealed by gravitation
wave detections. Cosmic rays and gamma rays cannot travel cosmic

distances without interacting. Identifying UHE neutrinos that originated from astrophysical sources would point to acceleration mechanisms that are hadronic rather than leptonic since only hadronic interactions would produce neutrinos through the decay of charged pions, their secondary muons, and neutrons. Identifying characteristics about the sources that differ from those observed from cosmic-ray measurements could mean that sources inside our local GZK horizon are not representative of the cosmic population.[17] The overall flux of cosmogenic neutrinos is an integral over the density of those sources over all distances, which is an important clue to the source types, although the level of the flux also depends on the cosmic-ray composition (heavier composition leads to lower flux predictions), so some care must be taken to separate the two effects.[18] The ultimate cutoff of UHE neutrinos will reveal the maximum energy that accelerators in the universe can reach. If that energy is beyond what is achievable through any acceleration mechanism, it could point to new physics.[19]

Since UHE neutrinos would be the highest energy neutrinos observed to date as well as the only particles at these energies that can travel cosmic distances, they will be important tests of fundamental physics. The center-of-mass energy E_{CM} of a neutrino–nucleon interaction is given by $E_{CM} = \sqrt{2m_N E_\nu}$, where $m_N \approx 1\,\text{GeV}$ is the nucleon mass, and so a $10^{18}\,\text{eV}$ neutrino will interact in a detection volume at approximately $E_{CM} = 45\,\text{TeV}$. Once a sample of order tens of high energy neutrinos are collected, a measurement of the neutrino–nucleon cross section can be made by observing the distance dependence of their absorption in the earth.[20,21] As shown in Fig. 1, first cross-section measurements at hundreds of GeV center-of-mass energies are already being made with the first astrophysical neutrino events measured by the IceCube observatory.[22,23] Once the measured neutrino spectrum is extended to the UHE regime, this measurement will be an important test of neutrino interactions at center-of-mass energies beyond those probed at the LHC.[24] High energy neutrinos already constrain models for Lorentz invariance violation (LIV) at the $\sim 10^{-27}$ level,[25,26] and current flux limits in the ultra-high-energy regime can even be in interpreted as constraints on LIV.[27] Neutrino measurements at higher energies will push these bounds even further.

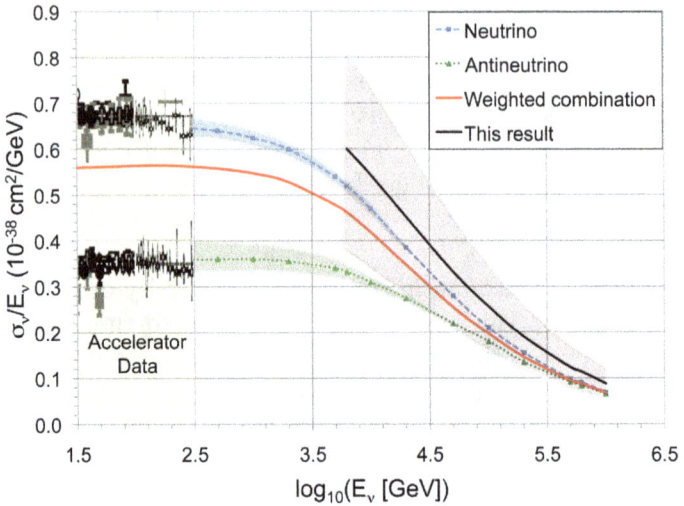

(a)

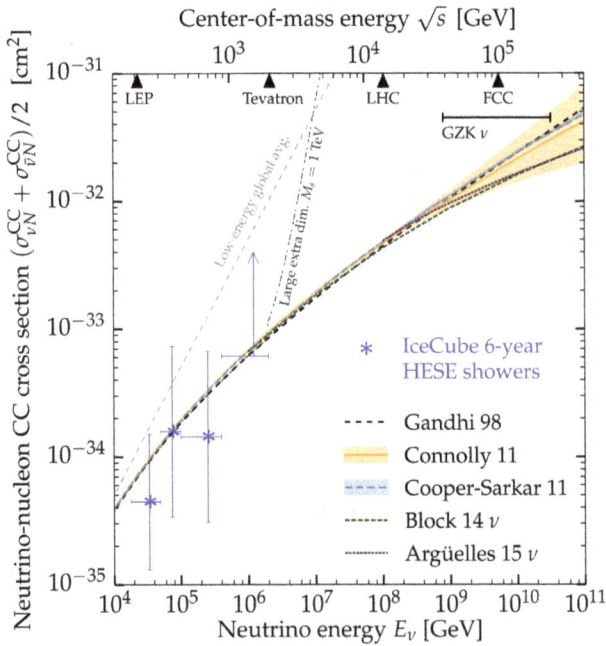

(b)

Fig. 1.   First glimpses of the neutrino–nucleon cross sections at hundreds of GeV center-of-mass energies. (Top) IceCube measurement,[22] the first of its kind, reporting the measured cross section across a single energy bin relative to the Standard Model expectation. (Bottom) Extraction of the energy-dependent cross section[23] using a public dataset of charged current electron neutrino events.

3. UHE Neutrino Interactions in Dense Media

A UHE neutrino interacts in a dense medium via one of the following interactions:

$$\nu + q \rightarrow \nu + X, \tag{2}$$
$$\nu + q \rightarrow \ell + X, \tag{3}$$

where q is a quark or antiquark from a neutron or proton in the medium and X is the remnant of the hadron, which subsequently develops a hadronic shower. Here, the ν could represent a neutrino or anti-neutrino, and in the latter case, the lepton would be an antiparticle. The fraction of the neutrino energy that goes to the initial hadronic recoil is called the inelasticity y. The inelasticities have a broad probability distribution that falls steeply with y, with a mean in the UHE regime near 0.2 with only a weak energy dependence.

These interactions produce different signatures depending on their end products. If a lepton is produced and is an electron, then this results in an electromagnetic shower consisting of electrons, positrons, and photons. A muon or tau would leave an ionizing track, and in the UHE regime, may also leave detectable showers due to secondary photonuclear interactions.[28,29] A tau lepton will decay over a length in the lab frame given approximately by

$$c\tau = 50\,\text{km}\frac{E}{10^{18}\,\text{eV}}, \tag{4}$$

where E and m_τ are the energy and mass of the tau, respectively, and τ_0 is the proper lifetime of the tau lepton, about 300 fs. The decay products of the tau lepton may induce an additional hadronic or electromagnetic shower depending on the decay channel.

The cross sections for interactions between UHE neutrinos and protons are in the range $10^{-32.5}$–$10^{-31}\,\text{cm}^2$ from 10^{17}–$10^{20}\,\text{eV}$.[24,30,31] The energy dependence of neutrino cross sections σ in this regime is approximately $\sigma \propto E^{0.3}$, softer than the linear dependence at energies below $\sim 10\,\text{TeV}$.

Recall that the cross section is related to the interaction length ℓ by

$$\ell = \frac{m_N}{\sigma\rho}, \tag{5}$$

where m_N is the nucleon mass, approximately 1.67×10^{-27} kg, or 1 GeV, and ρ is the density of the medium. An interaction length is a distance over which the probability that a neutrino will not have interacted is $1/e$. So, for example, in the Earth's crust, where the density is about 2600 kg/m^3, a 10^{18} eV neutrino will typically travel approximately 600 km before being absorbed. In ice, where the density is about 900 kg/m^3, that distance is about 1900 km.

4.   Radio Detection in Dense Media

Experiments using the radio technique for neutrino detection in dense media search for a characteristic impulse that arises from the Askaryan effect[32] originating from deep in the medium where other particles could not have reached before being absorbed. The signal power is predominantly in the hundreds of MHz, and polar ice is a common target medium because it is transparent over ~km distances in the relevant frequency band.

A neutrino interaction in a dense medium induces a particle cascade, and this cascade develops a charge asymmetry known as the Askaryan effect. The net current brought about by this charge asymmetry brings about the radio emission. The ~20% charge asymmetry arises from Compton scattering and annihilation of shower positrons with electrons in the medium, with Compton scattering being the dominant reason for the charge asymmetry. The evolution of the net current in the cascade as it increases, peaks, and then decreases along the axis of the shower produces the emission. The pulse observed near the Cherenkov angle is compressed in time due to the type of relativistic effects that come about in Cherenkov radiation. These effects are due to the speed of the particles exceeding c/n, where n is the index of refraction of the medium.

An important aspect of radio emission arising from the Askaryan effect is that it is coherent for frequencies up to about 1 GHz. An observer sees the field strengths emitted by particles in a particle shower summed *coherently* at long wavelengths. Coherence is maintained for wavelengths $\lambda \gg r_M$, where r_M is the characteristic transverse size of the shower called the Molière radius, which is approximately 10 cm.

The coherence of radio emission from the Askaryan effect is an important strength of the technique. While the power from incoherent optical Cherenkov emission is proportional to N, where N is the number of particles in a shower, for radio emission, the power $P \propto N^2$. This means that the power in the optical signal is proportional to the shower energy E_{sh}, and the radio emission is proportional to E_{sh}^2, and so once an energy threshold for detecting neutrino-induced cascades via the radio technique is reached, it can dominate over the optical technique.

Figure 2 demonstrates the dependence of the radio emission from the Askaryan effect on frequency and vantage point from which one views the cone. As is characteristic of Cherenkov emission, its power is linear in frequency. Askaryan emission, however, loses coherence at frequencies where wavelengths become shorter than the Molière radius at frequencies above approximately $1\,\mathrm{GHz}$. The emission is at peak strength when it is viewed at the Cherenkov angle with respect to the shower axis (which is aligned with the incident neutrino momentum), and the Cherenkov angle is related to the index

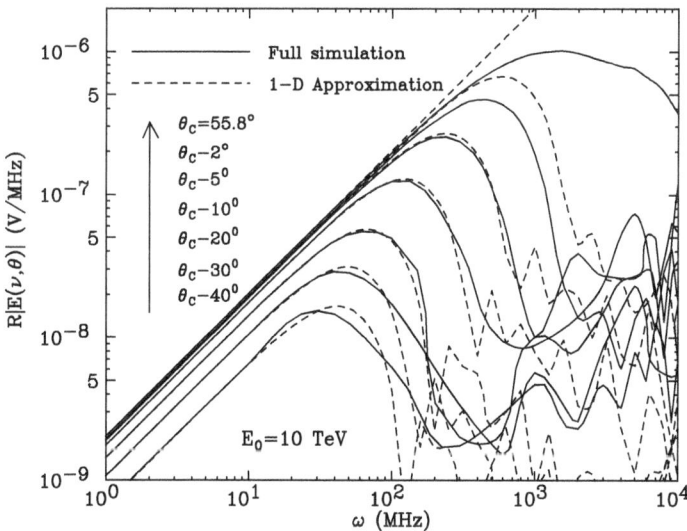

Fig. 2. Predicted frequency dependence of the Askaryan signal[33] for different viewing angles away from the Cherenkov angle. Away from the Cherenkov angle, the frequency at which coherence is lost is reduced. Equivalently, at lower frequencies, the width of the Cherenkov cone is broader.

of refraction in the medium by

$$\theta_C = \cos^{-1}\frac{1}{n}. \qquad (6)$$

In deep Antarctic ice, typically $n = 1.78$, giving $\theta_C = 56°$. The emission is in a cone that has a spread of $<10°$ about the Cherenkov angle above 100 MHz, being the broadest at the lowest frequencies.

The radio signal from the Askaryan effect is also linearly polarized. Considering a ray that propagates along wave vector $\hat{k}$ within that cone, the polarization of the signal points outward on the cone in the plane of the shower axis and $\hat{k}$ and perpendicular to $\hat{k}$.

The predicted Askaryan signal from high energy particle cascades in dense media was first observed in beam tests using sand and salt as targets in 2000.[35,36] Figure 3 shows later results from the ANITA experiment which, in advance of its first flight, measured the Askaryan signature with the properties expected of coherent radiation in an ice target bombarded with an electron beam at the then Stanford Linear Accelerator Center (SLAC).[34]

4.1. *Polar ice*

The most common target medium for neutrino detection via the radio emission from the Askaryan effect is polar ice, and there are two main reasons for this. First, as described in Section 1, due to their low flux and low cross sections, detection volumes ($\gtrsim 10\,\mathrm{km}^3$) are needed to measure tens of events, and such volumes can only be naturally occurring. Second, a detection volume of that size can only be sparsely instrumented because otherwise the cost becomes prohibitive. Since ice is in abundance in polar regions, particularly Antarctica and, as we will see later, it is also lossless over distances of about 1 km for frequencies in the hundreds of MHz range, it is a quite suitable medium for radio techniques.

It is important to understand the electromagnetic properties of the ice to enable proper modeling of experiments and interpretation of data for the detection of neutrinos with radio techniques. Here, we briefly review the most relevant properties and their effects on observations.

While it has long been known via remote sensing that radio signals can propagate long distances in polar ice,[39] several direct

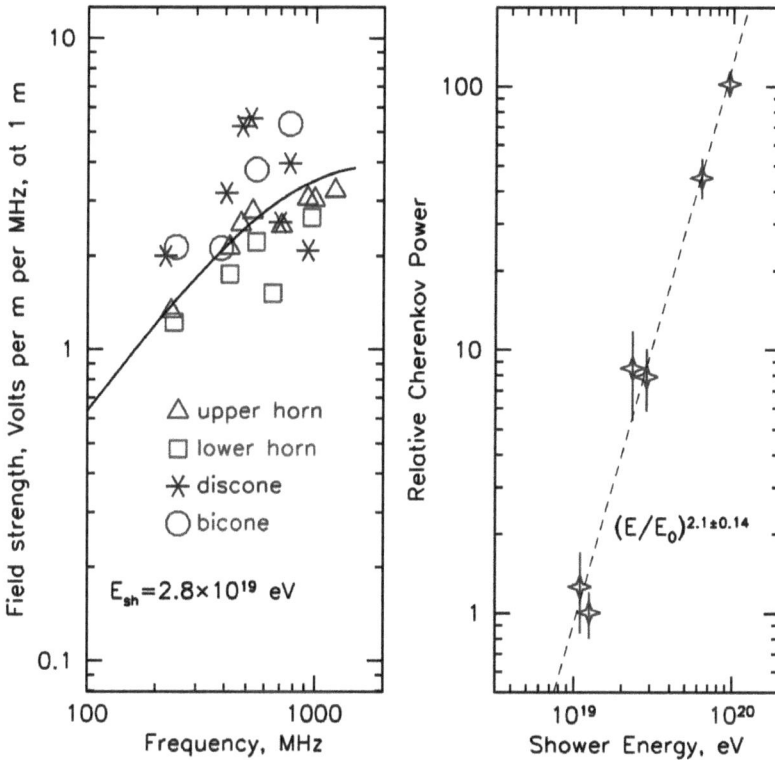

Fig. 3. Observation of the Askaryan effect in a beam test with an ice target by the ANITA experiment[34] shows (left) the expected linear dependence on frequency and (right) the quadratic dependence of the signal power on shower energy as expected from coherence.

measurements of in-ice attenuation lengths have been made over the past 20 years at the sites of different neutrino experiments.[37,40–42] The field attenuation length is the distance that a signal travels before the field strength drops to $1/e$ of its starting strength beyond the usual $1/r$ geometric loss. For this type of measurement, broad-band signals are transmitted and received ideally by antennas of the same design to simplify analysis. Sometimes, the signal is received after a reflection off of the bottom of the ice.

Figure 4 shows the results of one such field attenuation length measurements carried out by reflecting signals from the ice–bedrock interface in the South Pole ice. A simplified expression for the relationship between the attenuation length, L, and the transmitted and

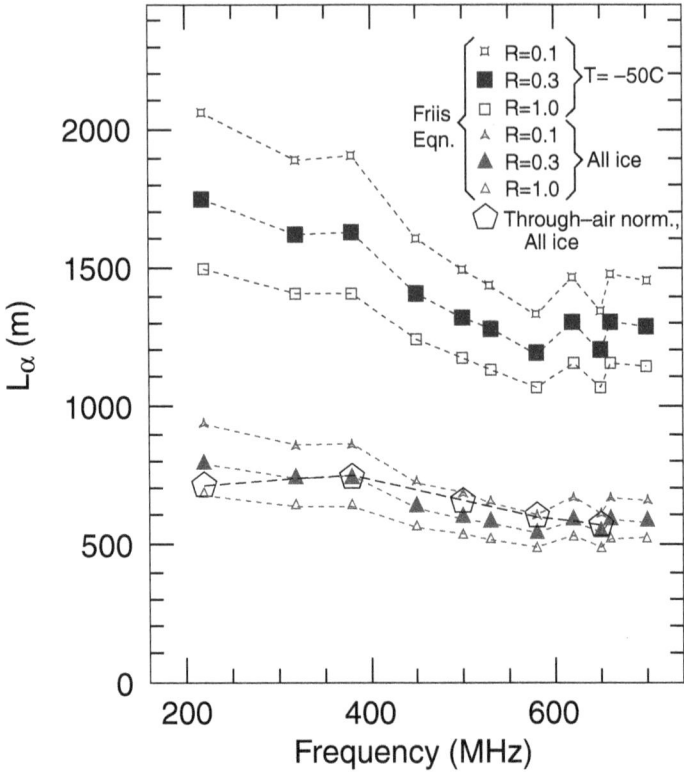

Fig. 4.　Measured effective attenuation lengths[37] at radio frequencies in the ice near South Pole for different choices of reflectivity R, which is the ratio of reflected power to incident power. Broadband pulses were transmitted into the ice from the surface and received after reflection from the ice–bedrock interface below.

received field strength is

$$rE_R = e^{-(r-r_0)/L}r_0 E_T, \tag{7}$$

where r is the propagation distance between the transmitter and receiver and r_0 is some reference distance close enough to the transmitter that attenuation is negligible. Due to the depth dependence of ice properties, what is measured directly is only an effective attenuation length; attenuation lengths decrease (loss increases) as temperature increases, and temperature increases with ice depth. Therefore, reported attenuation lengths rely on models for these dependencies.

　　Figure 5 shows a compilation of measurements of the depth-dependent index of refraction in the South Pole ice. The index of

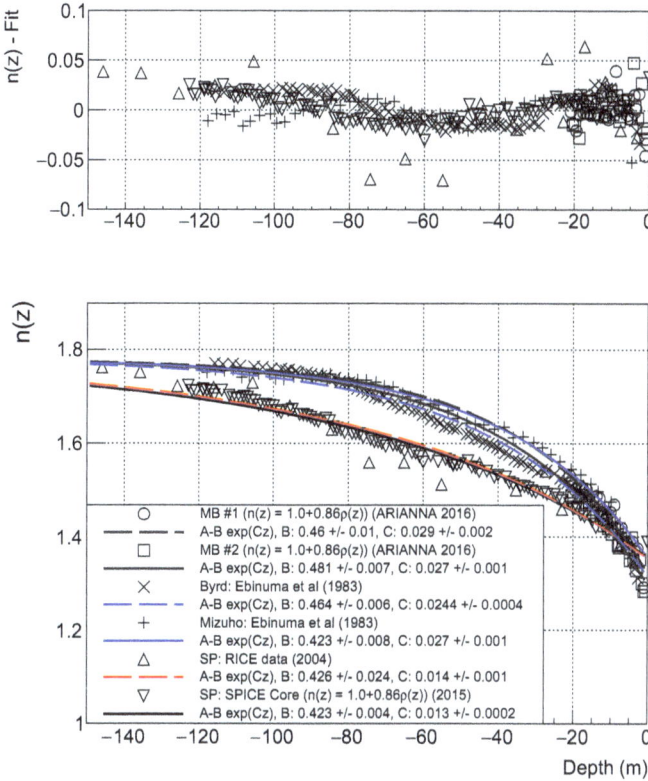

Fig. 5. Compilation of measurements of the depth-dependent index of refraction in the South Pole ice[38] compared to different models, each of exponential form. The top figure shows the residuals to the fits.

refraction in the South Pole ice varies from approximately 1.35 at the surface to 1.78 in ice $\gtrsim 200\,\mathrm{m}$.[43] This has important implications for detecting signals that have propagated over ∼km-scale distances in the ice. First, signals do not travel in straight trajectories but curve downward along their path. Second, there is often more than one ray-traced solution between the source of emission in the ice and an in-ice receiver. One solution is referred to as the "direct" ray and the other the "reflected" or "refracted" ray, depending on whether the signal reflects from the surface of the ice.

Both of these implications of depth-dependent indices of refraction are important for reconstructing properties of the incident neutrino. The ray bending is essential for reconstructing the neutrino direction. The two ray solutions together form a signature of an in-ice interaction and enable a reconstruction of the distance to an interaction. Where a double-pulse, "DR" event is observed, the distance to the interaction can be reconstructed to 15%.[44]

Additionally, ice is known to be birefringent, which means that the index of refraction depends on the direction of propagation. The birefringent properties of the ice come from that inherent to the ice crystals (a few mm in size) combined with the distribution of the crystal orientations being influenced by horizontal ice flow and vertical compression.[45] For each of the direct and reflected/refracted solutions, a signal emitted at the source propagates in the ice as two rays that have different polarizations and are separated in time by typically $\mathcal{O}(10\,\text{ns})$. This behavior has been measured in the South Pole ice in the ARA neutrino detector.[44,46] Effects due to birefringence have also been observed in the context of radar echoes in polar ice.[47]

4.2. *Experimental approaches*

Since Askaryan emission lies predominantly in the radio band, the detector elements for this technique are antennas, and the approach taken by different experiments begins with where to put them. Experiments seeking to detect neutrinos through radio emission from the Askaryan effect take on two main strategies. The first strategy is to detect the radio emission with instrumentation deployed in the ice itself, and the other is to view the ice from a high-altitude balloon.

The two strategies are complementary. In-ice detectors are sensitive to weaker signals and thus lower-energy neutrinos due to being in closer proximity to the interactions, but the volume of ice that they can view is limited by the extent of the array that can be built. From a high-altitude balloon, signals will need to be stronger, and thus neutrinos typically higher energy, to be detectable at the distance to the instrument. However, from a balloon, an immense volume of ice can be visible within its horizon.

Much like particle physics experiments at colliders, these projects use triggering systems to make decisions about when to record an event. Typically, the power of an incoming signal is split, with one half used to determine whether a trigger requirement is satisfied and the other to digitize the voltage traces from the different antennas once trigger conditions are met.[48,49a] Triggers use selection criteria based on properties that favor neutrinos and disfavor backgrounds, such as thermal noise and human-made noise, produced in the same band. Often, a signal is required to be consistent with being linear polarized and to exceed a power threshold in some number of antennas within some time coincidence window. The coincidence windows need to be long enough to allow the signals to cross the distances between antennas.

Due to the frequency range where the signal dominates (up to $f_{max} \sim 1\,\mathrm{GHz}$), the Nyquist–Shannon theorem[51] requires that the signals be sampled at a rate greater than $2f_{max}$. This means that sampling times must be less than 0.5 ns. This poses a technical challenge to digitize signals at an adequate rate while still operating at the limited power available in the remote locations of the experiments. The digitizers typically use switched capacitor arrays and write out $\mathcal{O}(\text{few } 100 \text{ ns})$–long waveforms that are sampled at a rate of $\sim 2\,\mathrm{GHz}$ when a trigger condition is satisfied.[52,53] Moving forward, some proposed projects are utilizing new streaming digitizers, where the trigger and digitization are integrated onto a single chip, and more complex requirements can be made at the trigger level.[54,55]

4.2.1. *From within the ice*

One approach to radio detection of in-ice neutrino interactions is to go where the interactions are and deploy antennas in the ice. Within this category itself, there are two approaches. One approach is to deploy antennas at depth, and another is to deploy them in the ice but near the surface. With a deep detector, a higher volume of ice can effectively be viewed for each detector element compared to at the surface. However, surface detectors do not need to drill into the ice and are not subject to the geometrical constraints that come with deploying down a hole.

The first experiment to boldly take this approach and deploy antennas in a borehole near South Pole was the Radio Ice Cherenkov Experiment (RICE), deployed in 1998–1999.[56] RICE deployed antennas up to 350 m deep along strings drilled for AMANDA, the

---

[a]An alternative approach taken by ARIANNA does away with the need to split the signal.[50]

predecessor to IceCube. The antennas were "fat dipoles" with bandwidth covering 200–1000 MHz. RICE placed strong constraints on the flux of UHE neutrinos, competitive with IceCube and the Pierre Auger Observatory (see Section 5) for about a decade.[57]

Following RICE, a series of prototyping efforts for in-ice detectors at South Pole[58] led to the deployment of the Askaryan Radio Array (ARA), illustrated in Fig. 6. Between 2010 and 2018, ARA deployed a prototype "Testbed" station[42] and five deep stations spaced by 2 km separation. Each station includes 16 receiver antennas at up to 200 m depths.[49] ARA deploys two types of antennas, separately targeting signal polarizations in two mutually perpendicular directions. As of this writing, ARA has published constraints on the diffuse neutrino flux based on a limited dataset of four years of livetime in two stations.[59]

The fifth ARA station operates with two different trigger systems. One is a traditional design and another uses a phased array (PA) trigger.[60] In the PA trigger, a set of directional hypotheses is considered. For each directional hypothesis, the signals arriving in seven vertically spaced antennas are corrected for their expected delays, and then the waveforms are summed. Then, the trigger criteria are imposed on that coherently summed waveform. The PA system reduces the trigger threshold on signal voltages by a factor

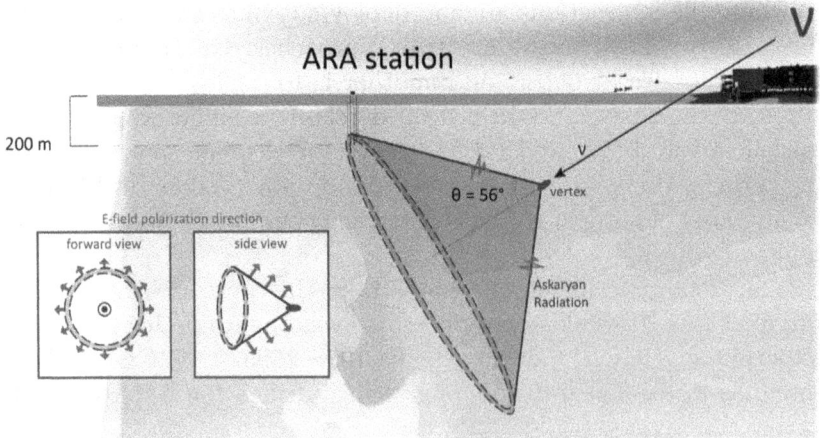

Fig. 6.   Illustration of the cone of radio emission induced by a neutrino interaction in the ice and visible by an ARA station with antennas ∼200 m deep.[59]

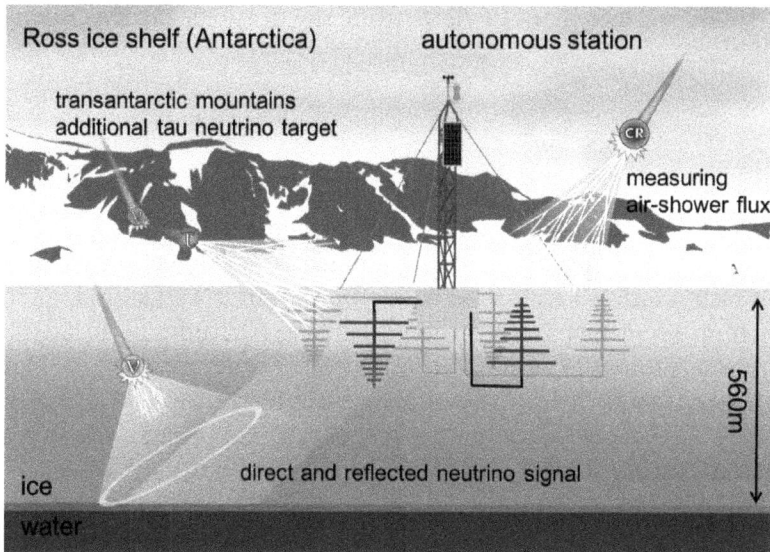

Fig. 7. Illustration of an ARIANNA station and the cosmic ray and neutrino signatures to which the detector is sensitive. ARIANNA uniquely searches for radio emission after reflection from the ice–water interface at the bottom of the ice.[61]

of approximately $\sqrt{N}$, where N is the number of antennas participating. This reduces the minimum detectable cascade energy. At any cascade energy, this leads to a greater trigger efficiency.

The ARIANNA experiment (Fig. 7) takes a different approach and deploys antennas at the surface rather than at depth, with nine stations of antennas on the Antarctic Ross Ice Shelf and another two stations at South Pole.[50] Each ARIANNA station consists of two pairs of log-periodic dipole antennas (LPDAs) aimed straight down into the ice, with the two pairs oriented orthogonal to one another for sensitivity to two polarizations. The LPDA gains are above 10 dBi from approximately 100 to 1300 MHz. On the ice shelf, ARIANNA is sensitive to signals reflecting from the ice–seawater interface below, allowing it to view neutrinos from solid angles spanning the sky above the detector. ARIANNA also deploys LPDAs pointed upward for detecting cosmic rays via geomagnetic emission from air showers (see Section 5 for more about that signature). Cosmic rays serve as a demonstration of observations of broadband radio impulses of natural origin.[62] ARIANNA has published constraints on the UHE neutrino flux based on 4.5 years of data from the stations on the Ross Ice Shelf.[63]

4.2.2.   *From a balloon*

The second main approach among experiments searching for radio emission from the Askaryan effect is to deploy antennas at an altitude above the polar ice, enabling a much larger detection volume to be in view. NASA's Antarctic balloon program is well suited for this, launching scientific payloads to stratospheric altitudes (~40 km) during austral summers with flight durations of order a month.[64] This is the approach taken by the ANtarctic Impulsive Transient Antenna (ANITA) experiment, having taken four flights under this program between 2006 and 2016. At altitude, ANITA can view as much as $1.5\,M\,km^2$ of ice within its horizon.

As shown in Fig. 8, the ANITA payloads consisted of antennas pointing outward and arranged with azimuthal symmetry. They point $10°$ below horizontal, where they can view the most ice. ANITA's quad-ridged horn antennas have nearly the same response in both polarizations, with gains of ~10 dBi in the 200–1000 MHz band. Over the four flights, the number of antennas increased from 32 to 48. This increase, modifications to the trigger, and other improvements to ANITA's design have allowed thresholds to be decreased, and thus sensitivity increased, for each successive flight.[65–68]

ANITA is most sensitive to UHE neutrinos at grazing incidence, where the Askaryan signal is viewed from the top of the cone of emission and the signal polarization is nearly vertical. Over its four flights, ANITA did not identify any neutrino candidates and set the best constraints on the UHE flux above 10^{19} eV.

After its first flight, ANITA detected geomagnetic emission (see Section 5) from cosmic-ray air showers.[69] Geomagnetic emission is nearly horizontally polarized because the magnetic field of the earth at Antarctic latitudes is nearly vertical.[69] While some cosmic rays are observed directly from the shower in the atmosphere, most are observed after the impulse is reflected from the ice surface. The two types of signals are observed with a relative inversion of signal polarity. ANITA's cosmic-ray observations[70] serve as a demonstration of detections of broadband radio impulses of natural origin and also serve as a calibration.

Fig. 8. Photograph of the ANITA experiment on ascent during the second flight in 2009–2010 austral summer season.[66] The inset shows a larger-scale picture including the ~100 m diameter balloon with the payload hanging underneath.

5. Neutrino Detection via Air Showers

Large arrays designed for the detection of cosmic-ray air showers are also sensitive to UHE neutrinos via an air shower signature. Here, I describe three techniques that are used to measure cosmic-ray air showers.

Cosmic-ray nuclei interact in the upper atmosphere to produce extended air showers. These are particle cascades composed of charged and neutral mesons, which decay to charged leptons, neutrinos, and photons, and these decay products leave various signatures. Charged particles in the shower interact with nitrogen in the atmosphere, leading to the emission of ultraviolet "fluorescence" light at ~300–430 nm wavelengths. High-energy muons from the shower reach the ground.

The charged particles in the air shower also produce what is called geomagnetic emission from a few to a few hundred MHz. This is due to the charged particles in the shower developing "transverse currents" perpendicular to the shower axis when they are accelerated by the Lorentz force in the Earth's magnetic field.[72] Like radio emission from the Askaryan effect, the geomagnetic emission adds coherently from the population of particles in the shower. Radio emission from the Askaryan effect is also present in air showers but at a much lower level than geomagnetic emission due to the low density of air. Thus, cosmic-ray air showers can be observed via fluorescence detectors, particle detectors on the ground, and radio arrays. The past two decades have seen improvements in our understanding of the latter signature through both theoretical and experimental developments.[69,70,73-75]

As discussed in Section 3, a tau neutrino that interacts via a charged current interaction will produce a tau lepton, which can decay after traveling tens of km and produce an air shower like the ones induced by cosmic rays. This means that air showers at grazing incidence can also be a signature of tau neutrinos for geometries where there is sufficient opportunity for the interaction, tau decay, and shower to occur before reaching the detector. The resulting shower can be detected through the same techniques, namely fluorescence, particle detection, and radio, that are used to measure cosmic-ray–induced air showers. In what follows, I discuss techniques for observing air showers for the detection of UHE neutrinos.

5.1. *Particle and fluorescence detection*

The Pierre Auger Observatory is a ground array in Argentina that has made the most precise measurement of the cosmic-ray spectrum while also setting competitive constraints on the diffuse flux of UHE neutrinos through the air shower technique for the detection of tau neutrinos. Auger is a hybrid array that utilizes two different detection mechanisms. One component is a surface detector (SD) of 1660 water tanks spaced 1500 m apart, filled with highly purified water, and instrumented with photomultiplier tubes to detect Cherenkov emission. The Cherenkov emission originates from muons that pass through the tank. The other component of the Auger hybrid detector is the fluorescence detector (FD) consisting of 24 fluorescence telescopes, with six at each of four elevated locations. All telescopes making up the FD face the interior array and cover a combined 180° in azimuth. At each telescope, a $10\,\text{m}^2$ mirror focuses the fluorescence light onto a camera consisting of 440 photomultiplier light sensors.

Tau-neutrino–induced air showers that are inclined downward can be detected by the SD. Other showers that are headed for higher inclinations can be detected by the FD, but due to the 10% duty cycle of the FD, the greatest sensitivity comes from the SD.

5.2. *Radio*

Several experiments are under development to use the geomagnetic technique for the detection of UHE neutrinos with arrays of antennas deployed to view landscapes where taus may emerge. Figure 9 illustrates the TAROGE experiment,[76] which searches for neutrinos at two sites, one in Taiwan overlooking the Pacific Ocean, and another on Mt. Melbourne overlooking Antarctic ice.

Among experiments utilizing the geomagnetic technique for tau neutrino detection, a key development has been the demonstrated capability to self-trigger on the radio signature from cosmic-ray air showers rather than relying on a particle detector for the trigger. This allows for greater efficiencies because the radio signal propagates a further distance than the particles in the shower and also leads to a lighter design for expansive arrays. However, radio frequency interference (RFI) poses a greater challenge with this type of design.[77] Several radio arrays have demonstrated this ability to self-trigger on the radio signature toward the aim of neutrino detection.[62,69,76–78]

ANITA could also be sensitive to Earth-skimming tau neutrinos exhibiting the same signature. In ANITA, radio impulses from tau

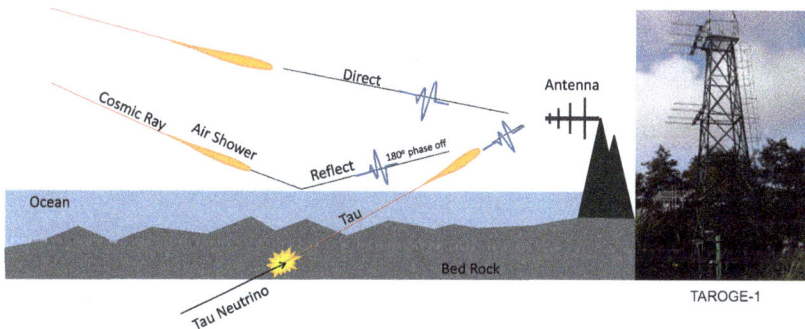

Fig. 9. Illustration[71] of the signatures sought after by the TAROGE experiment for (1) cosmic rays, similar to the signatures to which ANITA is sensitive, and (2) tau neutrinos, which is the same signature used by Auger and some other projects under development, such as GRAND.

neutrinos emerging from the ice would lack the polarity flip that impulses from cosmic rays undergo upon reflection at the ice surface but otherwise would be expected to have the same properties. Although ANITA has uncovered air showers with polarity consistent with the tau neutrino signature,[79-81] the events are in other ways difficult to square with a tau neutrino hypothesis[82] and so for now remain anomalous.

6. Future

Even with the vast detection volumes probed by current UHE neutrino observatories, typically $\mathcal{O}(10\,\mathrm{km}^3)$ water equivalent, the potential for probing the UHE neutrino flux could be characterized as "discovery level", i.e., if UHE neutrinos are uncovered, then they would likely be of order one to a few events. Even such a discovery of the first UHE neutrino events would be a landmark achievement and would lead to important first-ever glimpses of high energy astrophysics at cosmic distances and fundamental physics at center-of-mass energies never before probed.

To reach an era of precision measurements, neutrino events of order 100 UHE are needed. With a sample of this size, we will be able to tighten constraints on the properties of the most energetic cosmic sources and measure neutrino–nucleon cross sections at tens of TeV. This means that the detection volumes need to be scaled up by a factor of $\mathcal{O}(100)$ from where they are currently.

6.1. *In-ice arrays*

Proposed expansions of in-ice arrays searching for neutrinos via radio emission from the Askaryan effect build on the experience of existing deep and surface radio arrays, ARA and ARIANNA, with designs drawing from aspects of each type of detector. Deep antennas are desirable to view the most ice volume in a given station. The phased array triggers established in deep detectors keep thresholds low. Surface antennas supplement reconstruction capabilities and enable monitoring of backgrounds entering ice from the surface, both cosmic rays and anthropogenic noise. The most prominent proposed next-generation in-ice array is a radio component to the proposed expansion of IceCube called IceCube-Gen2. The reference design of the radio component takes a hybrid approach guided by these ideas.[83] First deployments of an IceCube-Gen2 array would take place as early as 2025. The full array would improve on the size of ARA and ARIANNA by a factor of $\mathcal{O}(100)$. In the meantime, an intermediate-sized radio array called RNO-G is being deployed in Greenland.[84]

6.2. *From a balloon*

Plans for the successor to the ANITA balloon-borne UHE neutrino array, the Payload for Ultrahigh Energy Observations (PUEO),[54] are underway, with a first flight planned for the 2025–2026 austral summer. PUEO will utilize a 16-channel delay-and-sum trigger, much like the phased array trigger design used for ARA, that will lower PUEO's threshold signal voltages by a factor of five compared to ANITA-IV thresholds. PUEO's high altitude keeps it the most competitive above 10^{19} eV through the Askaryan search channel. Its threshold-lowering triggering and a focus on the tau neutrino air shower channel expand its region of sensitivity below ANITA's, down to $\sim 2 \times 10^{18}$ eV. PUEO's ability to identify tau neutrinos via grazing air showers will make it unique in its ability to distinguish between flavors in this energy regime. In addition, for sources near its horizon, PUEO has an unmatched sensitivity to neutrinos from astrophysical sources that fall within that region of the sky.

6.3. *Ground arrays*

Efforts to detect neutrinos via air showers have expanded greatly in the past decade. Auger is currently the project that has the most potential to detect UHE neutrinos with this technique, and an upgrade named AugerPrime[85] is under construction. AugerPrime is designed to improve particle identification. AugerPrime will not expand on the Auger detection area, and so its projected sensitivity to UHE neutrinos will be a modest improvement over Auger because it is primarily from increased accumulated livetime.

The Auger Engineering Radio Array (AERA) is deployed within Auger and is the largest array of radio antennas for the detection of radio emission from air showers. It consists of 150 antenna stations spanning $17 \, \text{km}^2$. As part of the upgrade, the SDs will undergo a series of upgrades, including the addition of a radio antenna on the top of each detector. This radio component will target highly inclined showers. Auger and AugerPrime have laid/are laying the groundwork for other planned experiments using similar techniques that will be able to dig deeper into the range of UHE neutrino predictions.

Ground radio arrays are now being developed as dedicated UHE neutrino observatories. Giant Radio Array for Neutrino Detection (GRAND)[86] has ambitious plans for staged deployments that would reach 200,000 antennas operating in the 30–200 MHz frequency band over 200,000 square kilometers area in various mountainous locations around the world, with the mountains serving as the tau neutrino absorbers. Currently GRAND has deployed its first 300-antenna prototype array in China. The BEACON experiment[87] aims to reach the UHE flux with a compact array with low thresholds achieved through the phased array technique. BEACON is operating its first stations at the White Mountain Research Station in California. Both experiments plan to self-trigger on radio signals.

6.4. *The lunar technique*

Ground radio arrays designed for radio astronomy can be used to search for the Askaryan signature produced from neutrino interactions that occur near the surface of the moon. This is similar to the strategy used by ANITA/PUEO to view the detection medium from afar. At over 20 times the distance from which ANITA views the ice, the energy threshold for the lunar technique would be approximately 20 times higher for the same signal threshold.

The lunar technique has been pioneered by a few early experiments,[88,89] and the best limit from this technique so far was from the NuMoon project at the Westerbrook Synthesis Radio Telescope (WSRT).[90] At lower frequencies than previous Askaryan searches, the emission forms a wider cone, approaching isotropic. This allows for the ability to search over the entire 10^7 km^2 area of the moon facing the earth, instead of being restricted to the moon's limb, as has been done in searches at higher frequencies. The WSRT search was performed using eleven 25 m-diameter parabolic antennas forming two beams pointed at two halves of the moon. The search required an anticoincidence between the two beams since any signal should be observed in only one. This search resulted in the best limits on the flux of UHE neutrinos above 10^{23} eV.

There are plans for future searches for UHE neutrinos using the lunar technique. Low-Frequency Array (LOFAR)[91] is the world's largest radio array, consisting of 51 stations in eight countries with the largest number of stations being in the Netherlands. LOFAR includes low-band antennas (LBA) at 10–90 MHz and high-band antennas at 110–240 MHz. A short (one minute) test run was performed by the NuMoon at LOFAR in 2020 to establish the feasibility of a neutrino search using the lunar technique, demonstrating that backgrounds can be distinguished from simulated signals.[92] A future LOFAR search with reduced thresholds and 50 hours of data is expected to improve on ANITA's constraints on the flux of neutrinos above $10^{21.5}$ eV.

6.5. *New techniques*

There has been much development in a technique to detect neutrino-induced air showers via optical Cherenkov emission and fluorescence while viewing from an altitude so that larger volumes of the atmosphere are in view. Probe Of Extreme Multi-Messenger Astrophysics (POEMMA)[93] is a planned dual-satellite mission with two instruments, each designed to measure air showers induced in the atmosphere by cosmic rays or neutrinos above 20 EeV via the fluorescence signature. In addition, POEMMA will be sensitive to optical Cherenkov emission from air showers induced by upgoing tau neutrinos above 20 PeV. Both aims will be achieved using a wide field-of-view (45°) telescope with a focal surface that includes both near-ultraviolet and optical cameras. Both techniques will be validated in the Extreme Universe Space Observatory on a Super Pressure Balloon (EURO-SPB2) mission,[94] which will include cameras of both types and is set to fly in 2023. Both EURO-SPB2 and POEMMA will have the capability to slew the instrument in the direction of a candidate source in the event of an alert provided by gravitational wave, gamma ray, or other neutrino detectors. The Trinity experiment will also search for neutrinos via Cherenkov and fluorescence emission but from approximately 2 km altitude on Frisco Peak, UT, and due to its lower threshold, it will have a sensitivity to the astrophysical flux that overlaps that of IceCube.[95]

A new strategy to use radar for UHE neutrino detection is also under development. The energetic cascade induced by a neutrino interaction in ice leaves a plasma of electrons and ions that is dense enough to reflect a radio frequency signal. This radar signature was recently observed for the first time in a beam test in a dense medium.[96] The Radar Echo Telescope (RET) will deploy a prototype detector aiming to detect the radar signature from cosmic-ray air showers that reach ice at high altitude and then plans to follow with a full detector aimed at UHE neutrino detection.[97]

7. Conclusions

Figure 10 shows the current best constraints on the UHE neutrino flux from the different techniques described in the chapter. One can see that the ANITA balloon experiment holds the best constraints

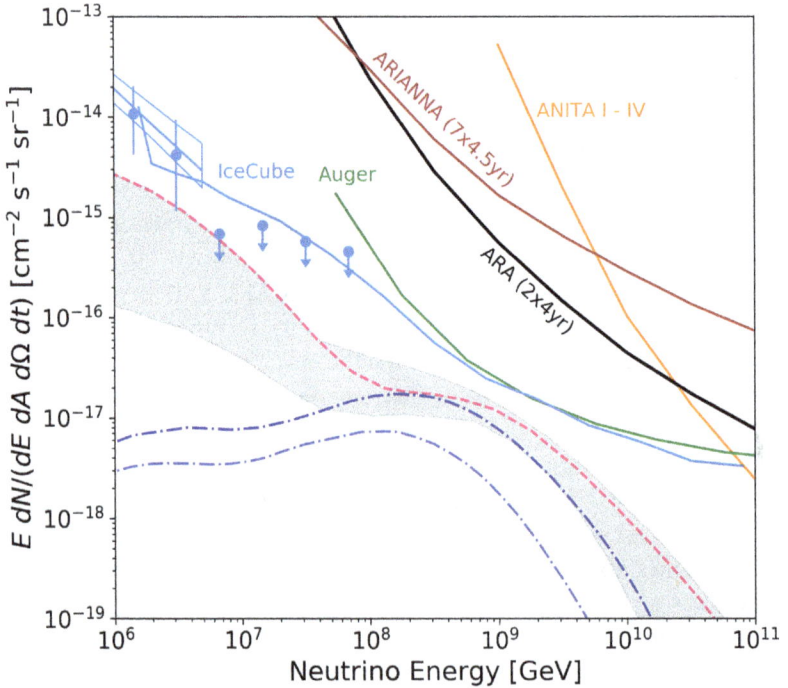

Fig. 10. Current constraints on the flux of UHE astrophysical neutrinos from experiments described in this chapter[59,63,68,98] compared to a range of models for the flux of neutrinos from the GZK process.[99–101] At the lowest energies, we can see the astrophysical flux measured by IceCube along with an estimated spectrum.[102] Extending to high energies, the figure includes the constraint set by IceCube on the UHE neutrino flux.[103]

above 10^{19} eV due to its large detection volume at a high threshold. At energies down to 10^{17} eV, the Auger ground array holds a competitive limit, and in-ice radio arrays are emerging competitors. Notably, the ARA result uses only a fraction of the existing dataset and will be expected to become competitive with an updated search. Below 10^{16} eV, the optical technique is dominant, and that is where the astrophysical neutrino spectrum has been measured by IceCube.

In the future, experiments described in this chapter will improve on these limits or reach a discovery of the UHE flux, as they expand, accumulate livetime, and improve designs, including by lowering thresholds. By increasing effective area and/or livetime, an experiment can bring their projected limits downward on the plot in Fig. 10. Lowering thresholds expands the region of sensitivity approximately to the left.

With so many techniques for UHE neutrino detection being employed, one might wonder why so many are needed. First, we have seen that different techniques complement each other in energy range and together span more than six decades in energy, which will lead to a broadband measurement of the UHE neutrino spectrum. Second, the different signatures probe different characteristics of the neutrinos with varying precision. The neutrino has a long history of surprising us and is the most elusive of the Standard Model particles. Having many different methods for scrutinizing them will be important for solidifying a discovery. The complementary of the techniques will also be essential for extracting new insights into astrophysics and fundamental physics at extreme energies that can only be revealed through measurements of UHE neutrinos, the final missing piece in the vibrant field of multi-messenger astrophysics.

Acknowledgments

The author would like to thank Constantine Sinnis and Pierre Sokolsky for the invitation to author this contribution and for their work to improve the drafts. The author would also like to thank the following people for providing their feedback on this chapter: Patrick Allison, Jaime Alvarez-Muniz, Tim Huege, Clancy James, John Krizmanic, Steven Prohira, Frank Schroeder, and Stephanie

Wissel. The author acknowledges support from the National Science Foundation under awards 1806923, 2013134, and 2012980.

References

1. M. G. Aartsen *et al.*, Evidence for high-energy extraterrestrial neutrinos at the IceCube detector, *Science.* **342**, 1242856 (2013). doi: 10.1126/science.1242856.
2. B. P. Abbott *et al.*, Observation of gravitational waves from a binary black hole merger, *Phys. Rev. Lett.* **116**(6), 061102 (2016). doi: 10.1103/PhysRevLett.116.061102.
3. M. Ackermann, M. Ahlers, L. Anchordoqui, M. Bustamante, A. Connolly, C. Deaconu, D. Grant, P. Gorham, F. Halzen, A. Karle, K. Kotera, M. Kowalski, M. A. Mostafa, K. Murase, A. Nelles, A. Olinto, A. Romero-Wolf, A. Vieregg, and S. Wissel, Astrophysics uniquely enabled by observations of high-energy cosmic neutrinos, *Bull. Am. Astron. Soc.* **51**, 185 (March, 2019).
4. M. Ackermann, M. Ahlers, L. Anchordoqui, M. Bustamante, A. Connolly, C. Deaconu, D. Grant, P. Gorham, F. Halzen, A. Karle, K. Kotera, M. Kowalski, M. A. Mostafa, K. Murase, A. Nelles, A. Olinto, A. Romero-Wolf, A. Vieregg, and S. Wissel, Fundamental physics with high-energy cosmic neutrinos, *Bull. Am. Astron. Soc.* **51**, 215 (March, 2019).
5. M. G. Aartsen *et al.*, Measurement of south pole ice transparency with the icecube led calibration system, *Nucl. Instrum. Meth. A.* **711**, 73–89 (2013). doi: 10.1016/j.nima.2013.01.054.
6. J. A. Aguilar *et al.*, Transmission of light in deep sea water at the site of the antares neutrino telescope, *Astropart. Phys.* **23**, 131–155 (2005). doi: 10.1016/j.astropartphys.2004.11.006.
7. K. Mannheim, The proton blazar, *Astron. Astrophys.* **269**, 67 (February, 1993).
8. A. Mucke and R. J. Protheroe, A Proton synchrotron blazar model for flaring in Markarian 501, *Astropart. Phys.* **15**, 121–136 (2001). doi: 10.1016/S0927-6505(00)00141-9.
9. S. Hummer, M. Ruger, F. Spanier, and W. Winter, Simplified models for photohadronic interactions in cosmic accelerators, *Astrophys. J.* **721**, 630–652 (2010). doi: 10.1088/0004-637X/721/1/630.
10. R. Engel, D. Seckel, and T. Stanev, Neutrinos from propagation of ultrahigh-energy protons, *Phys. Rev. D.* **64**, 093010 (2001). doi: 10.1103/PhysRevD.64.093010.

11. F. A. Aharonian, Proton-synchrotron radiation of large-scale jets in active galactic nuclei, *Mon. Not. R. Astron. Soc.* **332**, 215–230 (2002). ISSN 1365-2966. doi: 10.1046/j.1365-8711.2002.05292.x.

12. L. Maraschi, G. Ghisellini, and A. Celotti, A jet model for the gamma-ray emitting blazar 3C 279, *Astrophys. J. Lett.* **397**, L5–L9 (1992). doi: 10.1086/186531.

13. C. D. Dermer and R. Schlickeiser, Model for the high-energy emission from blazars, *Astrophys. J.* **416**, 458 (1993). doi: 10.1086/173251.

14. K. Greisen, End to the cosmic-ray spectrum? *Phys. Rev. Lett.* **16**, 748–750 (April, 1966). doi: 10.1103/PhysRevLett.16.748.

15. G. T. Zatsepin and V. A. Kuzmin, Upper limit of the spectrum of cosmic rays, *JETP Lett.* **4**, 78–80 (1966). [Pisma Zh. Eksp. Teor. Fiz.4,114(1966)].

16. V. S. Berezinsky and G. T. Zatsepin, Cosmic rays at ultrahigh-energies (neutrino?), *Phys. Lett.* **28B**, 423–424 (1969). doi: 10.1016/0370-2693(69)90341-4.

17. M. Ahlers, L. A. Anchordoqui, and A. M. Taylor, Ensemble fluctuations of the flux and nuclear composition of ultrahigh energy cosmic ray nuclei, *Phys. Rev. D.* **87**(2), 023004 (January, 2013). doi: 10.1103/PhysRevD.87.023004.

18. A. van Vliet, R. Alves Batista, and J. R. Hörandel, Determining the fraction of cosmic-ray protons at ultrahigh energies with cosmogenic neutrinos, *Phys. Rev.* **D100**(2), 021302 (2019). doi: 10.1103/PhysRevD.100.021302.

19. T. A. Thompson and B. C. Lacki, Upper bound on the energy of particles and their secondary neutrinos (March, 2011).

20. D. Hooper, Measuring high-energy neutrino nucleon cross-sections with future neutrino telescopes, *Phys. Rev. D.* **65**, 097303 (2002). doi: 10.1103/PhysRevD.65.097303.

21. E. Borriello, A. Cuoco, G. Mangano, G. Miele, S. Pastor, O. Pisanti, and P. D. Serpico, Disentangling neutrino-nucleon cross section and high energy neutrino flux with a km^3 neutrino telescope, *Phys. Rev. D.* **77**, 045019 (2008). doi: 10.1103/PhysRevD.77.045019.

22. M. G. Aartsen *et al.*, Measurement of the multi-tev neutrino cross section with icecube using earth absorption, *Nature* **551**(2017) 596–600 (November, 2017). doi: 10.1038/nature24459.

23. M. Bustamante and A. Connolly, Extracting the energy-dependent neutrino-nucleon cross section above 10 TeV using IceCube showers, *Phys. Rev. Lett.* **122**(4), 041101 (2019). doi: 10.1103/PhysRevLett.122.041101.

24. A. Connolly, R. S. Thorne, and D. Waters, Calculation of high energy neutrino-nucleon cross sections and uncertainties using the MSTW

parton distribution functions and implications for future experiments, *Phys. Rev. D.* **83**, 113009 (2011). doi: 10.1103/PhysRevD.83.113009.

25. R. Abbasi *et al.*, Search for a lorentz-violating sidereal signal with atmospheric neutrinos in icecube, *Phys. Rev. D.* **82**, 112003 (2010). doi: 10.1103/PhysRevD.82.112003.

26. M. G. Aartsen *et al.*, Neutrino interferometry for high-precision tests of lorentz symmetry with icecube, *Nature Phys.* **14**(9), 961–966 (2018). doi: 10.1038/s41567-018-0172-2.

27. P. W. Gorham *et al.*, Implications of ultra-high energy neutrino flux constraints for lorentz-invariance violating cosmogenic neutrinos, *Phys. Rev. D.* **86**, 103006 (2012). doi: 10.1103/PhysRevD.86.103006.

28. S. I. Dutta, M. H. Reno, and I. Sarcevic, Secondary neutrinos from tau neutrino interactions in earth, *Phys. Rev. D.* **66**, 077302 (2002). doi: 10.1103/PhysRevD.66.077302.

29. E. Bugaev, T. Montaruli, Y. Shlepin, and I. A. Sokalski, Propagation of tau neutrinos and tau leptons through the earth and their detection in underwater / ice neutrino telescopes, *Astropart. Phys.* **21**, 491–509 (2004). doi: 10.1016/j.astropartphys.2004.03.002.

30. R. Gandhi, C. Quigg, M. H. Reno, and I. Sarcevic, Ultrahigh-energy neutrino interactions, *Astropart. Phys.* **5**, 81–110 (1996). doi: 10.1016/0927-6505(96)00008-4.

31. A. Cooper-Sarkar, P. Mertsch, and S. Sarkar, The high energy neutrino cross-section in the Standard Model and its uncertainty, *JHEP.* **08**, 042 (2011). doi: 10.1007/JHEP08(2011)042.

32. G. A. Askar'yan, Excess negative charge of an electron-photon shower and its coherent radio emission, *Zh. Eksp. Teor. Fiz.* **41**, 616–618 (1961).

33. J. Alvarez-Muniz, R. A. Vazquez, and E. Zas, Calculation methods for radio pulses from high-energy showers, *Phys. Rev. D.* **62**, 063001 (2000). doi: 10.1103/PhysRevD.62.063001.

34. P. W. Gorham *et al.*, Observations of the Askaryan effect in ice, *Phys. Rev. Lett.* **99**, 171101 (2007). doi: 10.1103/PhysRevLett.99.171101.

35. D. Saltzberg, P. Gorham, D. Walz, C. Field, R. Iverson, A. Odian, G. Resch, P. Schoessow, and D. Williams, Observation of the Askaryan effect: Coherent microwave Cherenkov emission from charge asymmetry in high-energy particle cascades, *Phys. Rev. Lett.* **86**, 2802–2805 (2001). doi: 10.1103/PhysRevLett.86.2802.

36. P. Gorham, D. Saltzberg, P. Schoessow, W. Gai, J. G. Power, R. Konecny, and M. E. Conde, Radio frequency measurements of coherent transition and Cherenkov radiation: Implications for high-energy neutrino detection, *Phys. Rev. E.* **62**, 8590–8605 (2000). doi: 10.1103/PhysRevE.62.8590.

37. S. Barwick, D. Besson, P. Gorham, and D. Saltzberg, South Polar in situ radio-frequency ice attenuation, *J. Glaciol.* **51**, 231–238 (2005). doi: 10.3189/172756505781829467.
38. S. Barwick, E. Berg, D. Besson, G. Gaswint, C. Glaser, A. Hallgren, J. Hanson, S. Klein, S. Kleinfelder, L. Köpke, I. Kravchenko, R. Lahmann, U. Latif, J.-H. Nam, A. Nelles, C. Persichilli, P. Sandstrom, J. Tatar, and E. Unger, Observation of classically "forbidden" electromagnetic wave propagation and implications for neutrino detection, *J. Cosmol. Astropart. Phys.* **2018**, 055–055 (July, 2018). doi: 10.1088/1475-7516/2018/07/055.
39. D. M. Schroeder, R. G. Bingham, D. D. Blankenship, K. Christianson, O. Eisen, G. E. Flowers, N. B. Karlsson, M. R. Koutnik, J. D. Paden, M. J. Siegert *et al.*, Five decades of radioglaciology, *Ann. Glaciol.* **61** (81), 1–13 (2020). doi: 10.1017/aog.2020.11.
40. T. Barrella, S. Barwick, and D. Saltzberg, Ross ice shelf in situ radio-frequency ice attenuation, *J. Glaciol.* **57**, 61–66 (2011). doi: 10.3189/002214311795306691.
41. J. Avva *et al.*, Development toward a ground-based interferometric phased array for radio detection of high energy neutrinos, *Nucl. Instrum. Meth. A.* **869**, 46–55 (2017). doi: 10.1016/j.nima.2017.07.009.
42. P. Allison *et al.*, Design and initial performance of the askaryan radio array prototype EEV neutrino detector at the south pole, *Astropart. Phys.* **35**, 457–477 (2012). doi: 10.1016/j.astropartphys.2011.11.010.
43. S. W. Barwick *et al.*, Observation of classically "forbidden" electromagnetic wave propagation and implications for neutrino detection, *JCAP.* **07**, 055 (2018). doi: 10.1088/1475-7516/2018/07/055.
44. P. Allison *et al.*, Measurement of the real dielectric permittivity ϵ_r of glacial ice (December, 2017).
45. N. Hargreaves, The radio-frequency birefringence of polar ice, *J. Glaciol.* **21**, 301–313 (January, 1978). doi: 10.3189/s0022143000033499.
46. T. M. Jordan, D. Z. Besson, I. Kravchenko, U. Latif, B. Madison, A. Novikov, and A. Shultz, Modelling ice birefringence and oblique radio wave propagation for neutrino detection at the south pole (October, 2019).
47. K. Matsuoka, L. Wilen, S. P. Hurley, and C. F. Raymond, Effects of birefringence within ice sheets on obliquely propagating radio waves, *IEEE Trans. Geosci. Remote. Sens.* **47**(5), 1429–1443 (2009). doi: 10.1109/TGRS.2008.2005201.
48. P. W. Gorham *et al.*, The antarctic impulsive transient antenna ultra-high energy neutrino detector design, performance, and sensitivity for 2006–2007 balloon flight, *Astropart. Phys.* **32**, 10–41 (2009). doi: 10.1016/j.astropartphys.2009.05.003.

49. P. Allison *et al.*, Performance of two askaryan radio array stations and first results in the search for ultrahigh energy neutrinos, *Phys. Rev. D.* **93**(8), 082003 (2016). doi: 10.1103/PhysRevD.93.082003.

50. S. W. Barwick, E. C. Berg, D. Z. Besson, T. Duffin, J. C. Hanson, S. R. Klein, S. A. Kleinfelder, K. Ratzlaff, C. Reed, M. Roumi, T. Stezelberger, J. Tatar, J. Walker, R. Young, and L. Zou, Design and performance of the arianna hra-3 neutrino detector systems, *IEEE Trans. Nucl. Sci.* **62**, 2202–2215 (2015). ISSN 1558-1578. doi: 10.1109/TNS.2015.2468182.

51. H. Nyquist, Certain topics in telegraph transmission theory, *Trans. AIEE.* **47**, 617–644 (1928). ISSN 0096-3860. doi: 10.1109/t-aiee.1928.5055024.

52. J. M. Roberts, G. S. Varner, P. Allison, B. Fox, E. Oberla, B. Rotter, and S. Spack, LAB4D: A low power, multi-GSa/s, transient digitizer with sampling timebase trimming capabilities. *Nuclear Instruments and Methods in Physics Research Section A: Accelerators, Spectrometers, Detectors and Associated Equipment* **925**, 92–100 (2019). ISSN 0168-9002. doi: 10.1016/j.nima.2019.01.091.

53. S. A. Kleinfelder, E. Chiem, and T. Prakash, The SST fully-synchronous multi-GHz analog waveform recorder with Nyquist-rate bandwidth and flexible trigger capabilities. In *2014 IEEE Nuclear Science Symposium and Medical Imaging Conference and 21st Symposium on Room-Temperature Semiconductor X-ray and Gamma-ray Detectors* (May, 2015). doi: 10.1109/NSSMIC.2014.7431232.

54. Q. Abarr *et al.*, The Payload for Ultrahigh Energy Observations (PUEO): A White Paper (October, 2020).

55. S. Prohira *et al.*, The Radar Echo Telescope for Cosmic Rays: Pathfinder Experiment for a Next-Generation Neutrino Observatory (April, 2021).

56. C. Allen *et al.*, Status of the Radio Ice Cherenkov Experiment (RICE), *New Astron. Rev.* **42**, 319–329 (1998). doi: 10.1016/S1387-6473(98)00017-7.

57. I. Kravchenko, Status of RICE and preparations for the next generation radio neutrino experiment in Antarctica, *Nucl. Instrum. Meth. A.* **692**, 233–235 (2012). doi: 10.1016/j.nima.2012.01.032.

58. H. Landsman, AURA: Next generation neutrino detector in the South Pole, *Nucl. Phys. B Proc. Suppl.* **168**, 268–270 (2007). doi: 10.1016/j.nuclphysbps.2007.02.085.

59. P. Allison *et al.*, Constraints on the diffuse flux of ultrahigh energy neutrinos from four years of askaryan radio array data in two stations, *Phys. Rev. D.* **102**(4), 043021 (2020). doi: 10.1103/PhysRevD.102.043021.

60. P. Allison *et al.*, Design and performance of an interferometric trigger array for radio detection of high-energy neutrinos (September, 2018). doi: 10.1016/j.nima.2019.01.067.

61. A. Anker *et al.*, Targeting ultra-high energy neutrinos with the ARIANNA experiment, *Adv. Space Res.* **64**, 2595–2609 (2019). doi: 10.1016/j.asr.2019.06.016.

62. S. W. Barwick *et al.*, Radio detection of air showers with the arianna experiment on the ross ice shelf, *Astropart. Phys.* **90**, 50–68 (2017) (December, 2016). doi: 10.1016/j.astropartphys.2017.02.003.

63. A. Anker, S. W. Barwick, H. Bernhoff, D. Z. Besson, N. Bingefors, D. García-Fernández, G. Gaswint, C. Glaser, A. Hallgren, J. C. Hanson, S. R. Klein, S. A. Kleinfelder, R. Lahmann, U. Latif, J. Nam, A. Novikov, A. Nelles, M. P. Paul, C. Persichilli, I. Plaisier, T. Prakash, S. R. Shively, J. Tatar, E. Unger, S. H. Wang, and C. Welling, A search for cosmogenic neutrinos with the arianna test bed using 4.5 years of data, *JCAP03(2020)053* (September, 2019). doi: 10.1088/1475-7516/2020/03/053.

64. NASA scientific balloons https://www.nasa.gov/scientificballoons.

65. S. W. Barwick *et al.*, Constraints on cosmic neutrino fluxes from the anita experiment, *Phys. Rev. Lett.* **96**, 171101 (2006). doi: 10.1103/PhysRevLett.96.171101.

66. P. W. Gorham *et al.*, Observational constraints on the ultra-high energy cosmic neutrino flux from the second flight of the anita experiment, *Phys. Rev. D.* **82**, 022004 (2010). doi: 10.1103/PhysRevD.82.022004. [Erratum: Phys.Rev.D 85, 049901 (2012)].

67. P. W. Gorham, P. Allison, O. Banerjee, L. Batten, J. J. Beatty, K. Bechtol, K. Belov, D. Z. Besson, W. R. Binns, V. Bugaev, P. Cao, C. C. Chen, C. H. Chen, P. Chen, J. M. Clem, A. Connolly, L. Cremonesi, B. Dailey, C. Deaconu, P. F. Dowkontt, B. D. Fox, J. W. H. Gordon, C. Hast, B. Hill, S. Y. Hsu, J. J. Huang, K. Hughes, R. Hupe, M. H. Israel, K. M. Liewer, T. C. Liu, A. B. Ludwig, L. Macchiarulo, S. Matsuno, C. Miki, K. Mulrey, J. Nam, C. Naudet, R. J. Nichol, A. Novikov, E. Oberla, S. Prohira, B. F. Rauch, J. M. Roberts, A. Romero-Wolf, B. Rotter, J. W. Russell, D. Saltzberg, D. Seckel, H. Schoorlemmer, J. Shiao, S. Stafford, J. Stockham, M. Stockham, B. Strutt, M. S. Sutherland, G. S. Varner, A. G. Vieregg, S. H. Wang, and S. A. Wissel, Constraints on the diffuse high-energy neutrino flux from the third flight of anita, *Phys. Rev. D* **98**, 022001 (March, 2018). doi: 10.1103/PhysRevD.98.022001.

68. P. Gorham *et al.*, Constraints on the ultra-high energy cosmic neutrino flux from the fourth flight of ANITA, *Phys. Rev. D.* **99**(12), 122001 (2019).

69. S. Hoover, J. Nam, P. W. Gorham, E. Grashorn, P. Allison, S. W. Barwick, J. J. Beatty, K. Belov, D. Z. Besson, W. R. Binns, C. Chen, P. Chen, J. M. Clem, A. Connolly, P. F. Dowkontt, M. A. DuVernois, R. C. Field, D. Goldstein, A. G. Vieregg, C. Hast, C. L. Hebert, M. H. Israel, A. Javaid, J. Kowalski, J. G. Learned, K. M. Liewer, J. T. Link, E. Lusczek, S. Matsuno, B. C. Mercurio, C. Miki, P. Miočinović, C. J. Naudet, J. Ng, R. J. Nichol, K. Palladino, K. Reil, A. Romero-Wolf, M. Rosen, L. Ruckman, D. Saltzberg, D. Seckel, G. S. Varner, D. Walz, and F. Wu, Observation of ultra-high-energy cosmic rays with the anita balloon-borne radio interferometer, *Phys. Rev. Lett.* **105**, 151101 (May, 2010). doi: 10.1103/PhysRevLett.105.151101.

70. H. Schoorlemmer *et al.*, Energy and flux measurements of ultra-high energy cosmic rays observed during the first anita flight, *Astropart. Phys.* **77**, 32–43 (2016). doi: 10.1016/j.astropartphys.2016.01.001.

71. J. Nam, Taiwan Astroparticle Radiowave Observatory for Geo-synchrotron Emissions (TAROGE), *PoS.* **ICRC2015**, 663 (2016). doi: 10.22323/1.236.0663.

72. T. Huege and D. Besson, Radio-wave detection of ultra-high-energy neutrinos and cosmic rays, *PTEP.* **2017**(12), 12A106 (2017). doi: 10. 1093/ptep/ptx009.

73. W. D. Apel *et al.*, The wavefront of the radio signal emitted by cosmic ray air showers, *JCAP.* **09**, 025 (2014). doi: 10.1088/1475-7516/2014/ 09/025.

74. K. Belov *et al.*, Accelerator measurements of magnetically-induced radio emission from particle cascades with applications to cosmic-ray air showers, *Phys. Rev. Lett.* **116**(14), 141103 (2016). doi: 10.1103/ PhysRevLett.116.141103.

75. W. D. Apel *et al.*, Comparing LOPES measurements of air-shower radio emission with REAS 3.11 and CoREAS simulations, *Astropart. Phys.* **50–52**, 76–91 (2013). doi: 10.1016/j.astropartphys.2013.09.003.

76. S.-H. Wang *et al.*, TAROGE-M: Radio observatory on Antarctic high mountain for detecting near-horizon ultra-high energy air showers, *PoS.* **ICRC2021**, 1173 (2021). doi: 10.22323/1.395.1173.

77. R. Monroe, A. Romero Wolf, G. Hallinan, A. Nelles, M. Eastwood, M. Anderson, L. D'Addario, J. Kocz, Y. Wang, D. Cody *et al.*, Self-triggered radio detection and identification of cosmic air showers with the ovro-lwa, *Nucl. Instrum. Methods Phys. Res. Sec. A.* **953**, 163086 (February, 2020). ISSN 0168-9002. doi: 10.1016/j.nima.2019.163086.

78. D. Charrier *et al.*, Autonomous radio detection of air showers with the TREND50 antenna array, *Astropart. Phys.* **110**, 15–29 (2019). doi: 10.1016/j.astropartphys.2019.03.002.

79. P. W. Gorham *et al.*, Characteristics of four upward-pointing cosmic-ray-like events observed with ANITA, *Phys. Rev. Lett.* **117**(7), 071101 (2016). doi: 10.1103/PhysRevLett.117.071101.

80. P. W. Gorham *et al.*, Observation of an unusual upward-going cosmic-ray-like event in the third flight of ANITA, *Phys. Rev. Lett.* **121**(16), 161102 (2018). doi: 10.1103/PhysRevLett.121.161102.

81. P. W. Gorham *et al.*, Unusual near-horizon cosmic-ray-like events observed by ANITA-IV, *Phys. Rev. Lett.* **126**(7), 071103 (2021). doi: 10.1103/PhysRevLett.126.071103.

82. A. Romero-Wolf *et al.*, Comprehensive analysis of anomalous ANITA events disfavors a diffuse tau-neutrino flux origin, *Phys. Rev. D.* **99**(6), 063011 (2019). doi: 10.1103/PhysRevD.99.063011.

83. M. G. Aartsen *et al.*, IceCube-Gen2: The window to the extreme Universe, *J. Phys. G.* **48**(6), 060501 (2021). doi: 10.1088/1361-6471/abbd48.

84. J. A. Aguilar *et al.*, Design and sensitivity of the radio neutrino observatory in greenland (rno-g), *JINST.* **16**(03), P03025 (2021). doi: 10.1088/1748-0221/16/03/P03025.

85. A. Aab *et al.*, The Pierre Auger Observatory upgrade — preliminary design Report (Apr., 2016).

86. J. Álvarez-Muñiz *et al.*, The Giant Radio Array for Neutrino Detection (GRAND): Science and design, *Sci. China Phys. Mech. Astron.* **63**(1), 219501 (2020). doi: 10.1007/s11433-018-9385-7.

87. A. Zeolla *et al.*, Modeling and validating RF-only interferometric triggering with cosmic rays for BEACON, *PoS.* **ICRC2021**, 1072 (2021). doi: 10.22323/1.395.1072.

88. P. W. Gorham, C. L. Hebert, K. M. Liewer, C. J. Naudet, D. Saltzberg, and D. Williams, Experimental limit on the cosmic diffuse ultrahigh-energy neutrino flux, *Phys. Rev. Lett.* **93**, 041101 (2004). doi: 10.1103/PhysRevLett.93.041101.

89. T. R. Jaeger, R. L. Mutel, and K. G. Gayley, Project RESUN, a radio EVLA search for UHE neutrinos, *Astropart. Phys.* **34**, 293–303 (2010). doi: 10.1016/j.astropartphys.2010.08.008.

90. O. Scholten, S. Buitink, H. Falcke, C. W. James, M. Mevius, K. Singh, B. Stappers, and S. ter Veen, Ultra-high-energy cosmic ray and neutrino detection using the Moon, *Nucl. Phys. B Proc. Suppl.* **212–213**, 128–133 (2011). doi: 10.1016/j.nuclphysbps.2011.03.018.

91. M. P. van Haarlem, M. W. Wise, A. W. Gunst, G. Heald, J. P. McKean, J. W. T. Hessels, A. G. de Bruyn, R. Nijboer, J. Swinbank, R. Fallows, *et al.*, Lofar: The low-frequency array, *Astron. Astrophys.* **556**, A2 (July, 2013). ISSN 1432-0746. doi: 10.1051/0004-6361/201220873.

92. G. K. Krampah *et al.*, The NuMoon experiment: Lunar detection of cosmic rays and neutrinos with LOFAR, *PoS*. **ICRC2021**, 1148 (2021). doi: 10.22323/1.395.1148.

93. A. V. Olinto *et al.*, POEMMA: Probe of extreme multi-messenger astrophysics, *PoS*. **ICRC2017**, 542 (2018). doi: 10.22323/1.301. 0542.

94. J. Eser, A. V. Olinto, and L. Wiencke, Science and mission status of EUSO-SPB2, *PoS*. **ICRC2021**, 404 (2021). doi: 10.22323/1.395.0404.

95. A. Wang, C. Lin, N. Otte, M. Doro, E. Gazda, I. Taboada, A. Brown, and M. Bagheri, Trinity's sensitivity to isotropic and point-source neutrinos, *PoS*. **ICRC2021**, 1234 (2021). 37th International Cosmic Ray Conference (ICRC2021).

96. S. Prohira *et al.*, Observation of radar echoes from high-energy particle cascades, *Phys. Rev. Lett.* **124**(9), 091101 (2020). doi: 10.1103/ PhysRevLett.124.091101.

97. S. Prohira *et al.*, Toward high energy neutrino detection with the radar echo telescope for cosmic rays (RET-CR), *PoS*. **ICRC2021**, 1082 (2021). doi: 10.22323/1.395.1082.

98. A. Aab *et al.*, Probing the radio emission from air showers with polarization measurements, *Phys. Rev. D.* **89**(5), 052002 (2014). doi: 10.1103/PhysRevD.89.052002.

99. K. Kotera, D. Allard, and A. Olinto, Cosmogenic neutrinos: Parameter space and detectabilty from pev to zev, *J. Cosmol. Astropart. Phys.* **2010**(10), 013–013 (October, 2010). ISSN 1475-7516. doi: 10.1088/ 1475-7516/2010/10/013.

100. M. Ahlers and F. Halzen, Minimal cosmogenic neutrinos, *Phys. Rev. D.* **86**(8) (October, 2012). ISSN 1550-2368. doi: 10.1103/physrevd.86. 083010.

101. A. Olinto, K. Kotera, and D. Allard, Ultrahigh energy cosmic rays and neutrinos, *Nucl. Phys. B. Proc. Suppl.* **217**(1), 231–236 (August, 2011). ISSN 0920-5632. doi: 10.1016/j.nuclphysbps.2011.04.109.

102. M. G. Aartsen *et al.*, Observation and characterization of a cosmic Muon neutrino flux from the Northern Hemisphere using six years of IceCube data, *Astrophys. J.* **833**(1), 3 (2016). doi: 10.3847/ 0004-637X/833/1/3.

103. M. G. Aartsen *et al.*, Neutrino emission from the direction of the blazar TXS 0506+056 prior to the IceCube-170922A alert, *Science*. **361**(6398), 147–151 (2018). doi: 10.1126/science.aat2890.

Chapter 7

Very-High-Energy Gamma Rays

Felix Aharonian

Dublin Institute for Advanced Studies,
10 Burlington Road, Dublin 4, Ireland
Max Planck Institute for Nuclear Physics,
Saupfercheckweg 1, Heidelberg 69117, Germany
felix.aharonian@dias.ie

Over the last two decades, the impressive achievements of gamma-ray astronomy have raised the field to the level of a nominal observational discipline. The discovery of thousands of GeV and TeV gamma-ray emitters, representing over a dozen galactic and extragalactic source populations, revealed that cosmic-ray factories are distributed throughout the Universe in a wide diversity of forms and scales — from compact relativistic objects like neutron stars and stellar-mass black holes to large-scale cosmological structures like galaxy clusters. These days, we are witnessing a new revolution in gamma-ray astronomy, this time in the PeV band. With the arrival of the North and South Cherenkov Telescope Arrays (CTA), the Southern Wide-field Gamma-ray Observatory (SWGO), and the recently completed Large High Altitude Air Shower Observatory (LHAASO), one can anticipate groundbreaking discoveries in several research areas that would dramatically change the current concepts and paradigms of the most energetic and extreme phenomena in the non-thermal Universe. In this section, the major objectives and motivations of ground-based gamma ray observations for the coming years are discussed in the context of the origin of galactic and extragalactic cosmic rays.

1. Introduction

Extraterrestrial gamma rays carry crucial information about the most energetic and extreme phenomena in the Universe. The following three distinct features make them unique cosmic messengers: (i) copious production in both hadronic and electromagnetic interactions; (ii) free propagation over a substantial fraction of the Universe; (iii) effective detection by space-borne and ground-based instruments. Being a part of astroparticle physics, it remains a discipline in its own right addressing a broad range of high energy processes of particle acceleration, propagation, and radiation on all astronomical scales: from compact objects like pulsars (neutron stars) and microquasars (accreting stellar-mass black holes) to giant jets and lobes of radiogalaxies and galaxy clusters.[1]

The energy range covered by gamma-ray astronomy spans from low (LE; $E_\gamma \geq 0.1$ MeV) to extremely high (EHE, $E_\gamma \geq 100$ EeV) energies. The lower bound historically has been linked to the region of nuclear gamma-ray lines. From the opposite side, the upper bound is determined by the highest energy particles observed in cosmic rays (CRs); the formation of the GZK cutoff implies the presence of such energetic photons in the intergalactic medium. Because of different detection techniques and approaches applied to different energy bands, this enormously broad segment of the cosmic electromagnetic spectrum has been explored at different depths.

The past of gamma-ray astronomy was ambiguous and controversial. For decades, the very term "astronomy" has been used with reservations. However, the recent revolutions in the high-energy (HE; $E_\gamma \geq 0.1$ GeV) and very-high-energy (VHE; $E_\gamma \geq 0.1$ TeV) bands over the last two decades dramatically changed the status of the field. The detections of thousands of GeV and TeV γ-ray emitters representing more than a dozen galactic and extragalactic source populations[2–4] elevated the cosmic gamma-ray studies to the level of a modern astronomical discipline. These days, we are witnessing a new revolution, this time in the ultra-high-energy (UHE; $E_\gamma \geq 0.1$ PeV) domain. The recent fascinating discoveries of UHE γ-ray sources,[5–7] with energy spectra extending out of 1 PeV,[7] opened a new (UHE) window in the cosmic electromagnetic spectrum.

Except for the processes related to the decays of long-lived radioactive nuclei (leftovers from powerful cosmic events like supernovae explosions) and the dark matter, cosmic gamma rays are produced in interactions of nonthermal (accelerated) particles with the surrounding gas, radiation and magnetic fields. Gamma rays as products of the decay or annihilation of dark matter have been often invoked to explain different CR "anomalies". Although, so far, we do not have conclusive evidence of dark matter traces in gamma-ray data, these "exotic" but not entirely unrealistic interpretations give a particular cosmological and particle physics flavor to gamma-ray studies. In the following, the major objectives and motivations of VHE and UHE gamma-ray astronomy are discussed in the context of the origin of galactic and extragalactic CRs. Actually, the subject covers a broader range of astrophysical themes, particularly linked to supernova remnants, star formation, magnetospheres of pulsars and black holes, physics of relativistic outflows — pulsar winds, AGN jets, GRBs, etc.

2. Solving the "Century-Old Mystery"

In spite of the outstanding achievements in measurements of CRs and their interpretation within the framework of the current paradigm of galactic and extragalactic CRs, many aspects of CR studies remain unresolved. Therefore, the origin of CRs as a whole is still considered by many as a "century-old mystery". The main reason for such a pessimistic assessment is that we do not know yet which astronomical sources contribute to CR fluxes measured directly in the Earth's vicinity (local "CR fog"). The latter is supplied by many sources of different ages, time history, energy budget, acceleration spectra, etc. Because of deflections of charged particles in random interstellar and intergalactic magnetic fields, their original directions pointing to their production sites are lost. This makes the attempts of revealing the origin of local "CR fog" based on the "smell" (spectrum and chemical composition) of the "soup" of relativistic particles "cooked" over cosmological timescales a particular challenge.

The identification of contributors to the local CR flux with known astronomical source populations is one of the highest priorities of the field. However, the motivations of CR studies cannot be reduced merely to this objective. The term "cosmic rays" has indeed broader implications. After matter, radiation, and magnetic fields, the relativistic particles constitute the fourth substance of the observable Universe. Exploring physical conditions and processes in these CR factories, independent of their relative contributions to the "CR fog", is a fundamental issue in its own right. The high energy photons and neutrinos — the only stable and neutral secondary products of CR interactions with the ambient gas, radiation, and magnetic fields — are the two direct messengers of CR factories.

3.   Cosmic Messengers

Gamma rays of hadronic origin are produced through decays of π^0 mesons, the secondary products of inelastic pp and $p\gamma$ (photomeson) interactions. In standard scenarios, at GeV/TeV energies, the interactions with gas are more important, whereas in the EHE regime, the photomeson processes dominate both inside the sources and in the surrounding extended regions (e.g., the intergalactic medium). UHE gamma rays can be effectively produced in the interactions with gas (e.g., in SNRs) and radiation (e.g., in microquasars).

The inverse Compton (IC) scattering and synchrotron radiation provide the primary information channels about the directly accelerated electrons. The first mechanism results in the effective gamma-ray production from GeV up to PeV energies. The synchrotron radiation of VHE/UHE electrons is released typically in the X-rays band, but in the case of the so-called *extreme electron accelerators* like the Crab Nebula, the spectrum could extend up to the so-called synchrotron burn-off limit at $E_\gamma \approx 9/4\alpha_f^{-1}m_e c^2 \simeq 150$ MeV (in the source frame). Usually, in standard astrophysical environments, the proton synchrotron radiation can be safely ignored. However, in some scenarios, e.g., at the acceleration of EHE protons in a relativistic jet close to a supermassive black hole, the proton synchrotron radiation extending to $E_\gamma \approx 9/4\alpha_f^{-1}m_p c^2\delta \simeq 300\delta$ GeV could be used as a distinct signature of *extreme proton accelerators* (δ is the jet's Doppler factor).

As messengers of high-energy phenomena, neutrinos are similar to gamma rays, but there are also principal differences. Neutrinos are produced only in hadronic interactions and interact weakly with the surrounding matter and magnetic and radiation fields. Thus, neutrinos carry unambiguous information about the hadronic component of

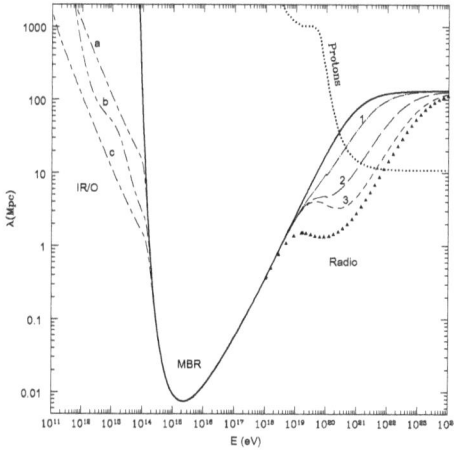

Fig. 1. Mean free path of gamma rays in the intergalactic medium.[8] Between 100 TeV and 10 EeV, gamma rays interact predominantly with 2.7 K MBR (sold line). Below 100 TeV, gamma rays interact more effectively with the infrared and optical photons of the extragalactic background light (EBL). Large uncertainties in these energy bands reflect the uncertainties of the EBL models. Curves a, b, and c correspond to three different EBL models. Above 10 EeV, gamma-rays interact with radio photons. Curves 1, 2, and 3 are calculated for three different positions of the low-frequency cutoff in the spectrum of the extragalactic radio background: at 5, 2, and 1 MHz, respectively. The triangles show the lower limit on the mean free path of the highest energy gamma rays, assuming that the observed radio background is entirely due to the extragalactic source populations. The thick dotted line shows the mean free path of protons (for details see Ref. [8]).

accelerated particles. Unlike fragile gamma rays, neutrinos can escape the hidden (opaque for gamma rays) sources. They freely propagate over cosmological distances, whereas the significant part of the Universe is not transparent for VHE and especially UHE gamma rays (see Fig. 1). This gives neutrinos a particular uniqueness and value. Unfortunately, the limited performance of the current high-energy neutrino detectors does not allow full implementation of these nice features of neutrinos. Recently, the IceCube collaboration reported the detection of dozens of high-energy neutrinos of extraterrestrial origin. Although the directions of some of the detected neutrinos point to several interesting objects in the sky, such as the blazar TXS 0506+056, their origin remains largely unknown. The upgrade of IceCube and the completion of ongoing new projects are aiming at the discovery of the first neutrino sources.

Finally, potentially great information about nonthermal hadronic processes is contained in the radiation component emitted not

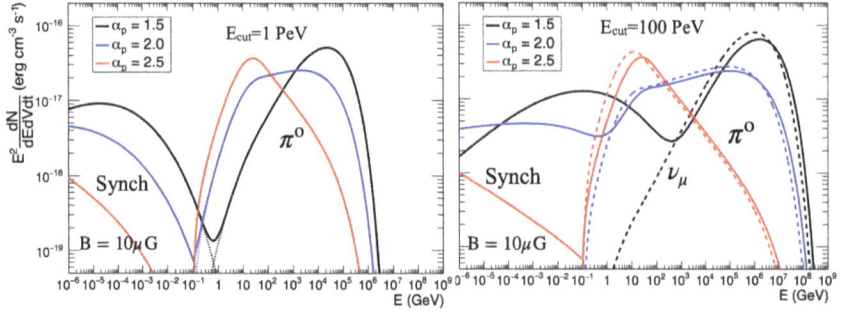

Fig. 2.   Emissivity of broadband radiation induced by interactions of relativistic protons with the ambient gas.[9] The spectra of gamma rays from the decay of π^0-mesons and the synchrotron radiation of the secondary electrons from the $\pi^\pm$-mesons are calculated for the power-law proton spectrum with exponential cutoff: $dN/dE \propto E^{-\alpha_p} \exp{-(E/E_{cut})}$. The power-law indices α_p and the cutoff energies E_{cut} are shown. The spectra of protons are normalized to the energy density above 100 GeV, $w_p = 1\,\mathrm{erg/cm^3}$. The gas number density $n = 1\,\mathrm{cm^{-3}}$ and the magnetic field $B = 10\mu G$. In the right figure (with $E_{cut} = 100\,\mathrm{PeV}$), the muon neutrino spectra are also shown (dashed lines).

directly but though the secondary electrons produced in the chain of reactions triggered by pp and $p\gamma$ interactions: $pp \to \pi^\pm \to e^\pm$ and $p\gamma \to \pi^\pm \to e^\pm$. In UHE and EHE sources, the synchrotron radiation of secondary electrons extending to X-rays and beyond is complementary to primary gamma rays and neutrinos.[1,9] In Fig. 2, the emissivities of the broadband radiation induced by pp interactions in a region with gas density $n = 1\,\mathrm{cm^{-3}}$ and magnetic field $B = 10\mu G$ are shown. It consists of the prompt gamma-ray and secondary synchrotron radiation components, calculated for different "power-law with exponential cutoff" proton spectra. For $E_{cut} = 100\,\mathrm{PeV}$, the spectra of high-energy neutrinos are also shown (dashed curves). One can see that the ratio of secondary 10 keV X-ray flux to 100 TeV gamma-ray flux varies between 0.1 and 1.

The secondary synchrotron radiation is delayed (relative to the primary gamma rays and neutrinos) by the synchrotron cooling time, $t_{\mathrm{synch}} \approx 1.5(B/1\mathrm{mG})^{-3/2}(\epsilon/1\mathrm{keV})^{-1/2}$ yr, where B is the magnetic field and ϵ is the energy of the emitted synchrotron photon. For various nonthermal sources, the cooling time of synchrotron radiation extending to X- and gamma-rays typically is shorter than other characteristic timescales. Thus, the secondary high-energy synchrotron emission is expected to be well synchronised (in space and time) with the prompt (primary) π-decay gamma rays and neutrinos. This makes the secondary synchrotron channel complementary and, in some scenarios, even advantageous compared to gamma rays and neutrinos. Namely, while the detectability of PeV gamma-ray sources is limited by objects within the Milky Way (see Fig. 1), the secondary synchrotron X-rays can be detected, like neutrinos, from cosmologically distant objects. Another advantage of secondary synchrotron radiation is the X-ray detectors' superior (10 arcseconds or better) angular resolution. With the arrival of eROSITA,[10] this potential can be facilitated for localization and identification of the galactic PeVatrons and search for UHE and EHE sources beyond the Milky Way.

4. Galactic Sources

4.1. *Supernova remnants*

The current paradigm of galactic cosmic rays (GCRs) assumes that SNe or supernova remnants (SNRs) — the results of these gigantic explosions — are responsible for GCRs up to at least 1 PeV. Over decades, this conviction has been based on sound phenomenological arguments and theoretical meditations. As early as in the 1930s, W. Baade and F. Zwicky envisaged a link between SN explosions and CRs based on the comparable energetics characterizing the SN explosions and CRs. Later, the Diffusive Shock Acceleration (DSA) theory emerged as a viable mechanism for CR acceleration in young SNRs (see e.g., Ref. [11]). The discovery of TeV gamma radiation from more than a dozen young SNRs is a remarkable achievement confirming the early theoretical predictions.[2-4] However, so far we do not have certain answers to the following two principal questions:

(i) On the origin of radiation of SNRs: "hadronic or leptonic?"
(ii) Whether the radiation continues effectively out to 100 TeV?

The current data do not allow us to distinguish between the hadronic and leptonic components of radiation even for the best-studied gamma-ray emitting SNRs, e.g., RXJ1713.7-3946. On the other hand, the detailed hydrodynamic treatment of the nonlinear

DSA applied to young SNRs provides conclusive predictions regarding the gamma-ray morphology for both leptonic and hadronic components of radiation in different energy bands (see e.g., Ref. [12]). One can predict that detailed study of the energy-dependent morphology over 3.5 decades of energy, from 30 GeV to 100 TeV by CTA, based on huge photon statistics, adequate angular (better than three arcminutes) and energy (better than 15%) resolutions, will resolve the radiation into the leptonic and hadronic components. Multiwavelength (MWL) data, especially on the gas distribution inside and outside SNRs, are required for derivation of the spatial distribution of CRs from the CTA data both within the remnant and outside. This information, based on observations of a large sample of SNRs, will help to understand the escape of accelerated particles at different epochs of evolution of SNRs. The study of the process of particle injection *in action* through gamma-ray observations would be a key contribution to the "SNR paradigm" of GCRs.

The clarification of the second question is of prime importance for the origin of the highest (PeV) energy GCRs. The unexpectedly steep (compared to the DSA's "benchmark" E^{-2} spectrum) energy distributions of all young SNRs measured at TeV energies (see Fig. 3) raised doubts in the CR community regarding the role of SNRs as the major suppliers of GCRs. It is premature, however, to draw a verdict. The recent studies of CR acceleration by strong shocks revealed new (not explored before) features of DSA allowing steep CR spectra. Moreover, the measured CR spectra combined with the realistic propagation scenarios in the Galactic plane (GP) also require steeper, close to $E^{-(2.3-2.5)}$, initial (acceleration) spectra. This could explain the observed steep TeV gamma-ray spectra from young SNRs such as Cas A and Tycho. Under the condition of amplification of the upstream magnetic field Ref. [33], the steep proton spectrum could continue to PeV energies implying small but not negligible UHE gamma-ray fluxes. For Cas A, the gamma-ray flux at 100 TeV could be as small as 3×10^{-14} erg cm^{-2}s^{-1}, which is significantly below the minimum detectable flux by CTA but still accessible by LHAASO after several years of observations.

LHAASO observations of young SNRs, in particular two historical remnants Cas A and Tycho, in the UHE regime should give a definite answer to the principal question whether the steep spectra have intrinsic origin and continue beyond 100 TeV or they are the result

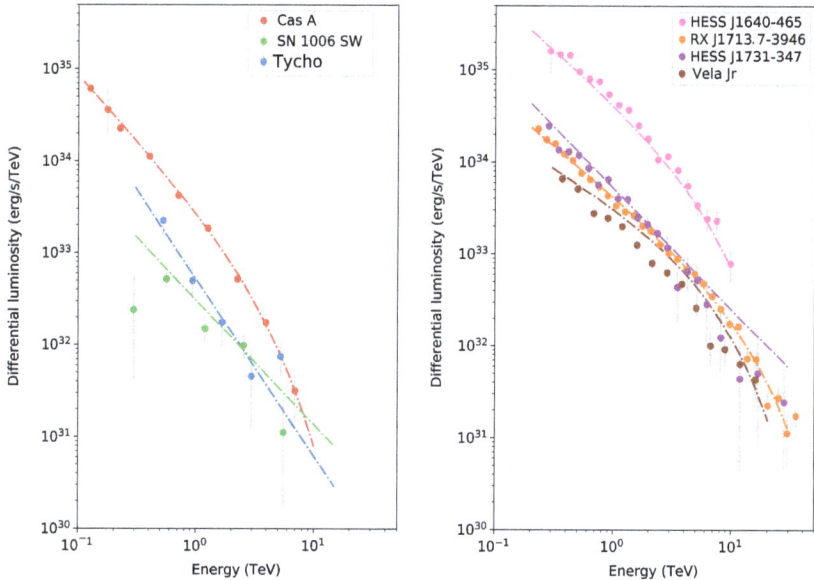

Fig. 3. Differential gamma-ray luminosities of historical SNRs Cas A, Tycho, SN1006 (left panel) and some other young TeV emitting SNRs (right panel).

of combination of hard power-law distribution imposed by "early" cutoffs in the proton spectrum well below 1 PeV.

The first outcome would confirm the recent theoretical studies that DSA, under realistic assumptions and careful treatment of the problem, does allow steep ($E^{-2.3}$ or steeper) acceleration spectra,[13–15] which actually is preferable for any reasonable diffusion coefficient in the interstellar medium (ISM) responsible for the formation of the locally observed ($\propto E^{-2.8}$) CR proton spectrum.

The second outcome could formally be interpreted as an argument against SNRs performing as PeVatrons. However, such a statement would be correct only for the present epoch; one cannot exclude that the production of PeV particles occurred during the first years after the SN explosion. Remarkably, for Cas A, a benchmark SNR-PeVatron candidate, this statement can be extended to any epoch of evolution. Assuming that PeV protons already have left the remnant, we should expect it to be detectable by LHAASO UHE gamma-ray flux from the dense ($\sim$10 cm^{-3}) environment surrounding Cas A. Even moving ballistically, protons cannot recede more than $\sim$100 pc from Cas A, corresponding to the angular size $\sim$1.5°. Non detection of UHE gamma rays, would imply a strong upper limit on the total energy in PeV protons, $W_p \leq 10^{46}$ erg, excluding the Cas A as a potential PeVatron.

The number of reported young SNRs emitting TeV gamma rays is limited, which does not allow unbiased conclusions regarding the entire SNR population. The sensitivities of the current detectors limit the detection of the hadronic component of gamma radiation from young SNRs located within a few kpc. The increase of the "detectability distance" by an order of magnitude would make visible all young SNRs in the Milky Way. Because of reduction of gamma-ray fluxes, $F_\gamma \propto 1/R^2$, the increase in the distance by an order of magnitude formally demands an improvement in the sensitivity for point-like sources by two orders of magnitude. However, in the case of young SNRs, an order of magnitude improvement in the sensitivity would be sufficient. Indeed, the sensitivity of IACT arrays for extended sources is proportional to the angular size of the source, $F_{min} \propto \Psi$, as long as Ψ is larger than the angular resolution of the IACT arrays (less than $0.1°$). On the other hand, the angular sizes of typical young SNR scales with the distance approximately as $\Psi \sim 0.5°(d/1 \text{ kpc})^{-1}$. Thus, the anticipated improvement in the sensitivity of CTA for point-like sources by an order of magnitude, compared to the current IACT arrays, will dramatically increase the number of potentially detectable young SNRs.

Moreover, the acceleration of protons in SNRs to multi-TeV and PeV energies can also be studied in an indirect way, using the gamma-ray "echos" of faded-away accelerators. In the presence of nearby dense clouds, particles which already left their accelerators, interacting with these clouds produce gamma rays potentially detectable even thousands of years after the epoch of active acceleration. Before being fully diffused away and integrated into the "sea" of galactic CRs, these particles produce gamma rays, the spectrum of which can significantly differ from both the radiation of the SNR shell and the diffuse galactic gamma-ray emission.[16] This scenario dramatically increases the number of SNRs detectable in gamma rays. Particularly promising are the middle-aged SNRs. Interestingly, some of the reported UHE sources,[5–7] e.g., LHAASO J1908+0621 and LHAASO J2226+6057, having nearby middle-aged (10–20 kyr) supernova remnants SNR G40.5-0.5 and SNR G106.3+2.7, respectively, could be interpreted within this scenario. The spectral energy distribution of

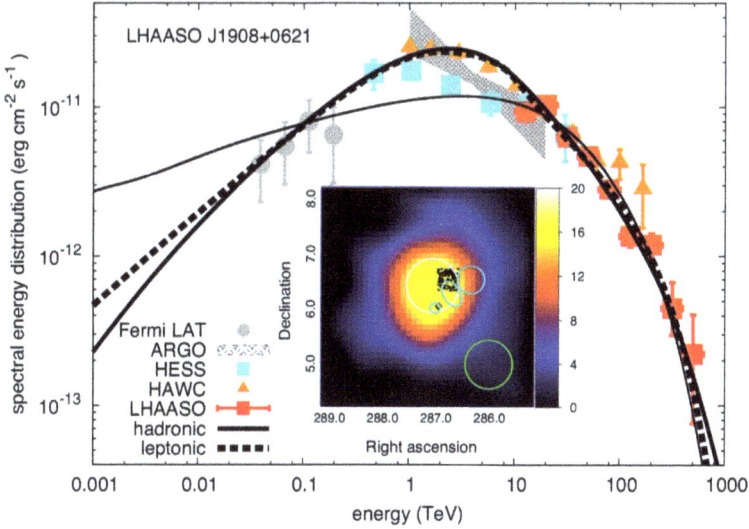

Fig. 4. Spectral energy distribution of LHAASO J1908+0621 and the potential counterparts reported by different groups.[7] The inset shows the significance map of UHE gamma rays. The dashed curve shows the leptonic model of radiation. The solid curves correspond to the following two hadronic models: (i) power-law spectrum of parent protons with exponential cutoff: $N(E) \propto E^{-1.85} \exp -E/0.38\,\text{PeV}$ (thin solid); (ii) broken power-law spectrum with an exponential cutoff of parent protons, with indices 1.2 and 2.7 below and above 25 TeV, respectively, and a cutoff energy at 1.3 PeV (thick solid curve).

LHAASO J1908+0621 is shown in Fig. 4 together with measurements at lower energies reported by different gamma-ray groups. The spectrum extends up to 0.45 PeV, which implies that the energy of parent particles, either protons or electrons, should exceed 1 PeV. Formally, the spectrum can be explained by both the hadronic pp interactions or IC scattering of electrons. In the hadronic scenario, the proton spectrum should be hard with a power-law index close to 1.75 and an exponential cutoff at 0.380 PeV.[7] Since the SNR is old, it is unlikely that gamma rays are produced inside the SNR. A more realistic site for gamma-ray production are massive clouds within the 100 pc vicinity of the remnant. This can also naturally explain the very hard spectrum of gamma rays caused by the fact that the low-energy protons propagating slower, did not yet reach the clouds.[16]

The future observations at sub-TeV and multi-TeV energies using the potential of CTA deep morphological studies and broadband spectrometry up to 100 TeV and beyond using the superior sensitivity of LHAASO should provide decisive tests of this scenario applied to young and middle-aged SNRs. The continuous monitoring of the sky by the Southern Wide-field Gamma-ray Observatory (SWGO) and LHAASO as well the deep survey of the Galactic plane by CTA could lead to serendipitous finding of tens if not hundreds of middle-aged SNRs through their gamma-ray "echoes". This should allow compelling population studies — a crucial condition for the proof of the SNR origin of GCRs.

4.2. *Young massive clusters*

The collision of powerful stellar winds and the multiple shocks initiated by supernovae explosions in clusters of young stars have long been suggested as possible alternatives to SNRs as CR factories contributing to Galactic CRs, particularly at PeV energies.[17,18] Massive stars formed by the collapse of giant molecular clouds usually form compact clusters and remain close to each other throughout their lives. Thus, unlike SNRs, which effectively accelerate particles over thousands of years or less, young massive clusters (YMCs) continuously inject CRs into the Galactic disk over several million years, with a total input up to $W_{CR} \sim 10^{52}$ erg, two orders of magnitude more than the typical contribution of an individual SNR. The recent reports of the detection of giant (up to $L \sim 100$ pc) gamma-ray structures around several YMCs with distinct spectral and morphological features (see, e.g., Ref. [19]) initiated a renewed interest in nonthermal processes in these complex environments, in particular, in the context of the origin of Galactic CRs.[20]

Figure 5 demonstrates a remarkable similarity of the energy and radial distributions of multi-TeV cosmic rays extracted from gamma-ray observations toward the Galactic center (GC) and two prominent clusters of young massive stars, Cygnus OB2 and Westerlund 1. This resemblance could be interpreted as a hint that CRs responsible for the diffuse multi-TeV gamma-ray emission from the central molecular zone (CMZ) are accelerated in one or all three ultra-compact stellar clusters, nuclear, arches, and quintuplet, located within the inner part of the GC. Note that the gamma-ray morphologies of these extended sources are rather irregular. However, the distribution of parent protons, derived from comparing the spatial distributions

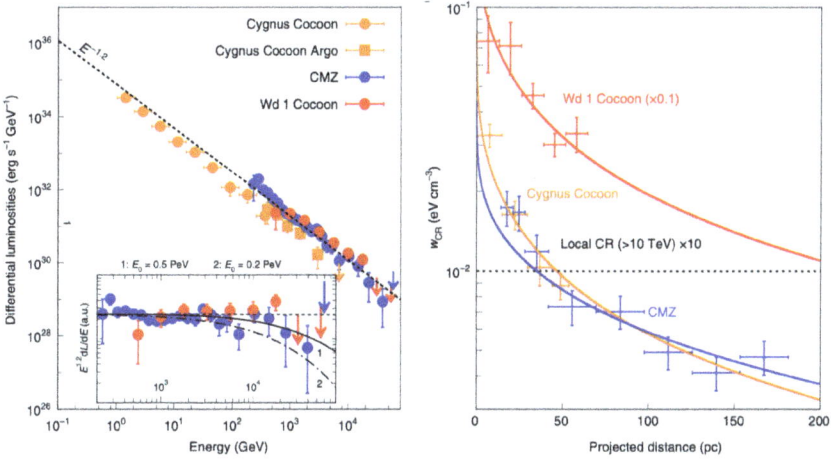

Fig. 5. Left: The differential gamma-ray luminosities of extended regions around the stellar clusters Cygnus OB2 (Cygnus Cocoon) and Westerlund 1 (Wd1 Cocoon) and in the CMZ of GC, assuming that the CMZ is powered by CRs accelerated in the arches, quintuplet, and nuclear clusters. The inset shows the differential luminosities of the CMZ and Wd1 multiplied by $E^{1.2}$. For protons, a power-law distribution (index = 2.3) imposed by an exponential cutoff at different energies is assumed. Right: The radial distributions of CR protons above 10 TeV in Cygnus Cocoon, Wd1 Cocoon, and CMZ derived from the gamma-ray and gas maps of these regions.

of gamma rays and the gas, appears with a characteristic $1/r$–type radial dependence indicating the continuous regime of particle injection into the circumstellar medium. The lack of a noticeable brightening of the gamma-ray images toward the stellar clusters makes the IC origin of gamma-ray radiation unlikely. The hard, $E^{-2.3}$–type power-law energy spectra of parent protons perfectly agree with the "source" function required by Galactic CR models (see, e.g., Ref. [21]). Remarkably, the recent reports of detection of UHE gamma rays by HAWC[22] collaboration with a spectrum extending to 1.4 PeV from Cygnus Cocoon containing Cygnus OB2, a host of some of the most massive and luminous stars in the Milky Way, implies that these objects initiate, in one way or another, operation of a super-PeVatron(s) boosting the energy of protons beyond 10 PeV. This is a result of extraordinary importance pointing to the likely source population responsible for the most problematic part of the CR spectrum between the "knee" and the "ankle".

4.3.   *Localizing PeVatrons*

The PeV photons also represent a special interest for another reason. For a typical scattering environment, the propagation of the parent $\geq 10\,$PeV protons, over large (up to tens of parsecs) initial distances, proceeds in the (quasi)ballistic regime, unless the diffusion coefficient in the Cygnus Cocoon is dramatically (by more than two orders of magnitude) suppressed compared to the standard value in the ISM. The spectral and angular distributions of gamma rays produced at the propagation of parent particles in the pre-diffusive stage provide unique information about (1) the initial (undistorted by energy-dependent diffusion) energy distribution of CRs and (2) location(s) of their accelerators. Although the radiation is produced in a large volume far from the accelerator, the angular size of the gamma-ray source decreases with energy. In the regime of ballistic (rectilinear) motion, they simply mimic the image of the accelerator. This effect, demonstrated in Fig. 6, offers a unique tool for localization of the currently operating accelerator. Based on the reported results obtained with the partly completed LHAASO,[7] in the coming years,

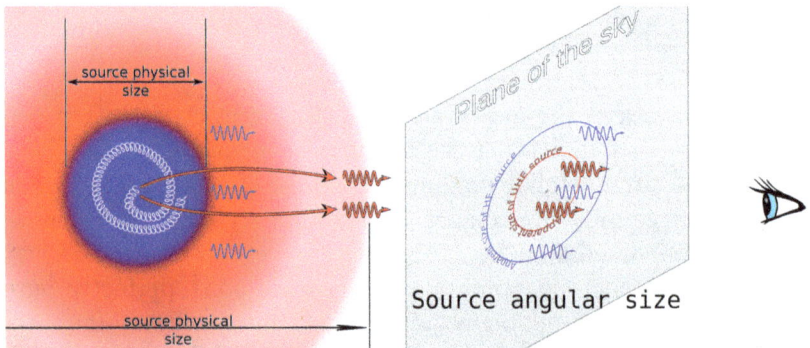

Fig. 6.   Sketch explaining the relation between the physical size of the emitter and the apparent angular size of the gamma-ray image in the diffusive-to-ballistic transition regime of propagation of parent charged particles (protons or electrons). At low energies (blue), particles propagate diffusively; therefore, for the observer, the angular size of the relevant gamma-ray source is determined as R/d, where R is the radius of the region occupied by low-energy particles and d is the distance to the source. At higher energies (red), particles move (quasi)ballistically and occupy more volume. Correspondingly, higher-energy photons are produced in a bigger volume than the low-energy gamma rays, but their apparent angular size is getting smaller (courtesy of Dmitry Khangulyan).

we anticipate more than several hundred photons above 100 TeV and at least ten PeV photons detected in the almost "background-free" regime. This should be sufficient to localize and identify the super-PeVatron. It would be interesting to analyze these data together with the observations of eROSITA of the Cygnus Cocoon. The "hadronic" synchrotron X-ray emission of the secondary electrons, the counterparts of $\geq$100 TeV gamma rays, should be sufficiently strong to be detected, although the contamination caused by different background components could be a serious issue. On the other hand, in the case of ballistic propagation of parent $\geq$1 PeV protons, the size of the X-ray emission would be significantly reduced making its detection feasible. And, of course, the mapping of gamma rays at lower ($E \leq 100$ TeV) energies with northern CTA telescopes with a few arcminute resolution would provide information about the broadband spectrum of accelerated protons, the energy-dependent diffusion coefficient, the total budget in accelerated particle, etc.

4.4. *Galactic center: SMBH vs. YMCs*

Several prominent stellar clusters in the Milky Way, e.g., Westerlund 1 and Westerlund 2, two powerful YMCs established by the HESS collaborations as prominent TeV sources,[23] are obvious targets for future broadband gamma-ray observations. Also of great interest are the arches, quintuplet, and nuclear stellar clusters in the Galactic Center (GC). These ultra-compact YMCs are unique in many aspects and generate a lot of interest because they can be responsible for the diffuse multi-TeV emission of the Central Molecular Zone (CMZ),[19] as alternative to Sgr A* — the supermassive black hole in the GC.[24] All three clusters are located within the central 20 pc region of GC. Therefore, to distinguish between Sgr A* and YMCs, we would need high-precision mapping of CMZ. Note that since these objects inject CRs continuously, the established $1/r$–type CR radial distributions from different points would result in a few "condensations" in the gamma-ray map. CTA with its $\approx$1 arcmin angular resolution and superb flux sensitivity should be be able to search for such condensations and localize their gravity centers with accuracy $\leq$1 pc, comparable to the linear scales of these ultra-compact clusters. Both CTA and SWGO should be able to measure the spectrum of diffuse gamma-ray emission up to 300 TeV and thus firmly establish the presence of a PeVatron(s) in the Galactic center. Any outcome of these observations ("SMBH vs. YMCs") will be fascinating and important for understanding the contribution of GC to Galactic CRs.

4.5.   *Microquasars as CR factories*

The history of gamma-ray observations of binary systems containing a black hole or neutron star is controversial. It is almost forgotten that these objects, particularly Cyg X-3, have been claimed as sources of TeV and PeV gamma rays decades ago. However, after the failure of confirming the early overoptimistic reports, these objects have not since been treated as relevant targets for gamma-ray observations. That stance has changed after the discovery of galactic sources with relativistic jets dubbed Microquasars. A clear message of this discovery was the important role of nonthermal processes and hence possible effective gamma-ray production in the accretion-driven objects.[1]

The interest in microquasars as sources of nonthermal radiation is multifaceted. For this chapter, they are attractive objects, primarily because of their ability to accelerate protons, potentially to hundreds of PeV. Despite the belief in the galactic origin of particles far beyond the knee, the astrophysics of this part of the spectrum remains unknown territory. Only a few galactic objects can accelerate, in principle, protons to such high energies. Microquasars can fill this gap. In this regard, the recent detection of multi-TeV gamma rays from SS 433 by HAWC[25] is a discovery of great importance. Although the authors prefer the leptonic origin of radiation, the hadronic interpretation cannot be excluded, given the uncertainties of the gas density in the jet termination region. The reported flux at 25 TeV is weak, detected at the margin of HAWC's sensitivity. For a firm conclusion concerning the origin of the radiation, comprehensive morphology of the source in different energy bands from 100 GeV to 100 TeV is needed. This will become possible soon with the WCDA and KM2A arrays of LHAASO. It is difficult to foresee the outcome of these observations. Still, from the point of view of the origin of Galactic CRs, the most exciting one would be the demonstration of the hadronic origin of radiation extending beyond 100 TeV.

LHAASO has great potential for the discovery of UHE sources with luminosities down to the level of $L_\gamma(\geq 0.1\,\mathrm{PeV}) \approx$

$10^{30}(d/1\,\mathrm{kpc})^2\,\mathrm{erg/s}$, where d is the distance to the source. It is orders of magnitude below the power of the jets in Cyg X-1, Cyg X-3, SS 433, and GRS 1915+105. This gives optimism that persistent UHE gamma-ray emission finally will be discovered from these objects. Note that the jets in all these objects are mildly relativistic with the bulk motion speed of a fraction of c, which is the most optimal one from the perspective of particle acceleration.

Several different scenarios of effective particle acceleration can also be realized through the internal shocks in the jet formed in the inner parts of the accretion disk. In this scenario, favorable conditions can also be established for gamma-ray production thanks to the presence of ample target material provided by the companion star or the accretion disk. This component is expected to be variable in the form of short (days) and ultra-short (hours and minutes) flares. The continuous monitoring of microquasars by LHAASO and SWGO, and follow-up observations with CTA for measurements of detailed light curves, is an obvious strategy which hopefully will lead to exciting discoveries. To the above list of prominent microquasars, some mysterious gamma-ray loud binaries could be added, in particular, LS 5039 and LSI +61 303. Despite some similarities with the binary pulsars, their origin is not yet established. The spectrum of LS 5039 continues up to 20 TeV. Within the binary pulsar model, the inverse Compton interpretation of the TeV emission requires extremely fast electron acceleration close to the theoretical limit. Even so, the accelerator should be located at the periphery of the binary system; otherwise, the severe Compton losses of electrons would prevent the acceleration of electrons to multi-TeV energies. The hadronic (pp or $p\gamma$) interpretations of the periodic TeV emission from LS 5039 are free from such constraints, although they are not free of other challenges. In any case, the detection of gamma rays by LHAASO with energy exceeding 20 TeV would imply a hadronic origin of radiation, making this and similar objects potential contributors to Galactic CRs.

4.6. *Pulsar wind nebulae: Extreme electron accelerators*

Nonthermal X-ray emission of young SNRs clearly indicates acceleration of electrons up to $\sim100\,\mathrm{TeV}$. Thus, it is natural to consider SNRs as essential contributors to the locally measured flux of CR electrons.

However, SNRs are not the only CR electron factories in the Galaxy. Pulsar wind nebulae (PWNs) constitute another class of powerful electron accelerators. Moreover, observations of UHE gamma rays from several PWNs, including the direct detection of PeV photons from the Crab Nebula,[26] tell us that we deal with electron PeVatrons. Because of synchrotron losses, the acceleration of electrons to PeV energies is a theoretical challenge. Understanding the operation of these perfectly designed machines accelerating particles at a rate close to the absolute theoretical limit is a major goal of future UHE gamma-ray observations.

The concept "Pulsar Wind" establishes the link between the energy reservoir (pulsar) and the extended MHD structure (PWN) through the formation and termination of an ultra-relativistic electron–positron wind.[27] UHE gamma rays contain the most critical information about these complex conglomerates. They are produced at inverse Compton scattering of electrons on 2.7 K MBR photons, thus carrying unambiguous information about the parent electrons. In particular, the extension of the Crab's spectrum to 1.1 PeV[26] implies an acceleration of electrons to 2.5 PeV. For the magnetic field $B \approx 100\mu G$ derived from the joint fitting of the IC and Synchrotron radiation components produced by PeV electrons, the acceleration rate should exceed 15% of the absolute theoretical limit. Complementary information about PeV electrons is carried also by synchrotron photons of energy ≈ 100 MeV. While the electrons responsible for UHE gamma rays are produced in a region with linear size $l \geq 10^{17}$ cm and magnetic field $B \leq 0.1$ mG, [26] the variability of synchrotron gamma-ray emission ("Crab flares")[28,29] requires compact ($l \leq 10^{16}$ cm) and strongly magnetized ($B \geq 1$ mG) blobs implying that the standard (one-zone) models should be revisited.

Moreover, the first LHAASO observations revealed a likely excess above 1 PeV,[26] indicating the presence of an additional radiation component (see Fig. 7). It can be interpreted as a result of a new radiation component but could also be caused by statistical fluctuations. Based on the first LHAASO results and the proper understanding of its performance, we anticipate an increase in photon statistics over the coming years by order of magnitude — more than 100 photons above 0.3 PeV detected at background-free conditions. This should allow accurate spectral measurements up to 1 PeV and confirm the presence of the second ≥ 1 PeV radiation component.

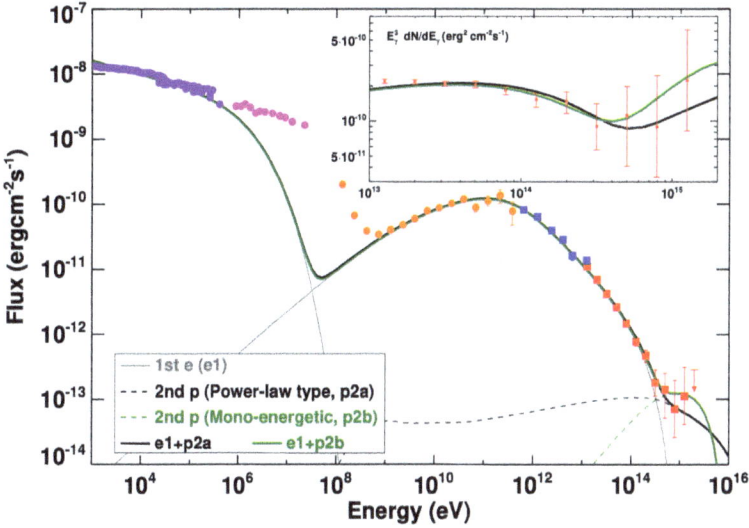

Fig. 7. Two zone radiation models of the Crab Nebula with the main (electron) and the second (proton) particle populations. Gamma rays from pp interactions are calculated for mono energetic (10 PeV) and power-law distribution with index = 2 and exponential cutoff at 30 PeV. Measurements: Fermi LAT (yellow), LHAASO WCDA (blue), and LHAASO KM2A (red). For details see Ref. [26].

The fit of the broadband emission of the Crab within the hybrid two-zone model is shown in Fig. 7. The overall radiation is caused by the main (electron) component responsible for the X-ray synchrotron and IC gamma rays up to ~300 TeV and, at higher energies, by the hadronic component produced at interactions of protons accelerated either in the pulsar's magnetosphere or at the termination shock. In this scenario, the cutoff energy in the spectrum of electrons is fixed at $E_0 = 450$ TeV; therefore, the synchrotron radiation appears well below the reported fluxes between 1 MeV and 1 GeV. It is assumed that the latter are produced in magnetized ($B \geq 1$ mG) compact ($\leq 10^{16}$ cm) region(s). A sudden increase in the electron injection rate in these blobs could be responsible for the synchrotron "Crab flares" in this energy band. The required acceleration power in PeV protons is substantial, about 10% of the pulsar's spin-down luminosity, which requires a harder than E_p^{-2} proton distribution. The continuation of Crab's spectrum beyond a few PeV would be a conclusive argument favoring the hadronic origin of the "excess" and a hint for a non-negligible contribution of pulsars to the Galactic CRs around the "knee".

While we cannot discard the hadronic origin of radiation around 1 PeV, below 300 TeV, it is robustly excluded — simply because of conflict with the available energy budget. Above 30 TeV, the 2.7 K CMBR photons represent the primary target for the IC scattering of electrons. Thus, the electron spectrum can be derived without any model assumption. After several years of observations, detection of more than 10,000 gamma rays by LHAASO is expected above 30 TeV with very low background contamination. With ~20% accuracy of energy reconstruction of individual UHE photons and given our perfect knowledge about the target (2.7 K MBR) and the Compton scattering cross section, the electron spectrum could be derived with unprecedented (even for the laboratory experiments) precision.

The Crab pulsar's spin-down luminosity significantly exceeds the rotational powers of other pulsars. On the other hand, the IC gamma-ray production efficiencies are dramatically enhanced compared to the Crab because of the smaller nebular magnetic fields. This compensates for the relatively modest rotational powers of other pulsars making their nebulae detectable in gamma rays. Two dozen extended VHE gamma-ray sources found in the H.E.S.S. Galactic Plane Survey,[23] and at least several UHE sources reported by HAWC[5] and LHAASO collaborations,[7] seem to have links to pulsars. These diffuse structures are composed of the following two components, both of IC origin: (i) produced by electrons confined in the pulsar wind nebula; (ii) linked to the "halo" of electrons and positrons that escaped the nebula and presently propagate diffusively through the ISM. As shown in Fig. 4, LHAASO J1908+0621 could be a candidate for this type of UHE emitter. However, this interpretation is not unique; the source could result from the delayed hadronic gamma-ray emission of the past activity of a middle-aged SNR (see Fig. 4). Significantly more extended sources surrounding pulsars are expected to be detected by CTA, SWGO, and LHAASO, allowing adequate population studies. More importantly, some bright representatives' comprehensive morphological and spectral studies should allow separating the two components, which is critical for proper interpretation of the results. While we anticipate detecting tens of PWNs and/or pulsar halos by SWGO and LHAASO-WCDA, their identification will be challenging. In-depth morphology by CTA and detailed spectrometry above 100 TeV by LHAASO-KM2A promise groundbreaking results. UHE gamma rays produced by electrons moving the initial paths of their propagation ballistically should show a tendency to reduce the angular size. This could be used to firmly associate these extended structures with an adjacent pulsar (see Fig. 6).

4.7. *Gamma-ray signatures of extragalactic CRs*

4.7.1. *Galaxies*

All types of gamma-ray emitters discussed above are well represented in other galaxies as well. Therefore, one can confidently claim that all galaxies are sources of high-energy gamma rays. The detectability of the overall gamma radiation consisting of the superposition of individual gamma-ray emitters as well as the diffuse component induced by interactions of CRs with ISM is determined by the following condition on the CR injection power: $\dot{W}_{CR} \geq 10^{40}(F_{min}/10^{-12}\,\text{erg/cm}^2\text{s})\,(\kappa/0.01)^{-1}(\text{d}/1\,\text{Mpc})^{-2}\,\text{erg/s}$, where F_{min} is the detector's sensitivity and κ is the fraction of total energy of CRs released in gamma rays in the given energy band and d is the distance to the source. For flat (E^{-2} type) spectra, κ may vary from 0.001 per energy decade, in the case of CRs propagating $\approx 1\,\text{g cm}^{-2}$ matter inside the sources and/or in the ISM, to 0.1, in the case of "full calorimetry". This simple condition tells us that LHAASO and SWGO can provide deep probes of gamma rays in the 1–100 TeV energy interval from nearby (within 1 Mpc or so) galaxies similar to our Galaxy. The derivation of the total CR acceleration power in these objects as well as detection of giant structures like Fermi Bubbles[30] in our Galaxy and $\sim$100 kpc halo around the Andromeda Galaxy[31] are essential for a better understanding of the propagation aspects of CRs and their possible acceleration in such structures in our own Galaxy. The potential of detection of ordinary galaxies drops with the distance as $1/d^2$. However, in gamma-ray emitting starburst galaxies such as M82[32] and NGC 253,[33] the flux reduction is compensated by much (at least an order of magnitude) higher CR acceleration power as well as by more effective confinement making gamma-ray production close to the calorimetric regime (see, e.g.,

Ref. [34]). In this regard, distant ultra-luminous starburst galaxies such as Arp 220 ($d \sim 100$ Mpc) are promising targets for observations with CTA and LHAASO up to ≈ 20 TeV.

4.7.2. *Cen A and M87*

Unfortunately, because of interactions of high-energy gamma rays with the intergalactic photon fields, only the nearby universe is transparent for UHE and EHE gamma rays (see Fig. 1). The mean free path of 1 PeV photons is only 8.5 kpc, thus all sources beyond the Milky Way are not visible in PeV gamma rays. The visibility range increases up to a few Mpc for $E_\gamma \sim 100$ TeV and $E_\gamma \sim 1$ EeV. This makes Centaurus A and M87, two nearby radiogalaxies at distances of 3.5 Mpc and 16 Mpc, respectively, unique objects for study of CR acceleration processes and exploring their contributions to the observed EHE CR fluxes. In principle, both objects can accelerate protons and nuclei up to 10 EeV or even higher. The straightforward and unbiased signature of the presence of such energetic particles in Cen A and M87 would be detection of 1 EeV gamma rays and neutrinos produced at interactions of protons and nuclei with the ambient gas and low-energy photons before they leave the host galaxies. The Pierre Auger Observatory, the Telescope Array, and the next-generation radio neutrino detectors are obvious tools for searching EHE gamma-ray and neutrino signals. However, given the fast escape of EHE particles, the efficiency of their interactions inside the source is likely to be quite low; therefore, one cannot be very optimistic concerning the detectability of these signals. Another signature of highest energy protons is their synchrotron radiation produced at the last stage of acceleration, presumably close to the supermassive black hole and released in the sub-TeV band (see the following). An effective search for variable TeV signals from M87 can be conducted by CTA during the strong flares of M87[35] which can be continuously monitored by the LHAASO detectors.

4.7.3. *Galaxy clusters*

We do not have many alternatives to M87 and Cen A as sources responsible for the observed EHE CRs. The choice is limited since they should be located at distances not exceeding significantly

~100 Mpc set by the mean free path of 100 EeV protons caused by interactions with 2.7 K CMBR. The nearby Virgo cluster and two other rich galaxy clusters in the 100 Mpc proximity — Coma and Perseus — have been proposed as potential production sites of the highest energy CRs. The speed of accretion shocks of about 2000 km/s formally, if one ignores dissipative processes, is sufficient for boosting the energy of protons over the Hubble time to 100 EeV. However, the proton acceleration rate is insufficient to overcome the energy losses caused by the (Bethe–Heitler) pair production at interactions with 2.7 K CMBR.[36–39] This results in a cutoff in the proton spectrum below 10 EeV (see Fig. 8). The secondary electron–positron pairs interacting with the ambient magnetic fields and 2.7 K CMBR produce broadband emission consisting of the synchrotron and IC components.[38] The spectral energy distributions of accelerated protons and secondary gamma rays calculated for the Coma galaxy cluster is shown in Fig. 8. The synchrotron radiation achieves its maximum in the hard X-ray and low-energy gamma-ray domain.

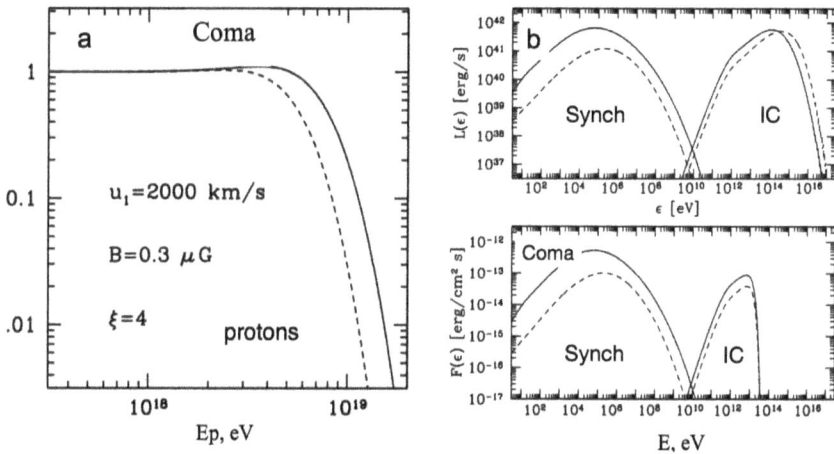

Fig. 8. Spectral energy distributions of protons and gamma rays in the Coma galaxy cluster. (a) Proton spectra at the shock location calculated within the model of diffusive shock acceleration of protons by accretion shocks for acceleration times 5 Gyr (dashed) and 10 Gyr (solid). The shock speed $v = 2000$ km/s, magnetic field upstream $B_1 = 0.3 \mu$G, and downstream $B_2 = 4B_1$. (b) The luminosity (top panel) and energy flux (bottom panel) of radiation consisting of the synchrotron and IC components from upstream (dashed) and downstream (solid). The intergalactic absorption of gamma rays is taken into account assuming 100 Mpc distance to the source. For details, see Ref. [38].

The IC component has a characteristic shape; it is very hard, reaches its maximum around 10 TeV, and then sharply declines because of the intergalactic γ–γ absorption. The detection of multi-TeV gamma rays from Coma with such a specific spectrum would imply the realization of the above scenario and prove the acceleration of protons and nuclei up to 10 EeV/nucleon. If 100 EeV CRs are dominated by heavy nuclei with negligible proton contribution, the galaxy clusters could be viable production sites of these particles. Otherwise, to explain the presence of 100 EeV protons in the CR spectrum, one would need more effective accelerators.

4.7.4. *Extreme proton accelerators*

Boosting the proton energy up to 100 EeV is a theoretical challenge. Even if acceleration proceeds at the maximum rate, $\nu = \eta q B c$ with $\eta = 1$, allowed by classical electrodynamics, the attainable energy is limited by two conditions as follows: (i) confinement — particles can stay in the acceleration region as long as their gyroradius remains smaller than the characteristic linear size L of the accelerator, i.e., $E_{\max} = 10^{20}(L/3 \times 10^{15}\,\text{cm})(B/100\,\text{G})\,\text{eV}$; (ii) synchrotron energy losses, $E_{\max} = 0.2 \times 10^{20}(B/100\,\text{G})^{-3/2}\eta^{-1/2}\,\text{eV}$.

These two conditions allow us to find the optimal size L and the magnetic field strength B that minimize the electromagnetic energy W stored in the accelerator.[40] The relativistic bulk motion reduces the energy requirements significantly but requires compact objects and large magnetic fields. For example, for a bulk motion Lorentz factor $\Gamma \sim 10$, the optimal size and magnetic field are $R \sim 10^{14}$ eV and $B \sim 300$ G, respectively. This corresponds to a quite reasonable total electromagnetic energy $\sim 3 \times 10^{47}$ erg for the inner jets of blazars. A reasonable combination of model parameters is also obtained for GRBs, for which a bulk motion Lorentz factor of about 300 is more typical.

While decreasing the accelerator size (and correspondingly increasing the magnetic field), synchrotron cooling plays an increasingly important role in the energy balance of accelerated particles. In compact objects, such as small-scale AGN jets or GRBs, the radiative losses become the dominant factor limiting the maximum attainable energy. If the acceleration in a jet with Lorentz factor $\Gamma \sim 10$ proceeds in the extreme regime, $\eta = 1$, the proton's energy could achieve 100 EeV. The proton acceleration in such objects is accompanied by hard synchrotron radiation extending to $\sim \alpha_f^{-1} m_p c^2 \Gamma \sim 1(\Gamma/10)$ TeV (in the observer's frame).[40] To some extent, here we have an analogue of the extreme electron accelerator in the Crab, characterized by synchrotron radiation shifted by the factor of $m_p/m_e \sim 2000$. This characteristic radiation can be used to search for extreme proton accelerators, the factories of 100 EeV protons.

In the context of the origin of extragalactic CRs, the nearby BL Lac objects, in particular, Mkn 501 and Mkn 42, well established TeV emitters, are of prime interest. The synchrotron cooling time of protons responsible for production of gamma rays in the cutoff region is $t_{\text{psynch}} \approx 2.4 \times 10^4 (B/100\,\text{G})^{-3/2}$ s. Thus, the Doppler-boosted TeV emission in the observer's frame is expected to vary on the hour or shorter timescales.

4.8. *Summary*

The ground-based gamma-ray observations of galactic and extra-galactic sources cover almost all topical areas of modern astrophysics. The sources and phenomena discussed above are only a fraction of these topics relevant to the origin of galactic and extragalactic cosmic rays. But even in this specific framework, one can see fascinating physics and astrophysics full of challenges.

References

1. F. Aharonian, L. Bergström, and C. Dermer, *Astrophysics at Very High Energies* (Saas-Fee Advanced Course, Springer, 2013).
2. F. Aharonian, J. Buckley, T. Kifune, and G. Sinnis, High energy astrophysics with ground-based gamma ray detectors, *Rep. Progr. Phys.* **71** (9), 096901 (September, 2008). doi: 10.1088/0034-4885/71/9/096901.
3. J. A. Hinton and W. Hofmann, Teraelectronvolt astronomy, *Ann. Rev. Astron. Astrophys.* **47**(1), 523–565 (September, 2009). doi: 10.1146/annurev-astro-082708-101816.
4. S. Funk, Ground- and space-based gamma-ray astronomy, *Ann. Rev. Nucl. Part. Sci.* **65**, 245–277 (October, 2015). doi: 10.1146/annurev-nucl-102014-022036.

5. HAWC Collaboration, Multiple Galactic sources with emission above 56 TeV detected by HAWC, *Phys. Rev. Lett.* **124**(2), 021102 (January, 2020). doi: 10.1103/PhysRevLett.124.021102.

6. Tibet AS γ Collaboration, First detection of photons with energy beyond 100 TeV from an astrophysical source, *Phys. ReV Lett.* **123**(5), 051101 (August, 2019). doi: 10.1103/PhysRevLett.123.051101.

7. LHAASO Collaboration, Ultrahigh-energy photons up to 1.4 petaelectronvolts from 12 γ-ray Galactic sources, *Nature.* **594**(7861), 33–36 (January, 2021). doi: 10.1038/s41586-021-03498-z.

8. P. S. Coppi and F. A. Aharonian, Constraints on the very high energy emissivity of the universe from the diffuse GeV gamma-ray background, *Astrophys. J. Lett.* **487**(1), L9–L12 (September, 1997). doi: 10.1086/310883.

9. S. Celli, F. Aharonian, and S. Gabici, Spectral signatures of PeVatrons, *Astrophys. J.* **903**(1), 61 (November, 2020). doi: 10.3847/1538-4357/abb805.

10. A. Merloni, K. Nandra, and P. Predehl, eROSITA's X-ray eyes on the universe, *Nature Astron.* **4**, 634–636 (June, 2020). doi: 10.1038/s41550-020-1133-0.

11. M. A. Malkov and L. O. Drury, Nonlinear theory of diffusive acceleration of particles by shock waves, *Rep. Progr. Phys.* **64**(4), 429–481 (April, 2001). doi: 10.1088/0034-4885/64/4/201.

12. V. N. Zirakashvili and F. A. Aharonian, Nonthermal radiation of young supernova remnants: The case of RX J1713.7-3946, *Astrophys. J.* **708**(2), 965–980 (January, 2010). doi: 10.1088/0004-637X/708/2/965.

13. M. A. Malkov and F. A. Aharonian, Cosmic-ray spectrum steepening in supernova remnants. I. Loss-free self-similar solution, *Astrophys. J.* **881**(1), 2 (August, 2019). doi: 10.3847/1538-4357/ab2c01.

14. A. R. Bell, J. H. Matthews, and K. M. Blundell, Cosmic ray acceleration by shocks: Spectral steepening due to turbulent magnetic field amplification, *MNRAS.* **488**(2), 2466–2472 (September, 2019). doi: 10.1093/mnras/stz1805.

15. D. Caprioli, C. C. Haggerty, and P. Blasi, Kinetic simulations of cosmic-ray-modified shocks. II. Particle spectra, *Astrophys. J.* **905**(1), 2 (December, 2020). doi: 10.3847/1538-4357/abbe05.

16. F. A. Aharonian and A. M. Atoyan, On the emissivity of pi0 decay gamma radiation in the vicinity of accelerators of galactic cosmic rays, *A& A.* **309**, 917–928 (May, 1996).

17. M. Casse and J. A. Paul, Local gamma rays and cosmic-ray acceleration by supersonic stellar winds, *Astrophys. J.* **237**, 236–243 (April, 1980). doi: 10.1086/157863.

18. C. J. Cesarsky and T. Montmerle, Gamma-rays from active regions in the galaxy — the possible contribution of stellar winds, *Space Sci. Rev.* **36**(2), 173–193 (October, 1983). doi: 10.1007/BF00167503.

19. F. Aharonian, R. Yang, and E. de Oña Wilhelmi, Massive stars as major factories of Galactic cosmic rays, *Nature Astron.* **3**, 561–567 (March, 2019). doi: 10.1038/s41550-019-0724-0.

20. A. M. Bykov, A. Marcowith, E. Amato, M. E. Kalyashova, J. M. D. Kruijssen, and E. Waxman, High-energy particles and radiation in star-forming regions, *Space Sci. Rev.* **216**(3), 42 (April, 2020). doi: 10.1007/s11214-020-00663-0.

21. R. Aloisio, E. Coccia, and F. Vissani, *Multiple Messengers and Challenges in Astroparticle Physics* (2018). doi: 10.1007/978-3-319-65425-6.

22. HAWK Collaboration, HAWC observations of the acceleration of very-high-energy cosmic rays in the Cygnus Cocoon, *Nature Astronomy* **5**, 465–471 (March, 2021). doi: 10.1038/s41550-021-01318-y.

23. HESS Collaboration, The H.E.S.S. Galactic plane survey, *A&A.* **612**, A1 (April, 2018). doi: 10.1051/0004-6361/201732098.

24. HESS Collaboration, Acceleration of petaelectronvolt protons in the Galactic Centre, *Nature.* **531**(7595), 476–479 (March, 2016). doi: 10.1038/nature17147.

25. HAWC Collaboration, Very-high-energy particle acceleration powered by the jets of the microquasar SS 433, *Nature.* **562**(7725), 82–85 (October, 2018). doi: 10.1038/s41586-018-0565-5.

26. LHAASO Collaboration, Peta–electron volt gamma-ray emission from the Crab Nebula, *Science.* **373**(6553), 425–430 (July, 2021).

27. M. J. Rees and J. E. Gunn, The origin of the magnetic field and relativistic particles in the Crab Nebula, *MNRAS.* **167**, 1–12 (April, 1974). doi: 10.1093/mnras/167.1.1.

28. R. Bühler and R. Blandford, The surprising Crab pulsar and its Nebula: A review, *Rep. Progr. Phys.* **77**(6), 066901 (June, 2014). doi: 10.1088/0034-4885/77/6/066901.

29. M. Tavani *et al.*, Discovery of powerful gamma-ray flares from the crab Nebula, *Science.* **331**(6018), 736 (February, 2011). doi: 10.1126/science.1200083.

30. M. Su, T. R. Slatyer, and D. P. Finkbeiner, Giant gamma-ray bubbles from Fermi LAT: Active galactic nucleus activity or bipolar Galactic wind? *Astrophys. J.* **724**(2), 1044–1082 (December 2010), doi: 10.1088/0004-637X/724/2/1044.

31. C. M. Karwin, S. Murgia, S. Campbell, and I. V. Moskalenko, Fermi-LAT observations of γ-ray emission toward the outer Halo of M31, *Astrophys. J.* **880**(2), 95 (August, 2019).

32. VERITAS Collaboration, A connection between star formation activity and cosmic rays in the starburst galaxy M82, *Nature.* **462**(7274), 770–772 (December, 2009). doi: 10.1038/nature08557.

33. HESS Collaboration, Detection of gamma rays from a starburst galaxy, *Science.* **326**(5956), 1080 (November, 2009). doi: 10.1126/science.1178826.

34. E. Peretti, P. Blasi, F. Aharonian, and G. Morlino, Cosmic ray transport and radiative processes in nuclei of starburst galaxies, *MNRAS.* **487**(1), 168–180 (July, 2019). doi: 10.1093/mnras/stz1161.

35. V. A. e. a. Acciari, Radio imaging of the very-high-energy γ-ray emission region in the central engine of a radio galaxy, *Science.* **325**(5939), 444 (July, 2009). doi: 10.1126/science.1175406.

36. C. A. Norman, D. B. Melrose, and A. Achterberg, The origin of cosmic rays above 10 18.5 eV, *Astrophys. J.* **454**, 60 (November, 1995). doi: 10.1086/176465.

37. H. Kang, J. P. Rachen, and P. L. Biermann, Contributions to the cosmic ray flux above the ankle: Clusters of galaxies, *MNRAS.* **286**(2), 257–267 (April, 1997). doi: 10.1093/mnras/286.2.257.

38. G. Vannoni, F. A. Aharonian, S. Gabici, S. R. Kelner, and A. Prosekin, Acceleration and radiation of ultra-high energy protons in galaxy clusters, *A&A.* **536**, A56 (December, 2011). doi: 10.1051/0004-6361/200913568.

39. V. N. Zirakashvili and V. S. Ptuskin. Cosmic ray acceleration in accretion flows of galaxy clusters. *J. Phys. Conf. Ser.*, **1181**, 012033 (February, 2019). doi: 10.1088/1742-6596/1181/1/012033.

40. F. A. Aharonian, A. A. Belyanin, E. V. Derishev, V. V. Kocharovsky, and V. V. Kocharovsky, Constraints on the extremely high-energy cosmic ray accelerators from classical electrodynamics, *Phys. Rev. D.* **66**(2), 023005 (July, 2002). doi: 10.1103/PhysRevD.66.023005.

Chapter 8

Particle Detection Arrays for Very-High-Energy Gamma-Ray Astrophysics

Petra Huentemeyer

Michigan Technological University, Physics Department,
1400 Townsend Drive, Houghton, MI 49931, USA
petra@mtu.edu

Remarkable improvements in the sensitivity and image resolution of γ-ray sky surveys provided by the current generation of ground-based particle detection arrays (PDAs) has led to groundbreaking and sometimes paradigm-shifting observations. In addition to the detection of an unprecedented number of PeVatron candidates, and multi-TeV γ-ray emission associated with particle acceleration in the jets of a microquasar, a new source class, γ-ray halos, has been discovered. These halos are produced by electrons and positrons from pulsars, accelerating in their vicinity, and then interacting with the ambient interstellar radiation field outside the classical pulsar wind nebulae. Following the discovery of γ-ray halos by the High-Altitude Water Cherenkov (HAWC) Observatory, the Fermi Large Area Telescope (LAT) confirmed their existence at the GeV scale. Shortly after, the High Energy Stereoscopic System (H.E.S.S.) telescope confirmed extended emission around the Geminga pulsar at the TeV scale using an adapted analysis. The high-duty-cycle, wide field-of-view, and unrivaled sensitivity of PDAs to ultra-high-energy γ-rays enables them to provide real-time alerts and archival data and makes them uniquely suited for multi-messenger astronomy and fundamental physics research. This chapter covers the current status and future plans for PDAs in the field of ground-based γ-ray astronomy. It includes descriptions of currently operating wide-field-of-view instruments, a discussion of their scientific achievements, and an outlook on

241

an exciting future facility in the Southern Hemisphere pursued by the Southern Wide-Field Gamma-Ray Observatory (SWGO) project.

Gamma ray astronomy is on the horizon[1]

— Kenneth Greisen, 1960

1. Introduction

The very-high-energy (VHE; 100 GeV–100 TeV) and ultra-high-energy (UHE; >100 TeV) γ-ray bands of the electromagnetic spectrum — despite major advancements in detector technology in recent years — remain the least explored. Since the first detection of a TeV γ-ray source with high significance (at about 9σ), namely the Crab pulsar wind nebula (PWN), in 1989,[2] more than 200 such sources have been found.[3] At the time of writing, about one-third of the sources have not been unambiguously identified or associated with known objects in the universe. Improved detection sensitivity, angular resolution, and energy resolution of γ-ray instruments are crucial in clarifying the origin of the TeV γ-ray emission. Other essential ingredients are coincidence alerts, studies, and global fits that include the spectral energy distribution, morphology, and time variability of emission structures and combine information from multiple astronomical wavelengths and multiple messengers, such as neutrinos, gravitational waves, and cosmic rays.

The astrophysical γ-ray flux measured near Earth decreases exponentially with increasing γ-ray energy, which poses challenges to achieving statistically sound measurements. For this reason, spaceborne telescopes with rather small detection areas become impractical for the study of TeV γ-rays. On the other hand, the Earth atmosphere is 100% opaque to VHE γ-rays and prevents them from reaching observers at the Earth's surface directly. It is however possible to indirectly detect γ-rays (and cosmic rays) by measuring the products of their interactions with atmospheric nuclei. When cosmic particles enter the Earth's atmosphere they can collide with air nuclei and the collisions result in what is known as *extensive air showers (EASs)*, "avalanches" of secondary relativistic particles. EASs penetrate deeply enough into the atmosphere for earthbound observatories to detect them. Generally, two classes of earthbound (often called *ground-based*) observatories may be distinguished: imaging atmospheric Cherenkov telescopes (IACTs) and particle detection arrays (PDAs). The former are discussed in the following chapter, the latter are the subject of this chapter.

This chapter covers basic ideas and concepts of indirectly detecting γ-rays with ground-based PDAs and their complementarity with respect to IACTs. We discuss performance drivers, technological features, and specifications of current major ground-based γ-ray PDA observatories, highlight key scientific results, and conclude with future perspectives.

2. Ground-Based γ-ray Observations with PDAs

A ground-based PDA in combination with the Earth's atmosphere above it is a detector concept that is well known from experiments at earthbound particle accelerators: the calorimeter. Calorimeters are used to measure the energy of particles through their interaction with the calorimetric material and can also provide directional information and particle identification. In the case of a ground-based PDA, the atmosphere serves as calorimetric material in which a particle shower is initiated by a cosmic particle. The detectors of the PDA are distinct from the atmosphere and sample the total energy deposit in the atmosphere by measuring the energy deposit of a particle shower at the ground level (at 5200 m a.s.l. $\sim$1/10 of the cosmic primary energy). Hence, PDAs fall into the category of sampling calorimeters.

The following sections explain basics of particle shower development in the atmosphere, main detector technologies, and resulting methods for directional and energy reconstruction.

2.1. *Extensive air shower development*

The Earth's atmosphere is opaque to high-energy cosmic particles. Figure 1 shows what happens when a γ-ray or a cosmic-ray proton enters the atmosphere as simulated with the widely used package CORSIKA (COsmic Ray SImulations for KAscade).[4] The γ-ray induces an *electromagnetic shower*, the proton a a *hadronic shower*. Because cosmic rays are the main background to detecting γ-rays — less than about 1 in 1000 EASs that trigger PDAs are γ-ray showers — it is worthwhile spending some time discussing both γ-ray– and cosmic-ray–induced EASs.

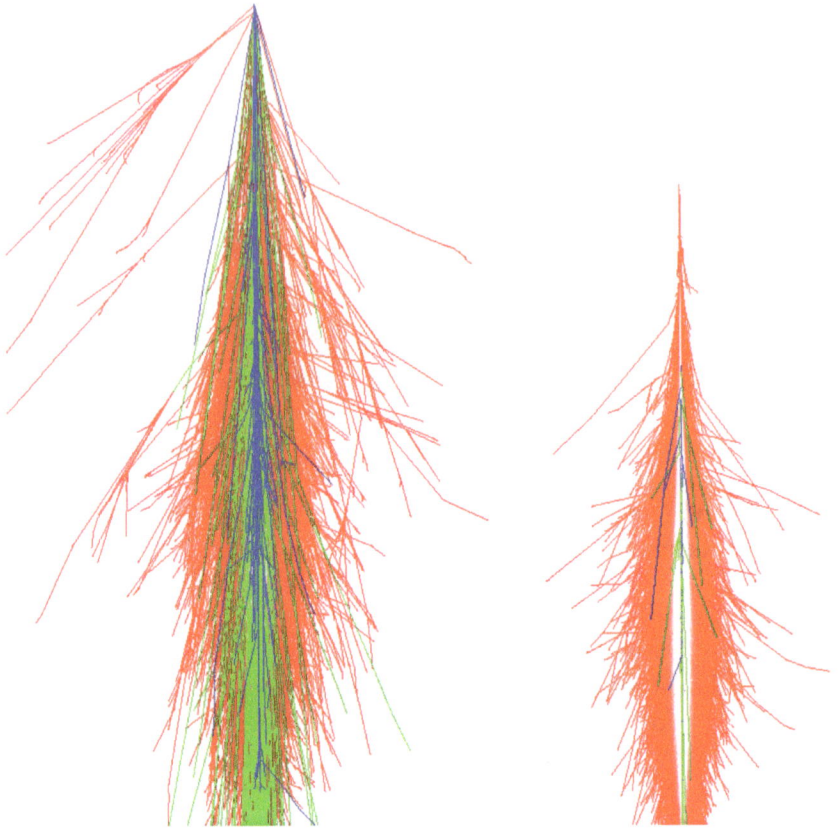

Fig. 1.   CORSIKA[4] simulation of a 10 TeV proton EAS (a hadronic shower, left) and a 10 TeV γ-ray EAS (an electromagnetic shower, right). The images show the secondary particle trajectories projected onto the (x, z) plane, where z measures the vertical atmospheric depth. The original γ-ray and cosmic-ray proton enter the Earth at a zenith angle of $0°$. The red trajectories represent photons, electrons, and positrons, the green muons, and the blue hadrons. To make the muons and hadrons close to the shower axis more visible, the photon, electrons, and positron are suppressed for the purpose of creating the image (courtesy of Hugo Ayala, Pennsylvania State University/HAWC Collaboration).[5]

A detailed and comprehensive description of EAS development can be found in Ref. [6]. Here, we limit ourselves to basic ideas that should enable the reader to understand the motivation for algorithms and parameters that are used to suppress the cosmic-ray background and to reconstruct the energy and direction of γ-rays.

If the energy of a γ-ray exceeds twice the electron-rest mass (1.02 MeV) the process of pair production becomes energetically possible and likely at several MeV. A VHE γ-ray travels a certain distance on average in the atmosphere, the mean free path, before it produces an $e^+ - e^-$ pair. These secondary particles then produce γ-rays again via *bremsstrahlung*. The distance over which these interactions occur is expressed in terms of radiation length, X_0, which is defined as the the distance over which an e^- loses all but $1/e$ of its energy by *bremsstrahlung*. Radiation length depends on the amount of matter traversed and is typically measured in units of $g\,cm^{-2}$. The Earth's atmosphere for instance is $\sim 30 X_0$ deep. The mean free path for pair production is equal to $7/9 X_0$. Therefore, to first approximation, one particle (which can be either a γ-ray or an e^-) transfers its energy to two particles of the next generation over a distance of one X_0. Repeating this process leads to an increasing number of photons, electrons, and positrons as the electromagnetic shower penetrates deeper and deeper into the atmosphere. The energy available in the shower is shared between the shower particles. Once the average energy per electron and positron drops below a value, the *critical energy*, at which the cross section for *bremsstrahlung* is less than that for ionization losses, the shower begins to dissipate because the electrons do not have enough energy to create another generation of shower particles.

When a cosmic-ray proton or heavier nucleus enters the Earth's atmosphere, it strikes a nucleus in the Earth's atmosphere and — in an inelastic collision — produces nuclear fragments and sub-atomic particles, including pions, kaons, and baryons. Neutral pions have a mean lifetime of the order of 10^{-16} seconds[7] and decay into γ-rays, which induce electromagnetic sub-showers, as explained above. Charged pions have a mean lifetime of the order of 10^{-8} and predominantly decay into muons and neutrinos.[7] The muons have a mean lifetime of $\sim 2.2\ \mu$s and due to relativistic time dilation can reach the Earth's surface. For instance, $\sim 20\%$ of muons with an energy of 1 GeV starting at an altitude of 10 km will make it to sea level.[b]

Muons have a mass that is ~ 200 times that of electrons so that they are not significantly affected by multiple Coulomb scattering. At the same time, the hadronic parent particles of the muons have a large transverse momentum, leading to the muon trajectories intersecting with the Earth's surface relatively far away from the shower axis.

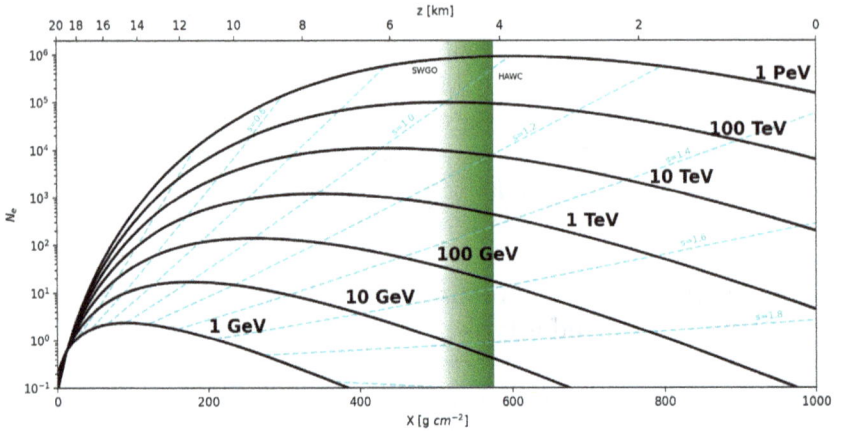

Fig. 2.   Longitudinal electromagnetic shower development for different γ-ray energies according to Greisen.[8–10] The solid lines indicate the number of electrons and positrons in the shower as function of atmospheric depth. The dashed lines correspond to equal-age curves. The altitude range of current and future PDA observatories is indicated by the shaded green band (courtesy of Xiaojie Wang, Michigan Technological University/HAWC Collaboration).

The resulting characteristics of the *footprint* of a hadronic shower at ground level in combination with their "muon richness" (see Fig. 1) is utilized in algorithms that reject the otherwise overwhelming cosmic-ray background (see the following).

Figure 2 shows the number of electrons and positrons in electromagnetic showers versus altitude as calculated using a semi-empirical model that was first proposed by Greisen.[9,10] It introduces a parameter called *shower age*:

$$s = \frac{3t}{t + 2y}, \quad \text{with} \quad y = \ln \frac{E_0}{E_c}, \tag{1}$$

where E_0 and E_c are the primary particle and critical energy, respectively, and $t = X/X_0$ is the reduced depth. Thus, the shower begins at an age of $s = 0$. The number of particles in the shower reaches a maximum at an age of $s = 1$, corresponding to a depth of $t_{max} = y = \ln \frac{E_0}{E_c}$, and the shower begins to dissipate at ages $s > 1$.

Greisen's approximation also provides an average for the number of electrons and positrons at depth t and specifically t_{max}, i.e., at shower maximum:

$$\overline{N_e}(t) = \frac{0.31}{\sqrt{y}} \exp\left[t\left(1 - \frac{3}{2}\ln s\right)\right] \quad \text{and} \quad \overline{N_e}(t_{max}) = \frac{0.31}{\sqrt{y}}\left(\frac{E_0}{E_c}\right). \quad (2)$$

A shower initiated by a γ-ray of 1 TeV for example reaches its maximum on average at $\sim$8000 m, while the shower maximum for a 1 PeV γ-ray occurs on average just above 4000 m. It should be noted that γ-rays outnumber electrons by between 5 and 6 to 1.

2.2. *Detector configurations & technologies*

The efficiency with which PDAs can detect γ-ray induced EAS is determined by their energy threshold and their cosmic-ray background rejection capabilities. The accuracy and sharpness of the γ-ray sky maps that can be produced from the raw data is limited by their pointing accuracy and angular resolution.

Considerations of the energy threshold are driven by the atmospheric depths above the array, by the density of sensitive detection units in the array, and the ability to detect γ-rays, electrons, and positron. Figure 2 gives an idea of the fraction of particles reaching a certain altitude and marks the location of current and future PDAs with respect to the number of electrons and positrons in a shower. Because γ-rays are the most abundant shower particles, their detection is a crucial consideration in lowering the energy threshold of a detector design. The front of the shower striking the ground appears to the detector stations like a convex pancake of $\sim$100 m diameter and a few meters thickness. The PDA density affects the energy threshold because the fraction of shower front particles falling within the *total* PDA area that *are* detected scales with the density of detector stations. Consequently, a dense array has a lower energy threshold than a sparse array.

There are different methods applied in PDAs to trigger on EAS. In the past, trigger logic was often implemented at the hardware level. A *hardware trigger* relies on signal thresholds and time windows defined by the electronics of the data acquisition system (DAQ). Nearly dead-time free data acquisition and sufficient computational power make it possible to operate PDAs with a *software trigger*. In this scheme, the digitized signals of all detector channels are transmitted to a computer system, where they are processed, noise filters are applied, and triggers are formed. An advantage of this approach is that it makes the development and maintenance of a dedicated hardware system to combine and process signals from potentially several thousand channels unnecessary. It also allows for more flexibility in case new or alternative trigger designs are desired. Both the HAWC detector array and the Large High Altitude Air Shower Observatory (LHAASO) successfully adopted software triggers.[11,12] In the case of the HAWC Observatory for example, the DAQ generates a data rate of ~500 MB/s, which is reduced to a data rate to disk of ~20 MB/s by the online computer system. This results in nearly 2 TB of data per day saved on portable disks that are then transported to the Universidad Nacional Autónoma de México for further distribution.

The detection ability of a specific PDA is typically expressed in terms of *effective area*, defined as the geometrical area over which events are detected, convolved with the detection efficiency. Its values versus primary particle energy are derived from simulations. Figure 3 shows an example of an effective area distribution computed for the specific configuration and detector elements of the original core array of the HAWC Observatory.[13]

The distribution is shaped by the shower physics discussed above. Due to the fluctuations in the depths of the first interaction initiating an EAS and its dissipation beyond the shower maximum, the energy threshold is not well-defined, as can be seen in the figure. While the rising edge shifts to higher energies with increasing zenith angle of the shower axis, its slope mainly depends on the spectral energy distribution of the primary particles and only to a lesser order on the zenith angle. Here, it is interesting to note that a change in zenith angle is equivalent to a change in altitude (the shower particles travel through more atmospheric matter). Specifically, a change in zenith from $0°$ to $60°$ corresponds to an altitude change of a factor of two.[6]

The effective area can exceed the geometrical size of an instrument if events are included in the computation for which the intersection point of the shower axis with the ground level, the so-called *shower core*, does *not* fall inside the area enclosed by the array. Requiring that the shower cores be reconstructed on the array and events within an angular radius around the true photon direction in which 68% of events are reconstructed, the effective area flattens at roughly half the physical area of the instrument, in this case the HAWC Observatory.[14]

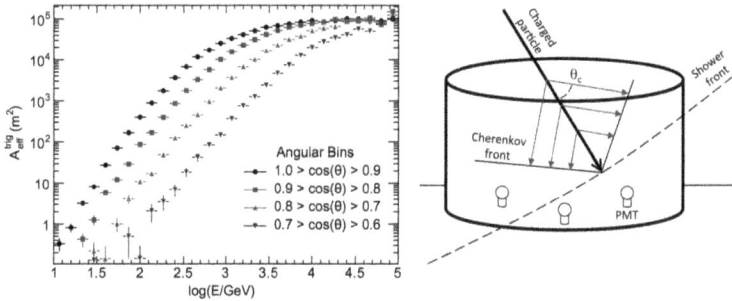

Fig. 3. Effective area and detector station design of the HAWC core array configuration on which it is based.[13] The left panel shows the effective area of the HAWC Observatory in its original configuration of 300 closely packed water-Cherenkov detectors (WCDs) as a function of γ-ray energy for four ranges of zenith angle. Requiring >30 photomultiplier tubes (PMTs) to record a significant signal results in a trigger rate of 17 kHz. The shower cores of the events are reconstructed within $1.1°$ of their true, originally, simulated direction. No cosmic-ray background rejection cut is applied. The right panel illustrates the principle of water-Cherenkov detection. Relativistic particles reach the ground in a shower front. The charged particles produce Cherenkov radiation as they travel in the WCDs. The Cherenkov light is emitted at an angle $\theta_c \approx 41°$ with respect to the particle trajectory, the Cherenkov angle characteristic of water, and detected by PMTs at the bottom of the WCD.

The angular resolution of a PDA configuration depends on the number of detected particles, their radial distance from the shower core, and the timing resolution of the signals in the detector units. The front of an EAS is curved because shower particles traveling at an angle with the shower axis have to travel a greater distance relative to the shower axis. Thus, particles detected near the shower core are observed to arrive earlier than particles far from the core. In addition, the arrival time of particles at the ground varies due to multiple scattering in the atmosphere. As a result, the shower front has a finite width of typically approximately several nanoseconds near the core and becomes wider at greater distances from the core. If the detector units of a PDA measure the arrival time of the earliest arriving particles, this time is more likely a measurement of the leading edge, not of the center of the shower front. The latter time would be the most probable if only one particle were detected. This effect coupled with the fact that the shower front is denser near the core than far away from it makes a position-dependent correction of the timing offset in the shower front necessary. This correction is called *sampling correction* because by making multiple measurements of a distribution of particle arrival times the detector units are "sampling" the shower front. The curvature and sampling effects are closely connected. The angular resolution of a PDA depends on the ability to first accurately reconstruct the shower core and then correct for both curvature and sampling.

Finally, considerations of background rejection capabilities play a critical role in optimizing PDA configurations. As previously mentioned, identifying muon signatures in the footprint of EASs is key. The detected muon content is almost independent of the observatory altitude, while the hadronic component decays more quickly as the EAS penetrates more deeply into the atmosphere (see green versus blue trajectories in the left panel of Fig. 1). The dominant hadronic component in the early stage of a cosmic-ray–induced shower in combination with the more deeply penetrating muons leads to a less smooth lateral distribution. The more efficiently a PDA configuration allows to distinguish between the *clumpiness* of a hadronic and the *smoothness* of an electromagnetic shower footprint, the better its detection sensitivity.

The following three basic detector technologies are used in the observation of TeV γ-ray induced EASs with PDAs on the ground: *scintillators*, *resistive plate chambers* (*RPCs*), and *water-Cherenkov detectors* (*WCDs*).

Early ground-based γ-ray PDAs, such as the CYGNUS[15] (2130 m a.s.l., Los Alamos, New Mexcio) and CASA-MIA[16] (1450 m a.s.l., Dugway Proving Grounds, Utah) observatories, covered a somewhat large area with a sparse array of scintillation counters, where <1% of the area was instrumented with detector material. A layer of lead was added on top of the scintillation material to convert the γ-rays, copious in an EAS, to electrons for which the detection efficiency in the scintillator is essentially 100%. Shielded detector units were added to identify muons, which penetrate material more deeply than e^-/e^+ or γ-rays, and these muon detector signals were used to reject the cosmic-ray background. The sparse coverage in combination with the low altitude relative to currently operated PDAs led to a very high energy threshold of $\sim$100 TeV.

A next-generation instrument, the Tibet ASγ Observatory, was designed to achieve a lower energy threshold by installing a relatively dense scintillator array at very high altitude (4300 m a.s.l.,

YangBaJing Observatory, Tibet). The array is able to detect γ-rays with an energy down to $\sim$3 TeV and the first PDA to successfully detect multi-TeV γ-ray emission from the Crab Nebula in 1999.[17] The Tibet ASγ Observatory has been upgraded repeatedly over its three-decades-long operation period and continues to produce pioneering scientific results. In its current configuration,[18] it consists of two sub-arrays: the air shower (AS) array, which encloses an area of 65,700 m^2, containing 597 plastic scintillators with a 0.5 m^2 detection area, and the muon detector (MD) array which consists of 64 WCDs buried 2.4 m beneath the AS array. The MD units are waterproof concrete chambers, 1.5 m in depth and $\sim$54 m^2 in area. They are filled with water which is viewed by a 20 in diameter downward-facing PMT installed at the top of each chamber. The angular resolutions (50% containment) of this configuration is $\sim$0.5° for 10 TeV γ-rays and $\sim$0.2° for 100 TeV γ-rays. For particle energies >100 TeV, 99.92% of the cosmic-ray background is rejected.

The Milagro experiment (2630 m a.s.l., Jemez Mountains, New Mexico)[19] was a 73$\times$53 m^2 central water pond instrumented with 750 eight-inch PMTs[20] arranged in two layers and surrounded by an "outrigger" array of 175 WCDs enclosing an area of $\sim$40,000 m^2. With its groundbreaking detection of galactic diffuse γ-ray emission >3.5 TeV,[21] the measurement of >10 TeV γ-ray emission from the direction of the Cygnus constellation,[22] and the consequential discovery of very extended TeV γ-ray emission around the Geminga pulsar,[23] Milagro proved the great potential of wide field-of-view observatories based on the water-Cherenkov technique for providing a vista of the sky that is complementary to that of IACTs.

Like the Tibet ASγ Observatory, the Astrophysical Radiation with Ground-based Observatory at YangBaJing (ARGO-YBJ)[24] was also operated at YangBaJing in Tibet until 2013. It used an array of 1836 RPCs arranged into 153 clusters enclosing a total area of 11,000 m^2. The central array of 130 clusters was closely packed on an area 74 $\times$ 78 m^2 and sensitive over 93% of its extent. Each RPC was composed of 10 pads divided into 8 strips of 6.7$\times$62 cm^2, providing pixelized spatial information with a time resolution about $\sim$1 ns. The central array of the ARGO-YBJ was surrounded by 23 additional clusters of the remaining RPCs, the "guard ring", to improve the accuracy of the shower core position reconstruction. The very dense central array ensured an energy threshold of a few hundred

GeV, significantly lower compared to other PDAs operational at the time.[25] The angular resolution of ARGO-YBJ ranged between 0.3° at the high end of detectable energies to ∼1.7° at the low end, with 0.6° for energies > ∼6 TeV.

The two most recently constructed ground-based γ-ray PDAs both makes use of the water-Cherenkov technique to detect EAS. The High-Altitude Water Cherenkov (HAWC) Observatory began science operation in 2015 at 4100 m a.s.l. in the Pico de Orizaba National Park, Veracruz/Puebla, with a completed densely packed core array of 300 large volume WCDs. The Large High Altitude Air Shower Observatory (LHAASO) began science operation in 2019 at 4400 m a.s.l. in the Mt. Haizi National Reservation in Sichuan with a partially completed array. Both observatories build on the pioneering design of the Milagro instrument.

A bird's-eye view of the the HAWC observatory is shown in Fig. 4. The large volume WCDs of the core array are steel tanks outfitted

Fig. 4. The site of the HAWC observatory at Pico de Orizaba National Park, Puebla, Mexico. The picture was taken with a drone from a height of 500 m above the observatory. The closely packed main array of 300 large volume water-Cherenkov detectors (WCDs) is surrounded by a sparse array of 345 small volume "outrigger" WCDs (courtesy of the HAWC Collaboration).

with a light-tide PVC bladder and a sand-color roof of military-grade opaque canvas. The bladder is instrumented at the bottom with three 8-inch Hamamatsu R5912 PMTs at 1.85 m distance from the center of the tank, and one 10-inch Hamamatsu R7081-MOD PMT at the center.[20] The tanks are 7.5 m in diameter and 5 m high. Water is filled to a depth of 4.5 m with 4.0 m of water above the upward-facing PMT. The core array is surround by a sparse outrigger array of 345 small WCDs, plastic tanks filled with water, 1.55 m in diameter and 1.65 m high, separated by 12–18 m and equipped with a single Hamamatsu R5912 8-inch upward-facing PMT.[26] The HAWC instrument is capable of detecting γ-rays from a few 100 GeV to a few 100 TeV, can achieve an angular resolution of $<0.2°$ (68% containment), and reject $\sim99.8\%$ of the cosmic-ray background at the highest energies.[14] As expected, but still worth noting, the trigger rate of the HAWC data acquisition system (DAQ) of $>20,000$ Hz significantly increases over that of the Milagro DAQ of $\sim1,700$ Hz.

The LHAASO combines lessons learned from both the Milagro and Tibet ASγ observatories and has three detector subsystems, known as the the Kilometer Square Array (KM2A), the Water Cherenkov Detector Array (WCDA), and the Wide Field-of-View Cherenkov Telescope Array.[27] The KM2A consists of 5,195 electromagnetic particle detectors (EDs) and 1,171 MDs. The EDs are plastic scintillators of one square meter, to which 1.5-inch PMTs are connected. As in the Tibet ASγ array, lead plates are added on top of the scintillation material. The MD units are also designed similarly to the Tibet ASγ MDs, buried 2.5 m deep, with a downward-facing 8-inch PMT viewing a 1.2 m deep water volume with a base area of 36 m^2. The KM2A surrounds the WCDA, which is composed of three Milagro-inspired water-filled ponds covering an area of about four times that of the compact HAWC core array. In total, the WCDA contains 3,120 optically isolated water volumes equipped with either 8-inch (900) or 20-inch (2,220) upward-facing PMTs at the bottom of each cell. Additionally, there are 900 1.5-inch and 2,220 3-inch PMTs installed in the cells to increase the dynamic range in measurements of the very particle-dense footprints of cosmic-ray–induced showers. The WFCTA is located on the edge of pond area and consists of 20 wide field Cherenkov light telescopes (WFCTs). All the components of a WFCT are contained in a shipping container on a chassis, allowing to adjust the elevation angle of the detector unit between 0° and 90°. The purpose of the WFCTs is to measure the Cherenkov and the fluorescence light produced in an EAS. The flexible WFCTA setup was conceived to allow to measure the energy spectrum of cosmic-ray components from $10^{13} - 10^{18}$ eV. The WCDA was optimized to survey the sky for γ-rays from 100 GeV to 30 TeV, while the KM2A provides unmatched sensitivity for γ-rays >30 TeV. The angular resolution and background rejection capabilities of the first operational pond of the WCDA — WCDA-1, which covers an area similar to that of the HAWC observatory — and of the half-completed KM2A were both independently confirmed by studying γ-ray emission from the Crab Nebula. For the WCDA-1, the angular resolution is measured to be 0.45° at 1 TeV and <0.2° above 6 TeV;[28] for the half-completed KM2A, the angular resolution (68% containment) is measured to be 0.5°–0.8° at 20 TeV and 0.24°–0.3° at 100 TeV.[29] With the WCDA-1, around 1 TeV 97.7% and around 6 TeV 99.8% of the cosmic-ray background are found to be rejected while keeping ~50% of the γ-ray–induced EAS; with the half-completed KM2A, above 100 TeV, a fraction of $<10^{-3}$ of the cosmic-ray events pass the background rejection cuts, while keeping >70% of the γ-ray–induced EAS. In its final configuration, the fully completed LHAASO is expected to have a γ-ray detection sensitivity above ~25 TeV that will be unmatched by any currently operational or constructed ground-based γ-ray observatory, including the HAWC Observatory or the Cherenkov Telescope Array (CTA)[30] (see also Fig. 13).

2.3. *Extensive air shower reconstruction methods*

The conventional reconstruction of EAS data collected with ground-based PDAs follows recipes that are overall very similar across different TeV instruments. The reconstruction steps are as follows: 1. data quality assurance via instrument calibration and selection of relevant EAS signals ("hit selection"); 2. shower core reconstruction, which may involve iterations, and subsequently allows to apply curvature and sampling corrections (see above); 3. directional reconstructions, which may involve iterations alternating with the previous step, core reconstruction; 4. cosmic-ray background rejection; finally, 5. γ-ray energy reconstruction.[14] In the following, instructive examples of

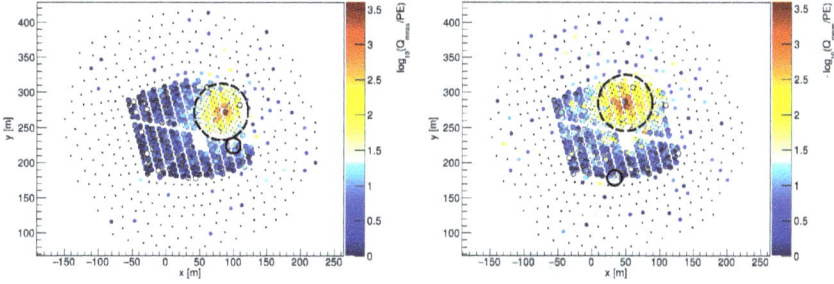

Fig. 5.   Footprint of EASs in the HAWC array induced by a 15 TeV photon (left) and 38 TeV proton (right). The orientation of the event display is the same as in Fig. 4, with the y-axis pointing north. The color scale indicates the measured PMT signal charge in terms of photoelectrons (PEs). The dashed black circle indicates a radial distance of 40 m from the shower core, and the smaller solid black circle represents the location of the PMT contributing the maximum PE signal to the shower footprint outside of the 40 m circle. This information is used in one of the hadronic background rejection algorithms (courtesy of Xiaojie Wang, Michigan Technological University/HAWC Collaboration).

algorithms are given that have recently been developed to suppress the cosmic-ray background and estimate the γ-ray energy.

Figure 5 shows the footprints in the HAWC WCD array of air showers initiated by a γ-ray and a cosmic-ray proton. The PMT contributing the maximum PE signal to the shower footprint is marked with a small solid circle. As can be seen, the maximum PE signal occurs farther from the shower core in the case of the proton shower, consistent with a muon that inherited a large transverse momentum from its hadronic parent particle in the EAS. This fact is used for one of the original parameters to distinguish electromagnetic showers caused by γ-rays from hadronic showers caused by cosmic rays:

$$C = \frac{N_{hit}}{CxPE_{40}}, \tag{3}$$

where N_{hit} is the number of PMTs with a significant signal produced as the EAS sweeps across the array, and $CxPE_{40}$ is the effective charge measured in the PMT with the largest effective charge measured beyond a radial distance of 40 m from the shower core (see caption of Fig. 6 for an explanation of effective charge). The parameter is called *compactness* and tends to smaller values for hadronic and larger values for electromagnetic showers. It was originally developed

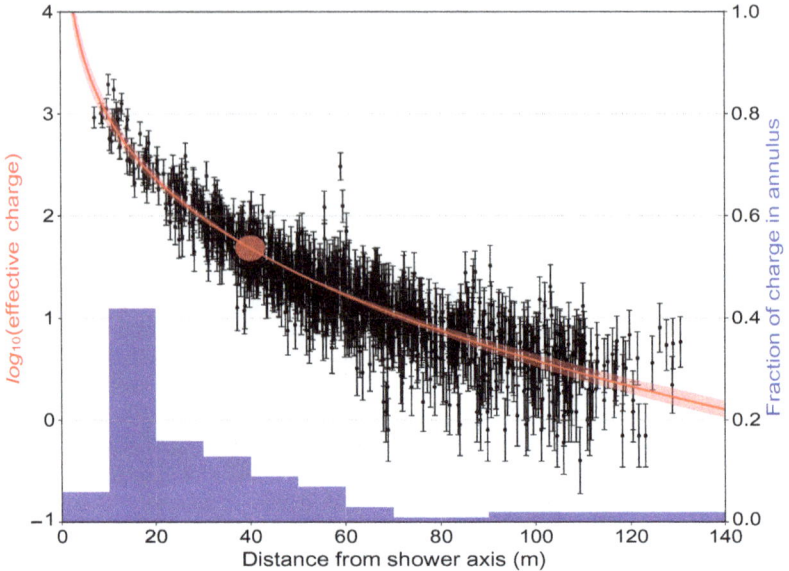

Fig. 6.   Information used by two algorithms developed by researchers from the HAWC Observatory to estimate the energy of γ-rays (visualized for a single EAS event). The black points represent the distribution of the log of the effective charge measured by the PMTs in the array as a function of the radial distance to the air shower core. The main array of the HAWC Observatory utilizes two sizes of PMTs, the 8-inch Hamamatsu R5912 and the 10-inch Hamamatsu R7081-02.[20] An effective charge is calculated by applying a scaling factor to adjust the measurements of the two PMT types. The red line is the best fit to an NKG-like function. The red dot marks the charge measurement at a distance of 40 m from the shower axis, ρ_{40}. The latter, together with the zenith angle, is used to infer the energy of the cosmic particle. The blue histogram represents the relative fraction of the charge in several annuli around the shower core. The width of the annuli up to an outer radial distance of 90 m is 10 m. The charge fraction outside of the 90 m radius is combined in one bin. The 10 charge fractions are a measure of the lateral shower distribution and serve as input to a neural network that is used to find the energy of the original cosmic particle energy in an alternative way to the ρ_{40} method.[33]

for the Milagro Pond,[31] has been optimized for the WCD configuration and altitude of the HAWC observatory, and adapted by the LHAASO in their WCDA-1 performance study (using an exclusion radius of 45 m instead of 40 m).[28]

An alternative parameter, fashioned after the common expression for χ^2, has been developed to measure the clumpiness of an

EAS footprint:[14]

$$P = \frac{1}{N} \sum_{i=0}^{N} \frac{(\zeta_i - \langle \zeta_i \rangle)^2}{\sigma_{\zeta_i}^2}, \tag{4}$$

where $\zeta_i = \log_{10}(Q_{eff,i})$ is the logarithm of the effective charge, $Q_{eff,i}$, measured by the ith PMT that participates in the shower fit. The term $\langle \zeta_i \rangle$ is an expectation value calculated by averaging the logarithms of the effective charges, ζ_j, measured by all the triggered PMTs in an annulus of 5 m width that is centered on the reconstructed shower core position and contains the ith PMT. The errors σ_{ζ_i} are derived from a study of a sample of γ-like events detected from the direction of the Crab Nebula. Thus, P tests for the evenness and radial symmetry of the lateral charge distribution of the EAS. The footprint of a hadronic EAS is expected to be clumpy and contain PMT charges that differ clearly from the expectation value leading to large values for P.

A third parameter that is currently being investigated to separate electromagnetic from hadronic showers is based on the Nishimura–Kamata–Greisen (NKG) function.[1,32] The NKG function was originally devised to describe the lateral and angular distribution of electromagnetic showers. Fitting this function to the EAS footprints on the ground tends to produce smaller values of χ^2 for electromagnetic showers than for hadronic showers, which can be used to discriminate between γ-rays and cosmic rays.

The NKG function also plays an important role in the estimation of γ-ray energies. The simplest way to reckon the energy of a VHE/UHE γ-ray is to measure the size of the footprint its EAS leaves in a PDA by counting the number of detector units that are triggered. This method has been used for example by the Milagro,[34] HAWC,[14] and ARGO-YBJ[35] instruments. The drawback of this method is that it does not take into account the location of the shower core and the shape of the shower profile, which leads to a somewhat weak correlation between this observable and the primary cosmic-particle energy. In addition, the dynamic range of this approach can be severely limited at the high end of the detectable energy range for an instrument because of saturation effects, especially when all the detection units in an array are triggered by an EAS. For this reason, researchers at the HAWC Observatory developed two alternative methods to independently estimate γ-ray energies. They are illustrated in Fig. 6.

The first method is inspired by a technique that was first proposed by Hillas in the 1970s[36] and is used by VHE and UHE cosmic-ray PDAs like the KASCADE-Grande[37] and Pierre Auger[38] observatories. It relies on fitting a modified NKG function to the charge profile of the shower footprint in the array and finding an optimal radial distance from the shower core at which the uncertainty in the density of the shower energy deposited in the array is minimized. For the HAWC configuration and altitude, this optimal radius is found to be 40 m. The charge deposit at this distance, ρ_{40}, is measured and combined with the reconstructed zenith angle of the EAS, θ, using the empirically determined Eq. (2) in Ref. [33] to determine the energy of the γ-ray primary. A similar implementation of the method has been used by the Tibet ASγ Observatory based on the measured particle density (instead of charge) at a radial distance of 50 m (instead of 40 m) from the shower core.[39] The energy resolution the Tibet ASγ Collaboration reports on the log scale is 16% at 100 TeV for EAS zenith angles of $<20°$ (compared to 27% when only the footprint size of the EAS is used).

The second approach utilizes the *toolkit for multivariate analysis*[40] to implement an artificial neural network. The input of the neural network represents three features of an EAS: the amount of energy deposited in the array, the containment of the shower footprint within the array, and the attenuation in the atmosphere. The specific input variables are the fraction of PMTs and WCDs triggered by the EAS and the logarithm of the normalization of a lateral distribution function that uses a smoothed approximation of the NKG function to fit the profile of the EAS footprint.[14] The primary energy resolution the HAWC Observatory achieves with these newly developed methods is $\sim$10% in log-energy space for γ-ray energies of around 100 TeV.

Over the past few years, the VHE γ-ray community has also increasingly started to turn to machine and deep learning methods for EAS reconstruction and γ-ray identification. The anticipation is that there is significant potential for improvement, especially at the low end of the energy range (<1 TeV) accessible to ground-based γ-ray PDA.[41]

Although new algorithms to measure γ-ray energies have led to major improvements in resolution of around a factor of two or more, it is worth pointing out that ground-based PDAs only take a snapshot of the shower development at a particular altitude. Even if located at a high altitude, they detect particles from the shower tail significantly below the shower maximum for primary particle energies below 10 TeV (see Fig. 2). Fluctuations intrinsic to the shower development combined with limited sampling at the ground level therefore present a fundamental boundary for the energy resolution.

3. Capabilities & Examples of Groundbreaking Results

Three main features of PDAs make them uniquely suited for certain research topics in the field of ground-based γ-ray astronomy. A *high duty cycle* (typical uptime >95%) enables monitoring transients (e.g., active galactic nuclei) over a long duration as well as sending and following up on alerts (e.g., for gamma-ray bursts and gravitational wave events), where the follow-ups can be immediate or archival. A *wide field of view* ($\sim$2 sr) lends itself to surveys of large swaths of the sky increasing the discovery potential and providing an advantage for the study of extended and large-scale structures. An *unparalleled detection sensitivity for γ-ray energies >10 TeV* offers the best prospects for the study of highest energy accelerators in our Galaxy and the discovery of PeVatrons that may explain the feature in the locally measured cosmic-ray spectrum at $\sim$10^{15} TeV known as the "knee". The direct study of cosmic rays and electrons cannot provide this insight because cosmic rays and electrons are charged and their trajectories are altered by interstellar magnetic fields. On the other hand, they can produce VHE/UHE γ-rays close to their accelerators in two ways: Cosmic rays may collide with interstellar matter that is for example present in nearby molecular gas clouds producing neutral pions that promptly decay into γ-rays or electrons may interact with interstellar radiation fields that permeate space producing γ-rays through inverse Compton scattering with low-energy photons. The γ-rays can carry about one-tenth of the energy of their progenitors and pass through space without being deflected. Hence, discovering locations of γ-ray emission exceeding 100 TeV is strong evidence for a PeVatron in the respective region.

There has been a significant increase in the number of publications reporting astrophysical results based on VHE/UHE γ-ray data collected with PDAs, a sign of maturing detector technologies and resulting improved instrument sensitivity. In the following, only a handful of high-impact results will be discussed briefly. More comprehensive publication lists can be found elsewhere.[42–44]

New VHE γ-ray sources and source classes: In 2017, the HAWC Observatory reported the detection of extended TeV γ-ray emission coincident with the location of two nearby middle-aged pulsars: Geminga and PSR B0656+14, which are located inside the Monogem ring (see Fig. 7).[45] Evidence for extended TeV γ-ray emission surrounding Geminga had previously been reported by the wide field-of-view instrument Milagro,[23] but the new HAWC measurements were significantly more sensitive allowing to study the emission profile. A fit of a simple leptonic progenitor model to the γ-ray profiles inspired by Ref. [47] was used to demonstrate that the diffusion of the leptons away from their accelerating pulsar sources is much slower than previously assumed. This put into question that the excess positron flux measured near Earth[48,49] could be explained by

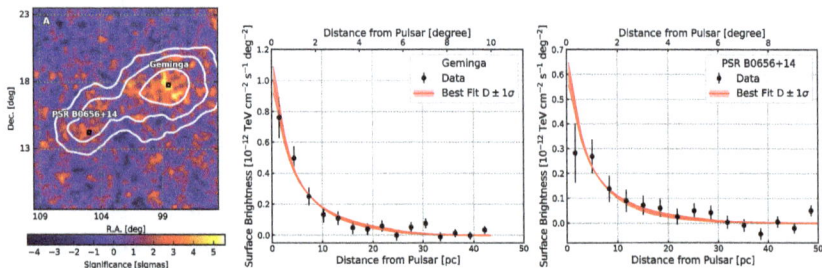

Fig. 7.   Surface brightness of Geminga and PSR B0656+14.[45] The color scale in the left-hand image indicates the statistical significance of the excess counts above the background of contaminating cosmic rays and γ-rays before accounting for statistical trials for γ-ray energies between 1 and 50 TeV convolved with the instrument point spread function. The middle and right-hand image show the surface brightness as a function of distance from the Geminga (A) and PSR B0656+14 (B) pulsars. The solid red lines are the best-fitting diffusion model. The shaded band represent the $\pm 1\sigma$ statistical uncertainty and the error bars are the statistical errors. The distance from each pulsar in parsecs is calculated based on nominal distances of 250 and 288 pc for Geminga and PSR B0656+14, respectively.[46]

local pulsars[50] instead of alternative hypotheses such as dark matter annihilation. Furthermore, the discovery of the Geminga/PSR B0656+14 halo by the HAWC Observatory and of several other TeV PWNe by the H.E.S.S. Telescope[46] led to the hypothesis that extended *halos* are a common feature of pulsars.[51–55] There are slightly differing approaches to interpret the halo phenomenon. One version posits that the observed γ-ray emission is due to inverse Compton up-scattering of cosmic microwave background photons by relativistic e^+/e^- that have escaped from the PWN but remain trapped in a larger region where diffusion is inhibited compared to the interstellar medium.[53] Another postulates that halos only form around pulsars that are older than $\sim$100,000 years and have either left their supernova remnant (SNR) shell or whose SNR shell already dissipated, allowing the relativistic e^+/e^- to diffuse freely in the vicinity of the pulsar.[55] Both interpretations have in common that they recognize halos as distinct objects since the e^+/e^- plasma has escaped from the classical X-ray PWN. Since their discovery by the HAWC Observatory, halos have also been detected by the LHAASO[56] and at lower γ-ray energies by the *Fermi*-LAT.[57]

In 2018, an analysis of HAWC sky maps revealed for the first time γ-ray emission of tens of TeV from structures in the lobes of the microquasar/SNR system SS433/W50 far away from the central compact object where the jets are formed.[58] A multi-waveband fit combining energy fluxes measured in the radio, X-ray, and γ-ray regime found the electromagnetic emission to be consistent with being caused by a single population of electrons previously accelerated to energies extending to at least hundreds of TeV in the jets of SS433. The result demonstrates the benefits of wide field-of-view observations coupled with good detection sensitivity for γ-ray energies above 10 TeV allowing to survey the sky without the need for a strategically designed campaign which has to take into account competing interests and may be rather time-consuming. In addition, the SS433 analysis is a great example of the power of a maximum likelihood formalism, in which a model — in this case, consisting of three separate emission sources with different extents and spectral energy distributions — is constructed to describe a particular region of the sky, convolved with the instrument response, and then compared to the corresponding data.[60] The succession of steps by which the SS433 lobe emission was revealed in this particular case are illustrated in Fig. 8.

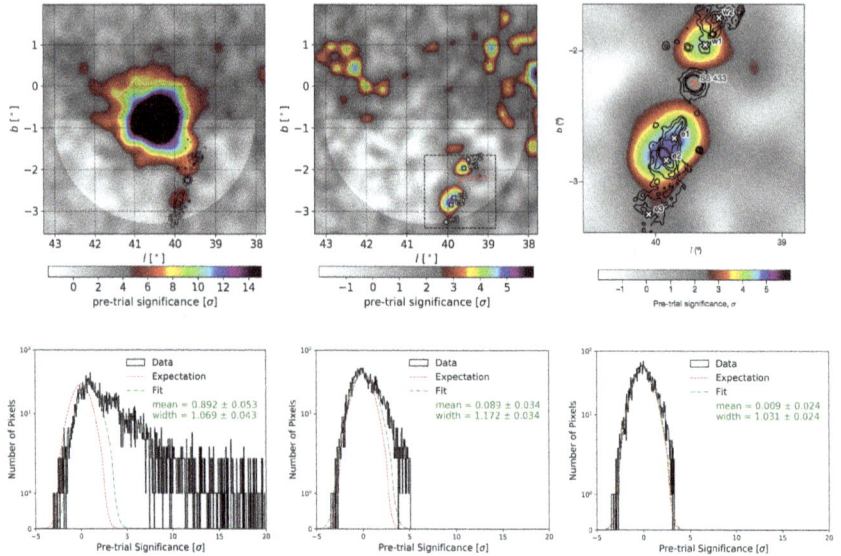

Fig. 8. Top panel: VHE γ-rays from MGRO J1908+06 and SS 433/W50.[58] The color scale indicates the statistical significance of the excess counts above the background of contaminating cosmic rays and γ-rays before accounting for statistical trials. The bright extended γ-ray source MGRO J1908+06 is shown at the center of the left-hand image with SS 433/W50 at the bottom. The dark contours show X-ray emission from SS 433 and its jets.[59] The semicircular area indicates the region of interest (ROI) used to fit the γ-ray observations. The middle image shows the γ-ray excess measured after the subtraction of γ-ray emission from the spatially extended source MGRO J1908+06. The dashed box indicates the region shown in the right-hand image. The jet termination regions e1, e2, e3, w1, and w2 observed in the X-ray data as well as the location of the central binary are indicated. Bottom panel: Distribution of pixel significance in the ROI of the fit. Shown are the distribution before fitting and subtraction of any source emission (left-hand image), after subtraction of emission from MGRO J1908+06 (middle image), and after subtraction of emission from the fully fitted model consisting of MGRO J1908+09 and emission associated with the two jet termination regions w1 and e1 (right-hand image). The left-hand and middle images correspond to the left-hand and middle image of the top panel. The distribution in the right-hand image is well described by a fit to a Gaussian with a mean of $\sim$0 and width $\sim$1, indicating that the remaining emission is consistent with background fluctuations.

***PeVatrons*:** In recent years, an increasing number of PeVatron candidates have been identified. In 2016, the H.E.S.S. Collaboration announced the detection of a significant flux of tens of TeV γ-rays from the Galactic center and interpreted them as tracers of PeV particles linked to past activity of Sagittarius A*.[67] The Tibet ASγ, HAWC, and MAGIC observatories detected γ-rays up to and above 100 TeV from the direction of the Crab Nebula.[18,33,68] In 2020, the HAWC Collaboration reported finding several more PeVatron candidates, with one maybe associated with an SNR.[69,70] In 2021, with data collected in less than a year by the half-completed KM2A, the LHAASO identified at least 12 PeVatron candidates.[71] About half of these sources had previously been detected by the HAWC observatory, but the LHAASO extended the measurement to higher γ-ray energies. Moreover, as the LHAASO sub-arrays are being completed and collect more data, the observation sensitivity, in particular, at extremely high energy, will increase. At the time of writing, the LHAASO has already reported the detection of γ-ray energies up to 1.4 PeV, coming from the direction of the constellation Cygnus. Both the Tibet ASγ[72] and HAWC collaborations[61] detected γ-ray emission exceeding 0.1 PeV from this region. Using a maximum likelihood approach similar to that applied in the SS433 analysis described above,[60] the HAWC researchers studied the morphology of the surface brightness of the region and separated out an extended emission component centered on the Cygnus OB2 star cluster (see Fig. 9). The γ-ray emission is well modeled as resulting from the interaction of protons colliding with ambient gas after the latter have recently been accelerated to PeV energies by powerful shock waves generated by strong stellar winds in the star cluster.

A recent result by the Tibet ASγ Collaboration provides indirect evidence for an undetected PeVatron population in the Milky Way Galaxy.[72] The observatory surveyed the Northern Hemisphere sky for γ-like events between 100 TeV and 1 PeV, that are not associated with any known VHE γ-ray source. The spatial distribution of the arrival directions of these events is shown in Fig. 10 for energies >0.4 PeV. As can be seen, the arrival directions tend to be clustered within a few degree of the Galactic plane. The authors of the study explain this distribution with γ-rays that are produced by PeV protons, which have moved away from their accelerators and then collide with the interstellar matter in the Galaxy. Thus, this sub-PeV

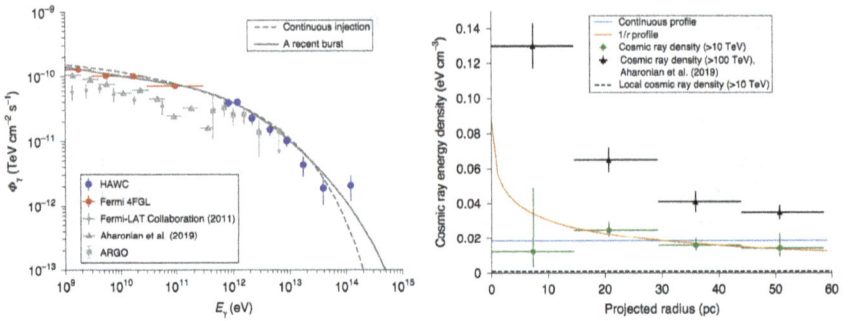

Fig. 9.   GeV–TeV γ-ray emission from the Cygnus Cocoon.[61] The left-hand plot shows the spectral energy distribution of the γ-ray emission from the Cygnus Cocoon measured by different instruments.[62–65] The errors on the flux points are the 1σ statistical errors. The gray solid and dashed lines are γ-ray spectra derived from modeling the emission as a result of proton progenitors. The right-hand plot shows the cosmic-ray energy density profile calculated for four rings centered on the OB2 association (0–15 pc, 15–29 pc, 29–44 pc and 44–55 pc). The green circles are the cosmic-ray densities derived from the γ-ray emission above 10 TeV as measured by the HAWC Observatory. The y errors are the statistical errors of the HAWC measurement. The x error bars denote the width of the rings. The orange and blue lines are the $1/r$ profile expected from continuous particle injection and constant profile expected from the burst injection, respectively. They are calculated by assuming a spherical symmetry for the γ-ray emission region and by averaging the density profile over the line of sight within the emission region. The black dashed line is the local cosmic-ray density above 10 TeV based on Alpha Magnetic Spectrometer measurements.[66] The black triangles are the cosmic ray densities above 100 GeV from Ref. [65].

diffuse γ-ray emission can be an indicator of potentially unidentified PeVatrons in our Galaxy.

Fundamental physics: The search for an excess flux of γ-rays that can be traced back to dark matter interactions is a topic in astrophysics that receives significant attention. PDAs with their wide field-of-view are well suited to perform such searches that may involve observing large-scale structures or stacking data collected from many astrophysical objects of a class that is expected to contain significant dark matter, like for example dwarf spheroidal galaxies.[75,76] Axion searches are also an area of interest.[77] Additionally, astrophysical sources provide unique test beds for possible signatures of LIV because of the high energies they accelerate particles to and their long

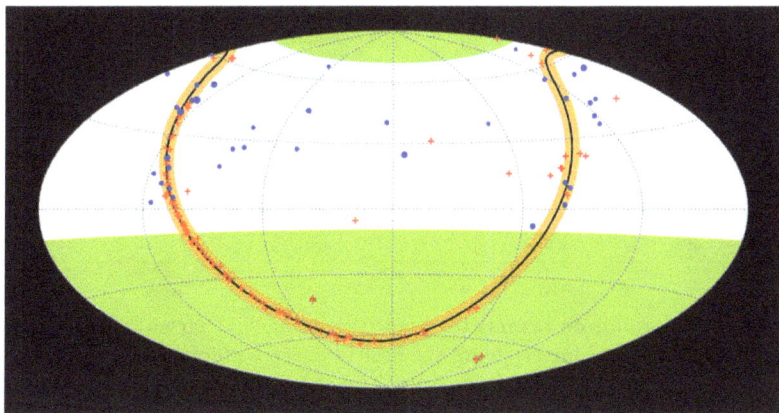

Fig. 10.   Arrival directions of γ-ray–like EAS events with energies between 0.4 and 1 PeV that are not associated with known TeV sources as measured by the Tibet ASγ Observatory (blue solid dots). Most events are clustered in the vicinity of the Galactic plane (yellow shaded area). The red crosses indicate the position of known TeV sources. The coordinate system is celestial and the green areas indicate the sky regions outside the field of view of the Tibet ASγ Observatory.[73] Reproduced with permission from Ref. [71] APS/Alan Stonebraker.

distances from the observer. Superluminal LIV for example allows for the decay of photons at very high energy. The currently unmatched sensitivity of PDAs above 10 TeV makes them the best instruments to measure LIV effects. The Tibet ASγ and HAWC observatories as well as the LHAASO are currently the most sensitive γ-ray instrument above 10 TeV, and all have found evidence for 100 TeV photon emission from several astrophysical sources. Figure 11 illustrates the method and shows results the HAWC Collaboration published in 2020.[74] The strongest limits derived from their observations exclude LIV to >1800 times the Planck energy. Studies with data from the LHAASO will improve these limits further.[78]

Source catalogs: The wide field-of-view coupled with background estimation methods like *direct integration*[15,31] form the basis for constructing VHE/UHE γ-ray sky maps. Based on these maps, the Milagro, Tibet ASγ, and ARGO-YBJ collaborations published surveys of the Northern Hemisphere, identifying new, previously undetected sources or following up on observations, for example, by the *Fermi*-LAT.[23,81–83] As the number of source detections increased along with instrument sensitivity, the HAWC Collaboration started

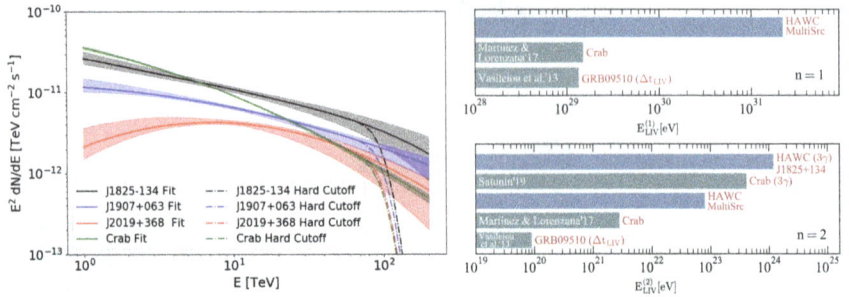

Fig. 11.   Constraints on Lorentz invariance violation (LIV) from HAWC observations.[74] The left-hand image shows the best fit spectra of four UHE γ-ray sources (including the Crab Nebula) as measured by the HAWC observatory. The overlaid dashed lines are the best fits that assume a maximum energy cutoff, E_c. In all cases, the fits without a cutoff in emission energy are statistically preferred. Setting a lower limit on E_c, beyond which there is weak or no emission, can be reinterpreted as setting an upper limit on observed photon energy, E_γ. A hard cutoff in photon energy can be caused by photon decays that are not kinematically allowed in classical relativity but may efficiently occur in LIV, and limits on E_γ can be translated into limits on the LIV energy scale. These are shown in the right-hand plot. It can be seen that HAWC measurements exclude the LIV energy scale of new physics, $E_{LIV}^{(1)}$ to greater than $> 10^{31}$ eV. This is over 1800 times above the Planck energy scale ($E_{Pl} \sim 10^{28}$ eV) and more constraining than prior limits.

releasing catalogs.[79,80,84] They are based on putative point and extended source searches in signal maps of arrival directions that are for example formatted in equatorial coordinates using the HEALPIX algorithm for pixelization.[85] An example of such a map made for the third HAWC catalog can be seen in Fig. 12. For each pixel in the celestial map (R.A./δ, Earth-centered inertial frame J2000), an independent fit is performed to the events recorded within a radius of (1°+source extent) around the pixel center. For the point (extended) source searches, the source model is composed of a point (extended, disk-like) source with a power-law energy spectrum of a predefined index (e.g., $\alpha = -2.5$), convolved with the instrument response function derived from simulations of the HAWC PDA. The only free parameter is the flux normalization. The combined source–background model that best fits the data (likelihood $\mathcal{L}_{s+b}$) is compared to a background-only model (likelihood $\mathcal{L}_b$) to calculate the test statistic, $TS = 2\ln \mathcal{L}_{s+b}/\mathcal{L}_b$. Then, a list of sources is compiled as locations of local maxima with $\sqrt{TS} > 5$. Maxima close to each

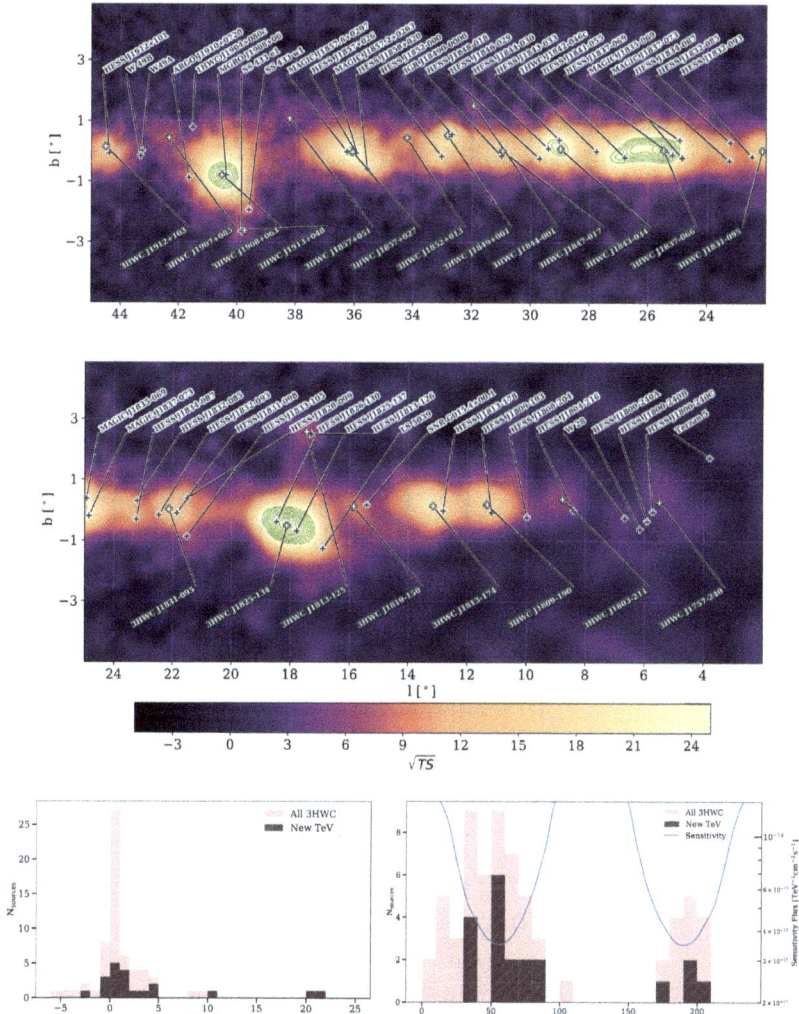

Fig. 12. The Third HAWC catalog (3HWC).[79] Top and middle panels: Maps in Galactic coordinates of γ-ray signal significance above the background in two selected regions along the galactic plane. The maps were calculated assuming point source emission. The green lines are significance contours starting at $TS = 26$, increasing in steps of $\Delta\sqrt{TS} = 2$. Top labels point to positions of known TeV sources (from TeVCat[3]); bottom labels point to positions of 3HWC sources. Bottom panel: 3HWC source distribution (excluding the known blazars) as a function of Galactic latitude (left) and longitude (right). The darker-shaded histograms represent 3HWC sources that were not present in the 2HWC catalog.[80] The blue solid lines in the right plot show the sensitivity of the HAWC Observatory at $b = 0°$. Due to its location, the observatory is most sensitive toward the Galactic anticenter ($l \approx -180°$) and the inner Galaxy ($l \approx 50°$).

other are identified as separate primary (secondary) sources if they are divided by a "valley" $\Delta\sqrt{TS} > 2$ ($2 > \Delta\sqrt{TS} > 1$) deep. Using this approach, 65 sources are detected in about four years of HAWC data. Twenty of these source are more than $1°$ away from previously detected sources and 14 have potential counterparts in the fourth *Fermi*-LAT catalog. Given the increase in sensitivity through improved reconstruction algorithms, additional data, and new instrumentation, it can be expected that future TeV catalogs constructed from PDA data will require more sophisticated approaches similar to those used for the *Fermi*-LAT catalogs involving multi-component fits to take into account point, extended, and diffuse emission.[86]

4. The Southern Wide-Field Gamma-Ray Observatory

Over the last decade, PDAs like the ARGO-YBJ as well as the Tibet ASγ and HAWC observatories have contributed a wealth of results that expand our understanding in astrophysics and astronomy, particularly regarding the nonthermal regime. Some results challenge prior assumptions and transform approaches to describing the energy and particle transport in the universe, like for example the discovery of γ-ray halos around pulsars.[45] The LHAASO, which came online with a partially completed array in 2019, already reports the detection of more than 12 PeVatron candidates, up from one just five years ago.[71] It has become obvious that PDAs provide a complementary view on the VHE/UHE γ-ray sky. This is supported further by a recent joint study of sections of the Galactic plane from the H.E.S.S. and HAWC collaborations that confirms four of seven HAWC sources previously undetected by IACTs with an adapted analysis of H.E.S.S. data in an overlapping region of the sky.[90] Likewise, research in multi-messenger astrophysics requires facilities that continuously survey the γ-ray sky in the space and time domains. Wide field-of-view observatories can meet these demands by providing prompt alerts about transient events and by storing archival information about γ-ray emission covering large regions of the sky. Currently, there exists no such VHE/UHE γ-ray survey observatory in the Southern Hemisphere, which would offer a clear view of the Galactic center. The Southern Wide-Field Gamma-Ray Observatory (SWGO) is a research & development project with the goal to fill

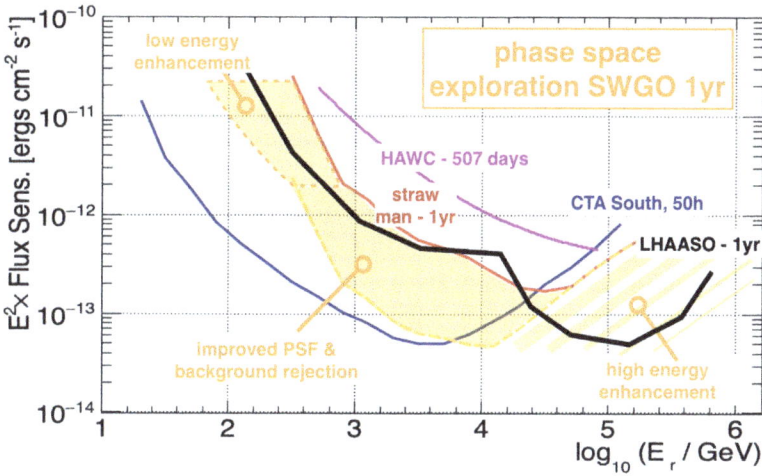

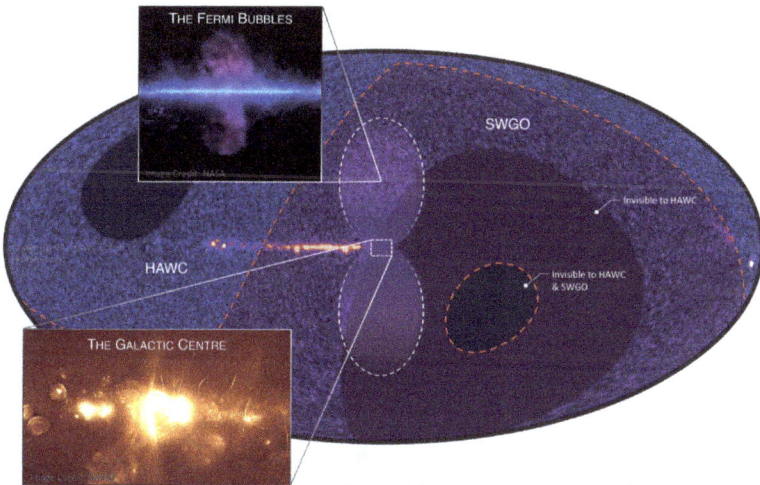

Fig. 13. Top panel: Differential point source sensitivity of different experiments.[14,30,87,88] Also shown are the sensitivity ranges that may be achievable for different SWGO designs.[89] Bottom panel: Sky coverage of the SWGO presented in the Galactic coordinate system. Overlaid is the HAWC significance map.[43]

this gap. Figure 13 shows the differential point source sensitivity in comparison with existing and emerging facilities (top)[89] and the sky coverage of the SWGO relative to the Galactic plane and center and the *Fermi* bubbles (bottom).[43,91] The shaded area in the upper plot indicates the phase space that will be explored in an optimization procedure guided by science potential and cost considerations for each design. The most conservative case is represented by the a red line and based on a scaled-up HAWC-like design of an earlier science case study.[88] There are three energy regimes where sensitivity enhancements are possible. The low-energy performance is for example tested against the ability to detect gamma-ray bursts and can be improved by lowering the energy threshold through detector unit design and configuration and building the observatory at higher elevation. The enhancements in the mid-energy range can be achieved through considerable improvements in angular resolution (by a factor of up to ~3) and background rejection (by a factor of up to ~10), where ideas include compact detector stations combined with machine learning algorithms to identify muons.[92–94] The recent LHAASO results at UHE γ-ray energies drive the exploration at the high energies, where the detection sensitivity can be enhanced by a large, low-density array with good background rejection (like the KM2A) surrounding the compact core array.

Several sites >4400 m in elevation are being evaluated in Argentina, Bolivia, Chile, and Peru for suitability for SWGO. A decision about the observatory design and location is expected by 2023/2024.

5.   Conclusion

Substantial advancement over the past three decades in detector technology, data reconstruction, and analysis methods (that latter made possible by a considerable increase in computing power) have lifted results from ground-based PDAs to another level of signal sensitivity and resolution in the space and time domains. The scientific contributions and discovery potential of PDAs make them crucial contributors to multi-messenger astronomy and fundamental physics. The SWGO, a next-generation PDA, promises to deliver similarly sensitive surveys of the Southern Hemisphere, expanding the search horizon for PeVatrons and other, possibly unexpected, signatures in the VHE and UHE *gamma*-ray sky toward the Galactic center.

References

1. K. Greisen, Cosmic ray showers, *Ann. Rev. Nucl. Part. Sci.* **10**, 63–108 (1960).
2. T. C. Weekes *et al.*, Observation of TeV gamma rays from the crab Nebula using the atmospheric Cerenkov imaging technique, *Astrophys. J.* **342**, 379 (1989).
3. S. P. Wakely and D. Horan, TeVCat: An online catalog for very high energy gamma-ray astronomy, *Int. Cosmic Ray Conf.* **3**, 1341 (2008). http://tevcat.uchicago.edu.
4. D. Heck, J. Knapp, J. N. Capdevielle, G. Schatz, and T. Thouw, CORSIKA: A Monte Carlo code to simulate extensive air showers (1998).
5. H. A. Ayala Solares, Search for high-energy gamma rays in the Northern Fermi Bubble Region with the HAWC Observatory, *Open Access Dissertation, Michigan Technological University* (2017).
6. F. Aharonian, J. Buckley, T. Kifune, and G. Sinnis, High energy astrophysics with ground-based gamma-ray detectors, *Rept. Prog. Phys.* **71**, 096901 (2008).
7. P. A. Zyla *et al.*, Review of particle physics, *PTEP.* **2020**(8), 083C01 (2020).
8. B. Rossi and K. Greisen, *Rev. Mod. Phys.*, **13**, 240 (1941).
9. R. W. Schiel and John P. Ralston, The Greisen equation explained and improved, *Phys Rev. D.* **75**, 016005 (2007).
10. M. de Naurois and D. Mazin, Ground-based detectors in very-high-energy gamma-ray astronomy, *Comptes Rendus Physique.* **16**, 610 (2015).
11. A. Abeysekara *et al.*, Data acquisition architecture and online processing system for the HAWC gamma-ray observatory, *Nucl. Instrum. Methods Phys. Res. A.* **888**, 138 (2018).
12. S. Wu *et al.*, Study of the trigger mode of LHAASO-KM2A, *Astropart. Phys.* **103**, 41 (2018).
13. A. Abeysekara *et al.*, On the sensitivity of the HAWC observatory to gamma-ray bursts, *Astropart. Phys.* **35**(10), 641 (2012).
14. A. U. Abeysekara *et al.*, Observation of the Crab Nebula with the HAWC gamma-ray observatory, *Astrophys. J.* **843**(1), 39 (2017).
15. D. E. Alexandreas *et al.*, Point source search techniques in ultrahigh-energy gamma-ray astronomy, *Nucl. Instrum. Meth. A.* **328**, 570 (1993).
16. M. C. Chantell *et al.*, Limits on the isotropic diffuse flux of ultrahigh-energy gamma radiation, *Phys. Rev. Lett.* **79**, 1805 (1997).

17. M. Amenomori *et al.*, Observation of multi TeV gamma-rays from the crab Nebula using the Tibet air shower array, *Astrophys. J. Lett.* **525**, L93 (1999).
18. M. Amenomori *et al.*, First detection of photons with energy beyond 100 TeV from an astrophysical source, *Phys. Rev. Lett.* **123**(5), 051101 (2019).
19. R. W. Atkins *et al.*, Milagrito: A TeV air shower array, *Nucl. Instrum. Meth. A.* **449**, 478 (2000).
20. Hamamatsu Photonics K.K. https://www.hamamatsu.com/resources/ pdf/etd/LARGE_AREA_PMT_TPMH1376E.pdf.
21. R. Atkins *et al.*, Evidence for TeV gamma-ray emission from the Galactic plane, *Phys. Rev. Lett.* **95**, 251103 (2005).
22. A. A. Abdo *et al.*, Discovery of TeV gamma-ray emission from the Cygnus region of the galaxy, *Astrophys. J. Lett.* **658**, L33 (2007).
23. A. A. Abdo *et al.*, Milagro observations of TeV emission from Galactic sources in the fermi bright source list, *Astrophys. J. Lett.* **700**, L127 (2009). [Erratum: *Astrophys. J. Lett.* 703, L185 (2009)].
24. A. Aloisio *et al.*, The ARGO-YBJ experiment in Tibet, *Nuovo Cim. C.* **24**, 739 (2001).
25. G. Di Sciascio, Physics results from the argo-YBJ experiment, *Italian Phys. Soc. Proc.* **98**, 555 (2009).
26. V. Marandon, A. Jardin-Blicq, and H. Schoorlemmer, Latest news from the HAWC outrigger array, *PoS.* **ICRC2019**, 736 (2020).
27. C. Zhen *et al.*, Introduction to large high altitude air shower observatory (LHAASO), *Chin. Astron. Astrophys.* **43**(4), 457 (2019).
28. F. Aharonian *et al.*, Performance of LHAASO-WCDA and observation of crab Nebula as a standard candle, *arXiv e-prints.* art. arXiv:2101.03508 (2021).
29. F. Aharonian *et al.*, The observation of the crab Nebula with LHAASO-KM2A for the performance study, *Chin. Phys. C.* **45**(2), 025002 (2021).
30. CTA Collaboration https://www.cta-observatory.org/science/cta-perf ormance.
31. R. Atkins *et al.*, Observation of TeV gamma-rays from the crab Nebula with MILAGRO using a new background rejection technique, *Astrophys. J.* **595**, 803 (2003).
32. K. Kamata and J. Nishimura, The lateral and the angular structure functions of electron showers, *Progr. Theor.Phys. Suppl.* **6**, 93 (1958).
33. A. U. Abeysekara *et al.*, Measurement of the crab Nebula at the highest energies with HAWC, *Astrophys. J.* **881**, 134 (2019).
34. A. A. Abdo *et al.*, Spectrum and morphology of the two brightest Milagro sources in the Cygnus region: MGRO J2019+37 AND MGRO J2031+41, *Astrophys. J.* **753**(2), 159 (2012).

35. B. Bartoli *et al.*, Galactic cosmic-ray anisotropy in the Northern Hemisphere from the ARGO-YBJ experiment during 2008–2012, *Astrophys. J.* **861**(2), 93 (2018).

36. A. Hillas, D. Marsden, J. Hollows, and H. W. Hunter, In *Proceedings ICRC (Tasmania)*, Vol. 12, p. 1001 (1971).

37. W. D. Apel *et al.*, Cosmic-ray energy reconstruction from the S(500) observable recorded in the KASCADE-grande air shower experiment, *Astropart. Phys.* **77**, 21 (2016).

38. D. Newton, J. Knapp, and A. A. Watson, The optimum distance at which to determine the size of a giant air shower, *Astropart. Phys.* **26**, 414 (2007).

39. K. Kawata, T. K. Sako, M. Ohnishi, M. Takita, Y. Nakamura, and K. Munakata, Energy determination of gamma-ray induced air showers observed by an extensive air shower array, *Exp. Astron.* **44**(1), 1 (2017).

40. A. Hocker *et al.*, TMVA — toolkit for multivariate data analysis, *arXiv e-prints.* art. physics/0703039 (2007).

41. I. J. Watson, Convolutional neural networks for low energy gamma-ray air shower identification with HAWC, *PoS.* **ICRC2021** (2021).

42. G. Di Sciascio, Ground-based gamma-ray astronomy: An introduction, *J. Phys. Conf. Ser.* **1263**(1), 012003 (2019).

43. P. Abreu *et al.*, The Southern Wide-Field Gamma-Ray Observatory (SWGO): A next-generation ground-based survey instrument for VHE gamma-ray astronomy (July, 2019).

44. P. Huentemeyer, Hunting the strongest accelerators in our galaxy, *Nature.* **594**, 30 (2021).

45. A. U. Abeysekara *et al.*, Extended gamma-ray sources around pulsars constrain the origin of the positron flux at earth, *Science.* **358**(6365), 911 (2017).

46. H. Abdalla *et al.*, The population of TeV pulsar wind Nebulae in the H.E.S.S. Galactic plane survey, *Astron. Astrophys.* **612**, A2 (2018).

47. A. M. Atoian, F. A. Aharonian, and H. J. Volk, Electrons and positrons in the Galactic cosmic rays, *Phys. Rev. D.* **52**, 3265 (1995).

48. O. Adriani *et al.*, An anomalous Positron abundance in cosmic rays with energies 1.5–100 GeV, *Nature.* **458**, 607 (2009).

49. M. Aguilar *et al.*, First result from the alpha magnetic spectrometer on the international space station: Precision measurement of the Positron fraction in primary cosmic rays of 0.5–350 GeV, *Phys. Rev. Lett.* **110**, 141102 (2013).

50. H. Yuksel, M. D. Kistler, and T. Stanev, TeV gamma rays from Geminga and the origin of the GeV positron excess, *Phys. Rev. Lett.* **103**, 051101 (2009).

51. T. Linden *et al.*, Using HAWC to discover invisible pulsars, *Phys. Rev. D.* **96**(10), 103016 (2017).
52. T. Linden and B. J. Buckman, Pulsar TeV Halos explain the diffuse TeV excess observed by Milagro, *Phys. Rev. Lett.* **120**(12), 121101 (2018).
53. T. Sudoh, T. Linden, and J. F. Beacom, TeV Halos are everywhere: Prospects for new discoveries, *Phys. Rev. D.* **100**(4), 043016 (2019).
54. H. Fleischhack *et al.*, Pulsars in a bubble? Following electron diffusion in the galaxy with TeV gamma rays, *BAAS.* **51**(3), 311 (2019).
55. G. Giacinti, A. M. W. Mitchell, R. López-Coto, V. Joshi, R. D. Parsons, and J. A. Hinton, Halo fraction in TeV-bright pulsar wind Nebulae, *Astron. Astrophys.* **636**, A113 (2020).
56. F. Aharonian *et al.*, Extended very-high-energy gamma-ray emission surrounding PSR J0622+3749 observed by LHAASO-KM2A, *Phys. Rev. Lett.* **126**(24), 241103 (2021).
57. M. Di Mauro, S. Manconi, and F. Donato, Detection of a γ-ray Halo around Geminga with the Fermi-LAT data and implications for the Positron flux, *Phys. Rev. D.* **100**(12), 123015 (2019).
58. A. U. Abeysekara *et al.*, Very high energy particle acceleration powered by the jets of the microquasar SS 433, *Nature.* **562**(7725), 82 (2018). [Erratum: Nature 564, E38 (2018)].
59. W. Brinkmann, B. Aschenbach, and N. Kawai, ROSAT observations of the W 50/SS 433 system, *Astron. Astrophys.* **312**, 306–316 (August, 1996).
60. G. Vianello *et al.*, The multi-mission maximum likelihood framework (3ML). p. arXiv:1507.08343 (2015).
61. A. U. Abeysekara *et al.*, HAWC observations of the acceleration of very-high-energy cosmic rays in the Cygnus Cocoon, *Nature Astron.* **5**(5), 465 (2021).
62. B. Bartoli *et al.*, Identification of the TeV gamma-ray source ARGO j2031+4157 with the Cygnus Cocoon, *Astrophys. J.* **790**(2), 152 (2014).
63. J. Ballet, T. H. Burnett, S. W. Digel, and B. Lott, Fermi large area telescope fourth source catalog data release 2, *arXiv e-prints.* art. arXiv:2005.11208 (2020).
64. M. Ackermann *et al.*, A Cocoon of freshly accelerated cosmic rays detected by fermi in the Cygnus superbubble, *Science.* **334**(6059), 1103 (2011).
65. F. Aharonian, R. Yang, and E. de Oña Wilhelmi, Massive stars as major factories of Galactic cosmic rays, *Nature Astron.* **3**(6), 561 (2019).
66. M. Aguilar *et al.*, Precision measurement of the proton flux in primary cosmic rays from rigidity 1 GV to 1.8 TV with the alpha magnetic spectrometer on the international space station, *Phys. Rev. Lett.* **114**, 171103 (2015).

67. A. Abramowski *et al.*, Acceleration of petaelectronvolt protons in the Galactic centre, *Nature.* **531**, 476 (2016).
68. V. A. Acciari *et al.*, MAGIC very large Zenith angle observations of the crab Nebula up to 100 TeV, *Astron. Astrophys.* **635**, A158 (2020).
69. A. U. Abeysekara *et al.*, Multiple Galactic sources with emission above 56 TeV detected by HAWC, *Phys. Rev. Lett.* **124**(2), 021102 (2020).
70. A. Albert *et al.*, HAWC J2227+610 and its association with G106.3+2.7, a new potential Galactic PeVatron, *Astrophys. J. Lett.* **896**, L29 (2020).
71. Z. Cao *et al.*, Ultrahigh-energy photons up to 1.4 petaelectronvolts from 12 γ-ray galactic sources, *Nature.* **594**(7861), 33 (2021).
72. M. Amenomori *et al.*, First detection of sub-PeV diffuse gamma rays from the Galactic disk: Evidence for ubiquitous Galactic cosmic rays beyond PeV energies, *Phys. Rev. Lett.* **126**(14), 141101 (2021).
73. P. Huentemeyer, Signs of PeVatrons in gamma-ray haze, *APS Phys.* **14**, 41 (2021).
74. A. Albert *et al.*, Constraints on Lorentz invariance violation from HAWC observations of gamma rays above 100 TeV, *Phys. Rev. Lett.* **124**(13), 131101 (2020).
75. A. U. Abeysekara *et al.*, A search for dark matter in the Galactic Halo with HAWC, *JCAP.* **02**, 049 (2018).
76. A. Albert *et al.*, Search for gamma-ray spectral lines from dark matter annihilation in dwarf galaxies with the high-altitude water Cherenkov observatory, *Phys. Rev. D.* **101**(10), 103001 (2020).
77. H. Vogel, R. Laha, and M. Meyer, Diffuse axion-like particle searches, *PoS.* **NOW2018**, 091 (2019).
78. Z. Cao *et al.*, Exploring Lorentz invariance violation from ultra-high-energy gamma rays observed by LHAASO, *arXiv e-prints.* art. arXiv:2106.12350 (2021).
79. A. Albert *et al.*, 3HWC: The third HAWC catalog of very-high-energy gamma-ray sources, *Astrophys. J.* **905**(1), 76 (2020).
80. A. U. Abeysekara *et al.*, The 2HWC HAWC observatory gamma ray catalog, *Astrophys. J.* **843**(1), 40 (2017).
81. A. A. Abdo *et al.*, TeV gamma-ray sources from a survey of the Galactic plane with Milagro, *Astrophys. J. Lett.* **664**, L91–L94 (2007).
82. M. Amenomori *et al.*, Observation of TeV gamma rays from the fermi bright Galactic sources with the Tibet air shower array, *Astrophys. J. Lett.* **709**(1), L6 (2010).
83. B. Bartoli *et al.*, TeV gamma-ray survey of the Northern sky using the ARGO-YBJ detector, *Astrophys. J.* **779**, 27 (2013).

84. A. U. Abeysekara *et al.*, Search for TeV gamma-ray emission from point-like sources in the inner Galactic plane with a partial configuration of the HAWC observatory, *Astrophys. J.* **817**(1), 3 (2016).

85. K. M. Gorski *et al.*, HEALPix — A framework for high resolution discretization, and fast analysis of data distributed on the sphere, *Astrophys. J.* **622**, 759 (2005).

86. A. A. Abdo *et al.*, Fermi large area telescope first source catalog, *Astrophys. J. Suppl.* **188**(2), 405 (2010).

87. S. Vernetto, Gamma ray astronomy with LHAASO, *J. Phys.* **718**, 052043 (2016).

88. A. Albert *et al.*, Science case for a wide field-of-view very-high-energy gamma-ray observatory in the Southern hemisphere, *arXiv e-prints*. art. arXiv:1902.08429 (2019).

89. H. Schoorlemmer, Simulating the performance of the Southern wide-view gamma-ray observatory, *PoS*. **ICRC2021**, 903 (2021).

90. H. Abdalla *et al.*, TeV emission of Galactic plane sources with HAWC and H.E.S.S, *arXiv e-prints*. art. arXiv:2107.01425 (2021).

91. M. Su, T. R. Slatyer, and D. P. Finkbeiner, Giant gamma-ray bubbles from Fermi-LAT: Active Galactic nucleus activity or bipolar Galactic wind? *Astrophys. J.* **724**(2), 1044–1082 (2010).

92. W. Hofmann, On angular resolution limits for air shower arrays, *Astropart. Phys.* **123**, 102479 (2020).

93. S. Kunwar, Double-layered water Cherenkov detector for SWGO, *PoS*. **ICRC2021**, 902 (2021).

94. R. Conceicao, Gamma/hadron discrimination using a small-WCD with four PMTs, *PoS*. **ICRC2021**, 707 (2021).

https://doi.org/10.1142/9781800612617_0009

Chapter 9

Networks of Atmospheric Cherenkov Telescopes for Very-High-Energy Gamma-Ray Astrophysics

Justin A. Vandenbroucke

Department of Physics, Wisconsin IceCube
Particle Astrophysics Center, University of Wisconsin–Madison,
Madison, WI 53706, USA
justin.vandenbroucke@wisc.edu

Networks of imaging atmospheric Cherenkov telescopes enable sensitive measurements of very-high-energy gamma rays through the Cherenkov light flash emitted by the particle showers that the gamma rays initiate in the atmosphere. Using a network of telescopes, rather than a single telescope, provides a larger effective gamma-ray–collecting area as well as stereoscopic imaging, resulting in improved signal reconstruction and background rejection. The power of the networked telescope technique has been established by the current generation of instruments, including H.E.S.S., MAGIC, and VERITAS. Construction of the next-generation instrument, the Cherenkov Telescope Array (CTA), is underway and will build on the success of the current generation to provide unprecedented sensitivity to TeV-scale gamma rays for astrophysics and multi-messenger astronomy.

1. Introduction

Following the initial detection of the first Galactic[1] and extragalactic[2,3] sources of very-high-energy (0.1 TeV–0.3 PeV) gamma rays by a single imaging atmospheric Cherenkov telescope (the Whipple 10 m telescope), the current generation of instruments (H.E.S.S., MAGIC, and VERITAS) have proven the potential of networked arrays of such telescopes for discoveries in physics and astrophysics. By viewing the same shower from slightly different angles (but still head-on), multiple telescopes detecting the same gamma-ray–initiated shower can much better determine its direction, as well as energy, and discriminate these signals from the more abundant cosmic-ray–initiated showers. The next-generation observatory, the Cherenkov Telescope Array (CTA), will build on this progress with two large arrays featuring large collecting area for detecting rare VHE photons. Gamma rays provide the key link from other messengers (neutrinos, gravitational waves, cosmic rays, and even dark matter) to the electromagnetic spectrum, and CTA will enable a range of scientific discoveries in multi-messenger, time-domain, and high-energy astrophysics.

2. Technique

Upon colliding with Earth's atmosphere, a very-high-energy (100 GeV–300 TeV) gamma ray produces an electron–positron pair, which in turn radiates *bremsstrahlung* photons that themselves pair produce, and the interactions continue in an electromagnetic cascade (air shower) in the atmosphere. Similar electromagnetic cascades are also produced by cosmic-ray electrons (or positrons) incident on the atmosphere. Energetic hadrons initiate analogous hadronic air showers, which present a large background that must be discriminated in order to identify air showers initiated by gamma rays (or electrons/positrons).

In addition to dense media, such as water, ice, and plastic, the atmosphere acts as a Cherenkov radiator for sufficiently energetic charged particles. The charged particles in air showers therefore emit Cherenkov radiation. Because the index of refraction, n, of the atmosphere is only a few parts per thousand above unity, the Cherenkov angle $\theta = \cos^{-1}\frac{1}{\beta n}$ is ~1°. Combined with the fact that TeV air showers are highly forward beamed, this means that Cherenkov emission is forward beamed in the same direction as the primary cosmic particle.

To illustrate the technique, consider a vertical, one-dimensional (pencil-beam) air shower initiated high in the atmosphere and propagating all the way to ground level (assume the detector is at very high altitude and/or the shower is very energetic). At the top of the atmosphere, where the density is small, the Cherenkov angle is barely greater than zero, and so any Cherenkov photons emitted early in the shower reach ground level at the center of the air shower core. As the shower develops into denser atmosphere, the Cherenkov angle grows and the photons illuminate a wider circle surrounding the core. Finally, the last emitted photons have the largest Cherenkov angle but zero distance to travel, so they again impact at the axis of the air shower. The maximum impact parameter of the photons determines the radius of the light pool, which is about 140 m for typical atmospheric density profiles and instrument altitudes. This implies that a single instrument can detect any air shower with impact point within $\sim$140 m and therefore has an effective area for detecting primary particles of $\sim 6 \times 10^4$ m^2.

The Cherenkov light pool radius also sets the approximate optimal spacing for a network of instruments. The precise spacing and array layout can be chosen to optimize the trade-off between spanning a larger footprint (and therefore providing a larger gamma-ray collection area) and having a larger number of telescopes detecting individual gamma rays (and therefore providing improved signal reconstruction and background rejection).

For imaging atmospheric Cherenkov instruments, the network detector unit is a focusing telescope that employs one or two large mirrors to focus the Cherenkov light onto a camera. The camera typically consists of hundreds to thousands of pixels of photodetectors (photomultiplier tubes or silicon photomultipliers) along with fast electronics to trigger on and read out the signals. Photodetector waveforms are typically read out. The Cherenkov flash is $\mathcal{O}(10)$ ns in duration and must be discriminated from the night-sky background produced by celestial objects (mostly stars) as well as both steady and variable anthropogenic light pollution. For these reasons, imaging atmospheric Cherenkov telescopes (IACTs) perform best at sites similar to those used for optical astronomy: those with low light pollution and good weather, with little cloud cover to obscure the Cherenkov light, and they perform best in dark night conditions. However, operation can be achieved in partial moonlight by using filters (blue pass, red block), reducing the photodetector gain, and/or using silicon photomultipliers rather than photomultiplier tubes in order to tolerate the large background light.

This telescope design enables imaging (and, with sufficient photodetector and electronics time resolution, even video capture with ~1 ns per frame) of each air shower. Typically, showers are viewed slightly off-axis and are therefore highly foreshortened. Since they are not viewed directly on-axis, their shape and orientation can be measured and used to determine the direction of the original gamma ray. If multiple telescopes image the same shower, better than 0.03° angular resolution can be achieved.[4]

The IACT technique is complementary to the ground-based particle detector array technique discussed in Chapter 8 of this book. The particle detector technique can view a larger fraction of the sky instantaneously and can observe with a larger duty cycle because it can operate in daylight. The particle detector technique is well suited for large-area sky surveys: For steady (or at least long-duration) sources, a large amount of exposure can be accumulated for a large number of sources by integrating many "transits" of each source overhead, each with a modest effective area. However, the IACT technique typically has superior background rejection, effective area, and angular resolution. These features combined imply that for deep observations of individual sources, morphological studies, or short durations (flaring or transient sources), the IACT technique is better suited.

Because they directly detect air shower particles, particle detector arrays must be located at high altitude. IACTs can be situated at lower altitude because they remotely sense the air showers via their Cherenkov light. For an IACT, lower altitude enables larger Cherenkov light pool, but this is overcome by atmospheric attenuation (especially for small showers produced by low-energy gamma rays) if the altitude is too low.

The IACT technique was established by the Whipple 10 m telescope, which detected the first source (the Crab Nebula) using the technique in 1989.[1] Subsequent instruments have demonstrated the importance of networking multiple telescopes together not only to increase the footprint (and therefore provide a larger effective area for collecting the rare VHE gamma rays) but also to improve background rejection and signal (direction and energy) reconstruction.

Because the Crab Nebula is the brightest and first source detected in this energy range, a convenient unit of VHE gamma-ray flux is the "Crab" (and "milliCrab"). Of course, the spectrum of the Crab Nebula is steeply falling (and curved), so the unit cannot be applied straightforwardly to different energy bands or spectral shapes but is convenient for approximate statements.

For the lowest detectable energy gamma rays of an IACT (near threshold), the showers are highest in the atmosphere, produce the lowest Cherenkov light intensity, and illuminate the fewest pixels in the camera. They are impacted by systematic effects including those of scattering and absorption in the atmosphere.

At TeV–PeV energies, the view of the universe by IACTs and other instruments is limited by absorption of gamma rays on the extragalactic background light (EBL). The EBL is the diffuse glow of optical, ultraviolet, and infrared light produced by the aggregate emission of galaxies and their contents, including reprocessing to longer (infrared) wavelengths. Collisions of the gamma rays with these low-energy photons results in electron–positron pair production. The process can continue cascading to lower energies. The resulting absorption of gamma rays is more severe at higher than lower energies in the VHE band. At the highest energies, sources must be especially bright (nearby and/or luminous) or instruments must be especially sensitive to overcome this absorption. Models quantifying the EBL can be used to estimate the amount of absorption as a function of energy and redshift. For a source with a known redshift, the model can then be used to convert the measured spectrum to the "intrinsic" spectrum that would have been measured without EBL absorption (the cosmological redshift of gamma rays between emission and detection must also be considered). EBL models are uncertain, and GeV–TeV measurements of source spectra at a variety of redshifts can be used to improve them. EBL absorption is quantified in Fig. 1.

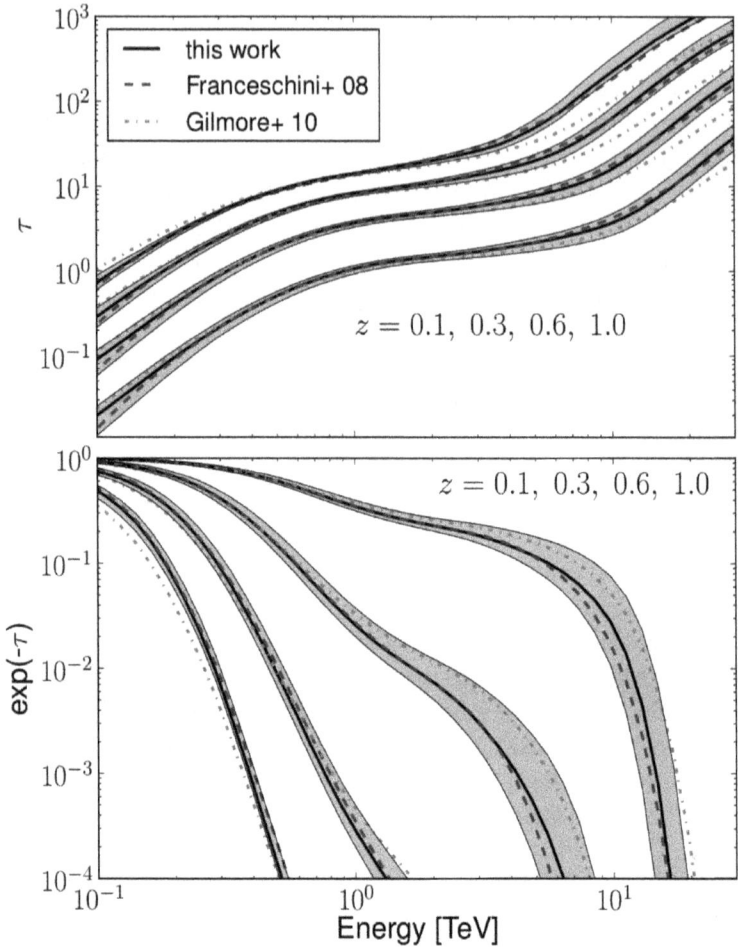

Fig. 1. Model of gamma-ray absorption by extragalactic background light (EBL). Taken from Ref. [5], to which "this work" refers. Gray bands indicate the estimated uncertainty of the model. Two additional models are also shown. Curves are included for four different redshifts z. Top panel: optical depth for absorption; bottom panel: resulting absorption coefficient (transmissivity). Both quantities are plotted as a function of the gamma-ray energy observed at Earth, after cosmological redshift.

3.   Current Instruments

3.1.   *H.E.S.S., MAGIC, VERITAS, and FACT*

Three IACT arrays have established the scientific power of networked IACTs. H.E.S.S is located in Namibia, providing excellent coverage of the southern sky, which enabled it to discovery a plethora of sources in the inner Galaxy. H.E.S.S. originally consisted of four telescopes and subsequently a fifth telescope with a very large mirror was constructed to provide a lower energy threshold. MAGIC is located on the island of La Palma in the Canary Islands and consists of two telescopes, designed with a low-mass (carbon fiber) design and fast slewing capabilities for detection of transient sources including gamma-ray bursts. VERITAS consists of four telescopes at the Fred Lawrence Whipple Observatory on the same mountain where the IACT technique was first established. These three instruments each use photomultiplier cameras. Another telescope design, FACT, was developed to establish the viability of silicon photomultipliers for IACTs.

3.2.   *Scientific results*

Since the detection of the Crab Nebula three decades ago, these instruments (in concert with the particle detector array technique pioneered by MILAGRO and further developed by HAWC) have established that there are a large number and variety of VHE gamma-ray sources. Over 200 individual sources have now been detected, spanning multiple Galactic and extragalactic source classes. Galactic VHE emitters include supernova remnants, pulsar wind nebulae (PWNs) as well as pulsars themselves, stellar binary systems including X-ray emitting binaries, massive star clusters, and star-forming regions. Extragalactic sources are mostly blazars, including both those of the BL Lac and flat-spectrum radio quasar (FSRQ) type. Detected non-blazar active galactic nuclei (AGNs) include the radio galaxies Centaurus A and IC 310 (a head-tail galaxy). In addition to AGNs, starburst galaxies including NGC 253 and M 82 have been detected. Since 2019, gamma-ray bursts have been established as an additional extragalactic source class detected in the VHE band by IACTs. Perhaps most intriguing, there are also dozens of unidentified sources.

After decades of searching, IACTs have recently detected four GRBs. In each case, the GRB was detected by following up a long-duration GRB first discovered by a satellite. MAGIC detected GRB 190114C (first discovered by Swift-BAT) between 0.2 and 1 TeV over a time range from 10^2 to 10^3 s.[6] The statistical significance is more than 50 sigma in the first 20 minutes. Its redshift was measured to be 0.4245. MAGIC also used the GRB observations to search for variation of the speed of light with energy, a form of Lorentz invariance violation predicted by many theories of quantum gravity.[7] H.E.S.S. detected VHE emission from GRB 180720B (initially discovered by both Fermi-GBM and Swift-BAT within 5 s of one another and measured to be at at $z = 0.653$) at 10 hours after the prompt burst.[8] H.E.S.S. also detected GRB 190829A, a nearby, low-luminosity burst at $z = 0.0785$, between 4 and 56 hours after the prompt burst, at energies between 0.18 and 3.3 TeV.[9] Finally, MAGIC detected GRB 201216C with greater than 5 sigma significance.[10] These measurements are enabling new insights into the nature of GRBs.

A notable exception in the list of detected source classes is galaxy clusters. Because they contain particle accelerators (cosmic-ray sources) and target gas (intra-cluster medium), they are expected to emit gamma rays but have so far eluded detection despite sensitive searches in both the GeV and TeV bands. In addition to cosmic-ray–induced emission, galaxy clusters are expected (and known, based on interacting clusters such as the Bullet Cluster) to be good repositories of dark matter. They are therefore excellent targets in the "indirect" search for dark matter via the Standard Model particles (including gamma rays) it is likely to produce if it self-annihilates or decays. In addition to galaxy clusters, good targets for indirect dark matter searches include the Galactic center halo and satellite galaxies of the Milky Way including dwarf galaxies which have a large mass-to-light ratio and are therefore believed to be rich in dark matter. Indirect dark matter searches with VHE gamma rays probe regions of WIMP parameter space inaccessible to direct detection or accelerator searches.[11]

In addition to VHE gamma-ray science, IACT networks are being used for a growing range of other exciting scientific applications. These applications exploit the unique advantages of IACTs compared to other telescopes that detect visible light: Although IACTs have worse imaging resolution, they have enormous mirror areas and excellent time resolution. These non-gamma-ray optical applications can use the main gamma-ray camera or additional instruments placed in the focal plane alongside the gamma-ray camera. They include searches for fast optical flashes that could be produced by distant technologies (optical searches for extra-terrestrial intelligence, OSETI)[12] and stellar intensity interferometry (SII),[13] in which the networked nature of IACTs is deeply exploited: interferometry (using light intensity rather than amplitude as is used in typical astronomical interferometers) with multiple telescopes can be used to measure the size and even shape of stars.

4. The Cherenkov Telescope Array

The next-generation IACT array, the Cherenkov Telescope Array (CTA), is now under construction. CTA promises up to an order of magnitude improvement in sensitivity compared to current IACT arrays, resulting in detection of $\sim 10^3$ VHE sources as well as studies of Galactic and extragalactic sources with excellent spectral, morphological, and temporal resolution. This will enable new discoveries concerning the sources of cosmic messengers including cosmic rays, gravitational waves, and neutrinos, and fundamental physics including searches for axion-like particles, Lorentz invariance violation, and the nature of dark matter.

In order to cover the entire celestial sphere, CTA will be located at two sites, one in the Northern Hemisphere and one in the Southern Hemisphere. CTA will include three sizes of telescopes. Large-sized telescopes (LSTs) have the largest mirror surfaces in order to achieve a low energy threshold of 20 GeV. Medium-sized telescopes (MSTs) cover the core energy range of CTA, from 0.1 TeV to 10 TeV. Small-sized telescopes (SSTs) cover the highest energy range, from 10 to 300 TeV. Because the highest energy gamma rays are the least abundant, SSTs need to span the largest footprint and can do so because they are small enough to be built in large quantity. Conversely, it is acceptable for a small number of LSTs to span a small area, and an intermediate number and footprint of MSTs is appropriate. According to this logic, there are few LSTs, many SSTs, and an intermediate number of MSTs.

The northern array will be located on La Palma at the Roque de Los Muchachos Observatory in the Canary Islands, Spain (near the two MAGIC telescopes). The southern array will be located in Paranal in the Atacama Desert, Chile. Funding is in place for construction of an initial ("alpha") array configuration. This array consists of 4 LSTs and 9 MSTs in the northern array (spanning a footprint of ~0.25 km^2), along with 14 MSTs and 37 SSTs in the southern array (spanning a footprint of ~3 km^2). The southern array is particularly well suited for Galactic sources, with a good view of the inner Galaxy and Galactic center and with SSTs providing excellent sensitivity at the 100 TeV energy scale. The northern array will be especially well suited for extragalactic sources, for which no SSTs are necessary due to absorption of the highest energy gamma rays by the EBL.

The first CTA telescope, LST-1, has been constructed and is performing scientific observations. It has already detected several VHE gamma-ray sources.

CTA will be operated as an observatory, the CTA Observatory (CTAO), featuring a time allocation committee that will prioritize among observing proposals submitted by scientists. The CTA Consortium (CTAC) has priority for observing time in the initial years of operation, gradually ramping down as more time is made available for the general scientific community. CTAC has organized its observing priority into several key science projects (KSPs) with observing targets matched to scientific objectives.[4]

4.1. *CTA telescope technologies*

Figure 2 illustrates the telescopes that have been designed for CTA, and Table 1 summarizes their characteristics. The LST and MST use single mirror dishes and photomultiplier cameras, similar to the current generation of IACTs. The SST design uses a Schwarzschild–Couder optical system consisting of two mirror dishes and a silicon photomultiplier camera. An additional design for the MST, known as the Schwarzschild–Couder telescope (SCT), is also based on a Schwarzschild–Couder optical system and silicon photomultiplier camera.

Fig. 2. CTA telescope designs. From left to right: SST, SCT, MST, and LST. Note the person and vehicle for scale.

Table 1. CTA telescope technologies.

Telescope	Energy range (TeV)	Mirror diameter (m)	FOV (deg)	Pixels	Photodetectors
LST	<0.1	23.0	4.3	1855	PMT
MST	0.1–10	11.5	7.5–7.7	1764–1855	PMT
SST	>10	4.3	10.5	2368	SiPM
SCT	0.1–10	9.7	8	11328	SiPM

Note: Large-sized telescope (LST), medium-sized telescope (MST), small-sized telescope (SST), and Schwarzschild–Couder telescope (SCT). For the MST, two distinct camera designs are planned, differing primarily in their electronics but also with slightly different pixel count and field of view. For dual-mirror telescopes, the primary mirror diameter is listed. FOV is the diameter of the field of view.

A prototype of each of the telescopes has been constructed: SST in Sicily, Italy; MST in Adlershof, Germany; SCT in Arizona, USA; and LST on La Palma (at the CTA northern site next to the MAGIC telescopes). The LST prototype is also considered the first telescope of CTA, while the others serve primarily as prototypes and are not planned to be incorporated into CTA itself. In addition to the SSTs at the northern site, an array of nine telescopes of a similar but different design (the ASTRI Mini Array) is planned to be constructed on Tenerife, Canary Islands.

In the core energy range of CTA, the SCT design enables excellent imaging resolution thanks both to the dual-mirror configuration and the small (0.067°) SiPM pixels. The excellent shower imaging resolution translates directly into improved background rejection and gamma-ray imaging resolution. A prototype SCT (pSCT) has been constructed and operated at the Fred Lawrence Whipple Observatory, next to the VERITAS array. The pSCT has established the viability of the telescope design concept, including detection of the Crab Nebula.[14] An excellent optical point-spread function has been measured, and the camera is being upgraded. The upgrade will replace the readout electronics with an improved version with lower noise and completely instrument the focal plane (increasing the instrumented region from 1600 pixels spanning 2.7° square to 11,328 pixels spanning 8° diameter). Based on the performance of the upgraded camera, characterization of the optical system, and experience operating the pSCT, an SCT design for CTA will be finalized. SCTs are intended for a second phase of CTA construction, for example by adding 11 SCTs to the southern array to bring the total number of MSTs to 25.

4.2.   *CTA performance*

The projected sensitivity of the initial (alpha) CTA configuration is shown in Fig. 3, and the angular resolution is shown in Fig. 4.

4.3.   *CTA science*

CTA will build on the success of the current generation of IACTs and complement modern observatories focused on other wavelengths and messengers to pursue a wide range of multi-messenger, high-energy, and time-domain astrophysics. As illustrated by GW 170817/GRB 170817A[8] and IC 170922/TXS 0506+056,[15,16] gamma-ray telescopes are essential for identifying the sources of gravitational waves and neutrinos and linking them to the wealth of information available across the spectrum of the traditional messenger, the photon.

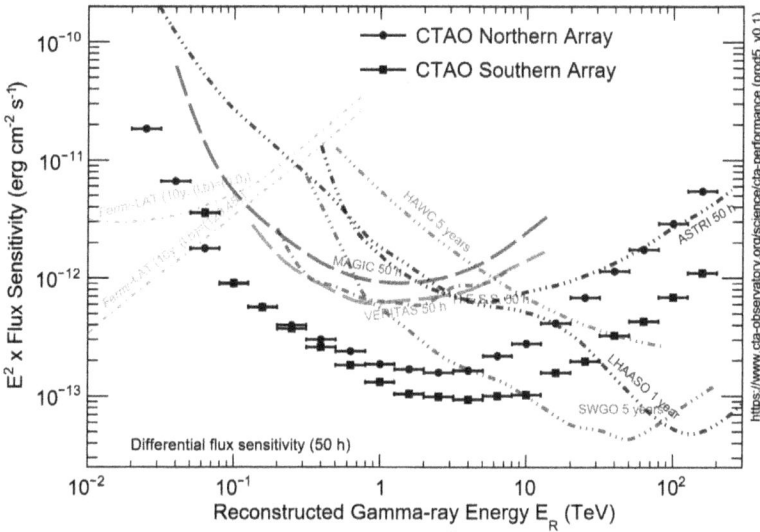

Fig. 3. Differential sensitivity of each of CTA's arrays to point-like sources of gamma rays, achieved in 50 h of observing time[a]. For comparison, existing IACTs (H.E.S.S., MAGIC, and VERITAS) are shown, as are current (HAWC and LHAASO) and future (SWGO) particle detector arrays as well as the Fermi LAT and ASTRI mini-array. CTA will provide approximately an order of magnitude improvement in sensitivity compared to the current generation of IACTs, lowering the threshold from $\sim 10^{-12}$ to $\sim 10^{-13}$ erg cm^{-2} s^{-1}.

4.3.1. *Time-domain and real-time astrophysics*

As shown in Fig. 3, the Fermi LAT is competitive with CTA at the low-energy range (with a cross-over energy of ~ 50 GeV for CTA-North) for observations that can be integrated over many years. This includes steady sources and studies of variable sources for which measuring their average behavior is sufficient. However, in the cross-over region of this figure, the Fermi-LAT sensitivity is achieved by integrating a decade of small effective area, compared to the same sensitivity achieved for CTA by integrating only 50 h of a much larger effective area. For short-duration scientific topics including fast variability and transient sources, long integration times are not possible and the large instantaneous sensitivity of CTA is essential. This is quantified in Fig. 5.

---

[a]https://www.cta-observatory.org/science/ctao-performance.

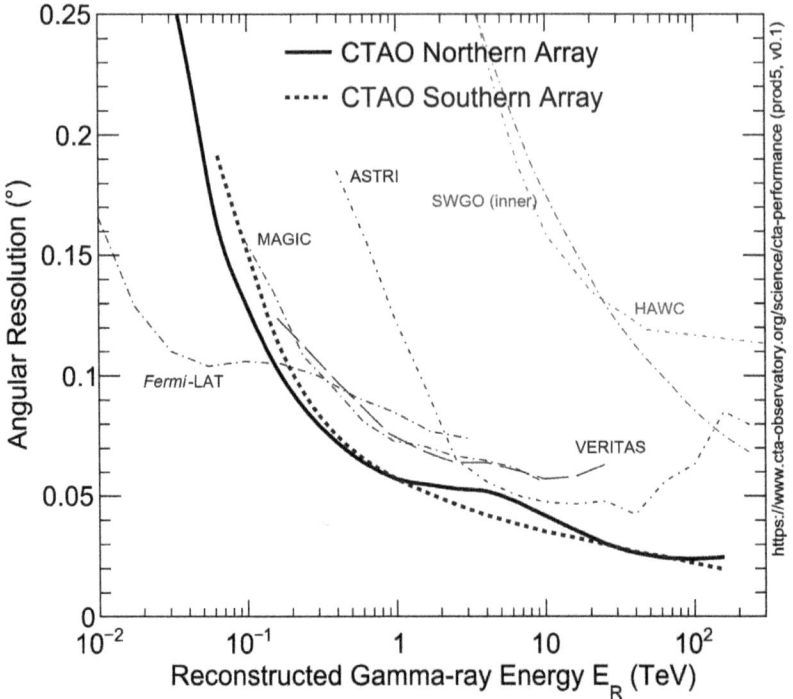

Fig. 4. Gamma-ray angular resolution (68% containment) of each of the CTA arrays compared to other instruments. CTA will achieve angular resolution better than any other instrument above 0.2 TeV and better than 0.03° above 100 TeV. Because of this, it will be especially well suited for identifying extended sources, resolving sources that are still confused in existing observations, and using morphology studies to determine source emission mechanisms (including distinguishing hadronic from leptonic emission).

Many of the most important topics in modern astrophysics occur in the time domain, sometimes on short (hour, minute, or even subsecond) timescales. Because of this, CTA is designed to both generate and respond to external alerts quickly. In response to external alerts such as GRB detections, for example, CTA can respond by repointing its telescopes within $\sim$30 s. CTA is also developing software pipelines both to analyze known targets quickly in order to generate alerts (e.g., in response to a flaring blazar) and also to identify new transients that occur serendipitously within the field of view during observation. Both types of phenomena will be announced publicly through rapid alerts.

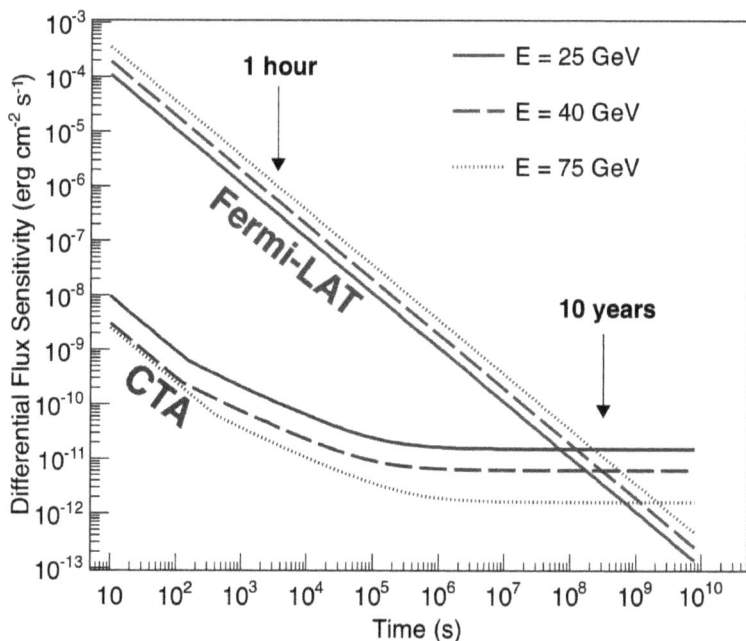

Fig. 5.   Comparison of Fermi-LAT to CTA sensitivity as a function of observing time. For steady sources and time-averaged studies, decade-scale LAT integration times provide excellent sensitivity. For short observation times, including for studying transients and variable sources, CTA will provide sensitivity superior to any other ground- or space-based instrument for energies above 20 GeV.

4.3.2.   *Astrophysical neutrinos*

The IceCube Neutrino Observatory has discovered a flux of high-energy astrophysical neutrinos (see Chapter 5 "Neutrinos: Scientific Motivation" of this book). Within existing observational constraints, the neutrino flux is consistent with being isotropic and diffuse. These characteristics imply that (1) the neutrino sources are predominantly extragalactic and (2) there is a large number of sources. To date there is evidence for one identified source, TXS 0506+056, which was identified thanks to GeV and TeV (IACT) gamma-ray observations.[15,16]

Beyond this, there is so far a surprising lack of correlation between high-energy neutrinos and gamma-ray sources. The high-energy neutrinos are likely produced in or near astrophysical particle accelerators by hadronic processes (interactions of protons with either photons or other protons) to produce both neutral and charged pions.

The neutral pions decay to gamma rays, while the charged pions decay to neutrinos (among other products). Because of this, high-energy gamma-ray production is generically expected to accompany high-energy neutrino production. The lack of a clear correlation between the two indicates either that gamma rays are absorbed within or near the sources (gamma-opaque sources) or that the neutrino sources are much more distant than the known gamma-ray sources, beyond the horizon of current gamma-ray instruments.

Future observations in the directions of high-energy neutrinos will likely identify additional sources, as has been demonstrated by the case of TXS 0506+056. Systematically aggregating many such follow-up observations, each with or without a gamma-ray detection, will constrain the relationship between neutrino-emitting sources and gamma-emitting sources, including whether most neutrino sources are gamma opaque or beyond the gamma-ray detection horizon. It is likely that the neutrino sky has a wide variety of sources, as is true of the sky viewed in gamma rays and other electromagnetic bands as well as gravitational waves. At the highest energies, the gamma-ray horizon is set by the universe itself due to EBL absorption. This exponential absorption is difficult to overcome with more sensitive instruments. At intermediate and lower energies, however, the new generation of instruments including CTA will expand the detectable volume of gamma-ray emitters, potentially including neutrino sources.

4.3.3. *Gamma-ray bursts and gravitational waves*

Now that four GRBs have been detected by IACTs (each in the afterglow phase by following up initial discoveries by satellites), the prospects for CTA detection of GRBs are bright. Population simulations estimate a CTA detection rate of ∼2 per year (with large uncertainty). Compared to the X-ray and gamma-ray satellites that discover GRBs thanks to their large field of view and duty cycle, the challenge for IACTs is that the fraction of observable GRBs (given Sun, Moon, and weather constraints) is smaller, and the probability of a burst occurring within the field of view is also smaller. IACT detection of a GRB in the prompt phase therefore remains a challenging (but worthwhile) goal, achievable through continuous monitoring for serendipitous bursts detected (either with or without a known component detected via satellite) in the field of view.

A complementary strategy, now proven, is to follow up satellite-detected bursts as quickly as possible and to continue observations to late times (multiple days!) to study the afterglow phase. Although the rate of GRB detections in the VHE band will remain small compared to the satellite-detected bursts, the bursts that are detected will be measured precisely with excellent sensitivity at the highest energies and will yield important insights concerning the emission mechanisms and surrounding environments of these brightest explosions in the Universe. They will also be used for tests of fundamental physics including searches for Lorentz invariance violation to test the nature of spacetime.[7]

One GRB in particular, GRB 170817A, proved the long theorized link between short-duration gamma-ray bursts and compact-object mergers, thanks to its detection in coincidence with GW 170817.[17] Gamma rays (in this case, those detected by Fermi-GBM) provided the essential link from gravitational waves to the electromagnetic spectrum. Thanks to the ongoing preparations of LIGO, Virgo, and KAGRA for observing run O4, gravitational wave interferometers will soon deliver compact binary merger detections at a rate of $\sim$2 per week.[18] Especially important for pointed, narrow-field instruments including IACTs, some of the GW sources will be much better localized on the sky than during previous observing runs. CTA can search for GW counterparts using either the standard observing mode (all telescopes of an array pointing at the same target and then tiling observations over the localization region) or the "divergent pointing" mode in which the pointing of individual telescopes is offset from one another during a single observing run, covering a wider region of sky instantaneously albeit with worse sensitivity at each point.

4.3.4. *Lorentz invariance violation*

Some quantum theories of gravity predict that the speed of light varies with frequency (energy). A sensitive way to test this is using variable or transient, distant sources of a wide range of photon energies, i.e., cosmic VHE sources. Gamma-ray bursts, blazars, and pulsars can all be used for such tests. The sensitivity of such searches increases with the maximum detected gamma-ray energy and the distance to the source, in order that there is sufficient distance for a detectable time delay to accumulate during propagation. If a significant energy-dependent time delay is measured, it must be determined whether this is intrinsic to the emission process or introduced during propagation. However, if no such effect is detected and a large span (from low to high energy) of arrival times is used, a conservative upper limit can be set on the variation with energy which includes both intrinsic and propagation effects. Moreover, multiple sources with different characteristics, such as distance or period, can be used to break the degeneracy between time delays intrinsic to the emission and those introduced during propagation, if and when a possible signal is detected.

4.3.5.   Galactic sources

CTA will survey the Galactic plane to a depth of 2–4 milliCrab. This will enable new source discoveries, population studies of individual source classes (supernova remnants, pulsar wind nebulae, and the TeV haloes that surround some PWNs), and detailed temporal, spectral, and morphological studies of individual bright sources. Morphological studies, for example, can identify whether TeV emission is correlated with molecular clouds and therefore is likely produced by protons colliding with the clouds as opposed to through a leptonic mechanism.

4.3.6.   Active galactic nuclei

Figure 6 shows the estimated flux distribution (source count distribution) of VHE blazars. Because IACTs have a limited field of view, deep and wide extragalactic surveys have not yet been executed, so this distribution is uncertain not only in the low-flux region below the sensitivity of current instruments but also at larger fluxes above their sensitivity. CTA will perform a deep extragalactic survey on a fraction of the sky, resulting in discoveries of hundreds of VHE blazars and a systematic measurement of the source count distribution. The extragalactic survey will cover 25% of the sky to a depth of 1 milliCrab.[4] Observations of these extragalactic sources will enable improved EBL modeling, searches for axion-like particles, constraints on the highly uncertain extragalactic magnetic fields and measurement of the Hubble constant.[20]

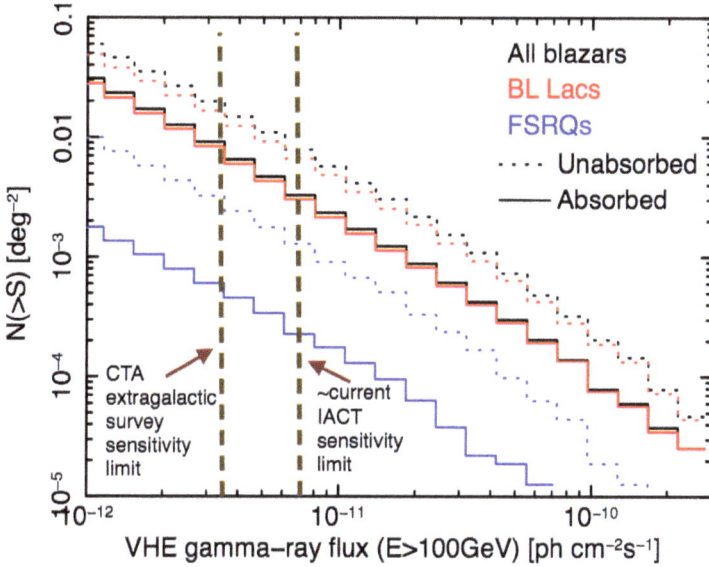

Fig. 6.   Source count distribution (cumulative distribution of source fluxes) of VHE (>0.1 TeV) blazars, including both BL Lacs and flat-spectrum radio quasars. Solid lines show the distribution as detected at Earth, and dotted lines show the hypothetical distribution that would be detected if there were no EBL absorption. CTA's extragalactic survey will produce an unbiased measurement of this distribution and push the flux limit lower by a factor of a few compared to current instruments, resulting in a factor of a few more detected VHE blazars. Figure taken from Ref. [4], where the figure was produced by modifying a figure from Ref. [19].

4.3.7.   Dark matter

CTA will provide unique tests of dark matter candidates including weakly interacting massive particles (WIMPs) and axion-like particles (ALPs). WIMPs can be detected when they annihilate with one another or decay to Standard Model particles including photons. High-mass (greater than $\sim$100 GeV) WIMPs can produce a bright signal in VHE gamma rays. Good targets for CTA WIMP searches include the Galactic center halo, dwarf galaxies, and galaxy clusters. Figure 7 shows CTA sensitivity for several astrophysical targets, compared with existing limits by Fermi-LAT and H.E.S.S. For thermal-relic WIMPs, CTA will test the WIMP mass range between 0.2 and 50 TeV, an important region of WIMP parameter space inaccessible by other methods.[11]

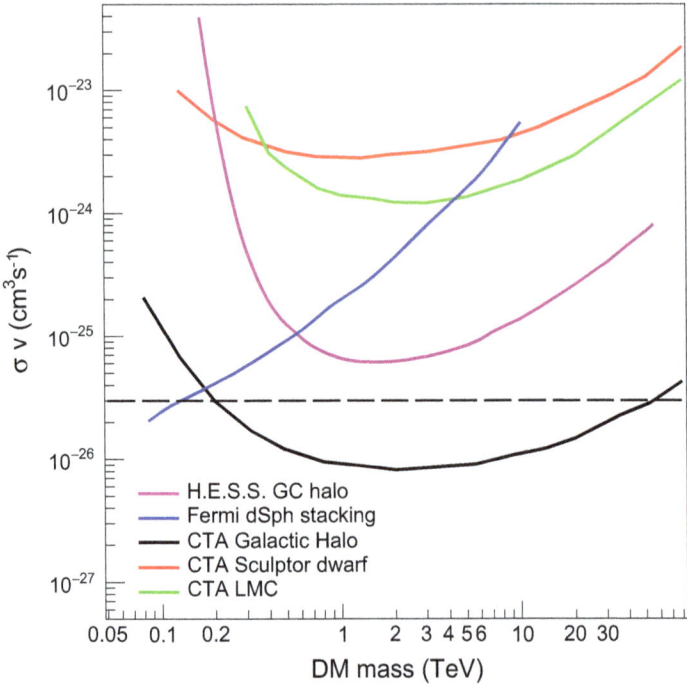

Fig. 7. CTA sensitivity to the WIMP dark matter self-annihilation cross section σ (multiplied by the relative velocity of WIMP particles v), for several astrophysical targets including the Galactic center halo, Sculptor dwarf galaxy, and Large Magellanic Cloud.[4] Sensitivities are compared to existing limits from Fermi-LAT stacking of dwarf spheroidal (dSph) galaxies and H.E.S.S. analysis of the Galactic center halo. If WIMPs are a thermal relic of the Big Bang, then their cross section is predicted to be at the scale of the dashed line, so CTA observations will provide a sensitive test of this scenario over a wide range of masses.

ALPs are another dark matter candidate. While axions would solve the strong CP problem in addition to serving as dark matter, ALPs are a generalization that do not solve the strong CP problem. Like axions, ALPs can convert to photons and vice versa. Such a signal can be detected in VHE gamma rays if EBL absorption is anomalously low or through signatures in the spectra of blazars.

4.3.8. *Charged particle measurements*

In addition to its primary purpose as a gamma-ray observatory, CTA can measure the flux of charged cosmic-ray particles including heavy-ion nuclei, electrons, and positrons. Heavy ions can be identified by the "direct Cherenkov" signal emitted by the original cosmic-ray nucleus when it first enters the atmosphere before showering. Such a signal can be identified by a bright spot at the beginning of the shower image, thanks to the Cherenkov light intensity scaling with particle charge squared. The diffuse "all-electron" (including e^+ in addition to e^-) cosmic-ray spectrum can also be measured using a sample of events with strict cuts to select electromagnetic showers in directions without known gamma-ray sources (such a diffuse $e^{\pm}$ measurement may include contamination by a diffuse gamma-ray signal, but such a signal is sub-dominant). Finally, it may be possible to perform a charge-separated measurement of the positron spectrum using the Moon spectrometer technique (the flux is shadowed by the Moon, with the e^+ and e^- signals offset in different directions due to the Earth's magnetic field).

5. Conclusion and Outlook

Since it was first proven over three decades ago by a single telescope, the imaging atmospheric Cherenkov technique has developed by exploiting the advantages of networked arrays of such telescopes. The current generation of such arrays (H.E.S.S., MAGIC, and VERITAS) consist each of a modest number (2–5) of telescopes, enabling detection of individual gamma-ray (and background) events by multiple different telescopes. The next-generation instrument, CTA, will build on the power of this networked detector method by using much larger arrays, enabling a substantially larger collecting area as well as improved signal reconstruction (including gamma-ray angular resolution) and background rejection.

Gamma-ray measurements are at the heart of some of the most exciting fields in contemporary astrophysics, including those of high-energy, multi-messenger, and time-domain astrophysics. CTA promises to make a variety of discoveries in these fields as well as in topics of fundamental physics, including the nature of space–time and the search for dark matter particles including axion-like particles and weakly interacting massive particles.

References

1. T. C. Weekes *et al.*, Observation of TeV gamma rays from the Crab Nebula using the atmospheric Cerenkov imaging technique, *Astrophys. J. Lett.* **342**, 379 (July, 1989). doi: 10.1086/167599.
2. M. Punch *et al.*, Detection of TeV photons from the active galaxy Markarian 421, *Nature.* **358**(6386), 477–478 (August, 1992). doi: 10.1038/358477a0.
3. J. Quinn *et al.*, Detection of gamma-rays with E > 300-GeV from Markarian 501, *Astrophys. J. Lett.* **456**, L83–L86 (1996). doi: 10.1086/309878.
4. *Science with the Cherenkov Telescope Array*. World Scientific (2019). doi: 10.1142/10986. https://www.worldscientific.com/doi/abs/10.11 42/10986.
5. A. Dominguez *et al.*, Extragalactic background light inferred from AEGIS galaxy-SED-type fractions, *Mon. Not. R. Astron. Soc.* **410**(4), 2556–2578 (January, 2011). ISSN 0035-8711. doi: 10.1111/j.1365-2966. 2010.17631.x.
6. V. A. Acciari *et al.*, Teraelectronvolt emission from the γ-ray burst GRB 190114C, *Nature.* **575**(7783), 455–458 (November, 2019). doi: 10.1038/s41586-019-1750-x.
7. V. A. Acciari *et al.*, Bounds on Lorentz invariance violation from MAGIC observation of GRB 190114c, *Phys. Rev. Lett.* **125**, 021301 (July, 2020). doi: 10.1103/PhysRevLett.125.021301.
8. H. Abdalla *et al.*, A very-high-energy component deep in the γ-ray burst afterglow, *Nature.* **575**(7783), 464–467 (November, 2019). doi: 10.1038/s41586-019-1743-9.
9. H. Abdalla *et al.*, Revealing x-ray and gamma ray temporal and spectral similarities in the GRB 190829A afterglow, *Science.* **372**(6546), 1081–1085 (June, 2021). doi: 10.1126/science.abe8560.
10. S. Fukami *et al.*, Very-high-energy gamma-ray emission from GRB 201216C detected by MAGIC, *PoS.* **ICRC2021**, 788 (2021). doi: 10.22323/1.395.0788.
11. M. Cahill-Rowley, R. Cotta, A. Drlica-Wagner, S. Funk, J. Hewett, A. Ismail, T. Rizzo, and M. Wood, Complementarity and searches for dark matter in the pMSSM. In *Community Summer Study 2013: Snowmass on the Mississippi* (March, 2013).
12. A. U. Abeysekara *et al.*, A search for brief optical flashes associated with the SETI target KIC 8462852, *Astrophys. J. Lett.* **818**(2), L33 (February, 2016). doi: 10.3847/2041-8205/818/2/L33.

13. A. U. Abeysekara *et al.*, Demonstration of stellar intensity interferometry with the four VERITAS telescopes, *Nat. Astron.* **4**, 1164–1169 (January, 2020). doi: 10.1038/s41550-020-1143-y.

14. C. B. Adams *et al.*, Detection of the Crab Nebula with the 9.7 m prototype Schwarzschild-Couder telescope, *Astropart. Phys.* **128**, 102562 (March, 2021). doi: 10.1016/j.astropartphys.2021.102562.

15. M. G. Aartsen *et al.*, Neutrino emission from the direction of the blazar TXS 0506+056 prior to the IceCube-170922A alert, *Science.* **361**(6398), 147–151 (July, 2018). doi: 10.1126/science.aat2890.

16. M. G. Aartsen *et al.*, Multimessenger observations of a flaring blazar coincident with high-energy neutrino IceCube-170922A, *Science.* **361** (6398), eaat1378 (July, 2018). doi: 10.1126/science.aat1378.

17. B. P. Abbott *et al.*, GW170817: Observation of gravitational waves from a binary neutron star inspiral, *Phys. Rev. Lett.* **119**(16), 161101 (October, 2017). doi: 10.1103/PhysRevLett.119.161101.

18. B. P. Abbott *et al.*, Prospects for observing and localizing gravitational-wave transients with advanced LIGO, advanced Virgo and KAGRA, *Living Rev. Rel.* **21**(1), 3 (2018). doi: 10.1007/s41114-020-00026-9.

19. P. Padovani and P. Giommi, A simplified view of blazars: The very high energy gamma-ray vision, *MNRAS.* **446**(1), L41–L45 (January, 2015). doi: 10.1093/mnrasl/slu164.

20. Domínguez *et al.*, A new measurement of the Hubble constant and matter content of the universe using extragalactic background light γ-ray attenuation, *Astrophys. J. Lett.* **885**(2), 137 (November, 2019). doi: 10.3847/1538-4357/ab4a0e.

Index

www.ingramcontent.com/pod-product-compliance
Lightning Source LLC
Chambersburg PA
CBHW050540190326
41458CB00007B/1860